New Approaches to the Scientific Study of Religion

Volume 3

Series editors
Lluis Oviedo, Pontifical University Antonianum, Roma, Italy
Aku Visala, Helsinki University, Helsingin Yliopisto, Finland

Editorial Boards
Helen de Cruz, Oxford Brookes University, UK
Nathaniel Barrett, University of Navarra, Spain
Joseph Bulbulia, Victoria University, New Zealand
Miguel Farias, Coventry University, UK
Jay R. Feierman, University of New Mexico, NM, USA
Jonathan Jong, Oxford University, UK
Justin McBrayer, Fort Lewis College, CO, USA

Introduction to the Series

This series presents new approaches to the scientific study of religion, moving from the first generation of studies that try to 'explain religion' towards a more critical effort to explore alternative paths in correspondence with this highly complex human and social feature. The series supports the development of new scientific models that advance our understanding of religious faith, emotions, symbols, rituals, meaning, and religions' anthropological and cultural dimensions, integrating them into more complex models.

Recent decades have witnessed a growing interest in the study of religious mind and behavior from new disciplinary fields, such as cognitive psychology, neuroscience and bio-evolutionary science. The amount of published research is impressive and has reached a level that now calls for evaluation and revision of current models and developments. This new series supports this fast-moving development, encouraging the publication of new books that move on from current research towards larger horizons and innovative ideas.

This series:

- Increases academic communication and exchange in this multi-disciplinary research area.
- Gives a new impetus to the science and religion dialogue.
- Opens up new avenues of encounter and discussion between more scientific and more humanistic traditions.

More information about this series at http://www.springer.com/series/15336

Gerrit Glas • Jeroen de Ridder
Editors

The Future of Creation Order

Vol. 1, Philosophical, Scientific,
and Religious Perspectives
on Order and Emergence

Editors
Gerrit Glas
Department of Philosophy
Vrije Universiteit Amsterdam
Amsterdam, The Netherlands

Jeroen de Ridder
Department of Philosophy
Vrije Universiteit Amsterdam
Amsterdam, The Netherlands

ISSN 2367-3494 ISSN 2367-3508 (electronic)
New Approaches to the Scientific Study of Religion
ISBN 978-3-319-89003-6 ISBN 978-3-319-70881-2 (eBook)
https://doi.org/10.1007/978-3-319-70881-2

© Springer International Publishing AG 2017, corrected publication 2018
Softcover re-print of the Hardcover 1st edition 2018
This work is subject to copyright. All rights are reserved by the Publisher, whether the whole or part of the material is concerned, specifically the rights of translation, reprinting, reuse of illustrations, recitation, broadcasting, reproduction on microfilms or in any other physical way, and transmission or information storage and retrieval, electronic adaptation, computer software, or by similar or dissimilar methodology now known or hereafter developed.
The use of general descriptive names, registered names, trademarks, service marks, etc. in this publication does not imply, even in the absence of a specific statement, that such names are exempt from the relevant protective laws and regulations and therefore free for general use.
The publisher, the authors and the editors are safe to assume that the advice and information in this book are believed to be true and accurate at the date of publication. Neither the publisher nor the authors or the editors give a warranty, express or implied, with respect to the material contained herein or for any errors or omissions that may have been made. The publisher remains neutral with regard to jurisdictional claims in published maps and institutional affiliations.

Printed on acid-free paper

This Springer imprint is published by Springer Nature
The registered company is Springer International Publishing AG
The registered company address is: Gewerbestrasse 11, 6330 Cham, Switzerland

Acknowledgments

The editors are grateful to the following organizations for their generous support which helped to make possible both the 2011 conference "The Future of Creation Order" at Vrije Universiteit Amsterdam and the publication of this and its companion volume: C.J. de Vogel Stichting, Dimence Groep, Stichting Bijzondere Projecten Vereniging voor Reformatorische Wijsbegeerte, Stichting Dr. Abraham Kuyperfonds, Stichting Pro Religione et Libertate, Stichting Zonneweelde, Subsidiecommissie Verenigingsfondsen VU, and two anonymous private sponsors.

We also wish to thank the many anonymous referees who provided insightful, constructive, and often extensive feedback on the individual chapters in this volume, and two anonymous referees who gave helpful suggestions for improvement of the volume as a whole.

Finally, we would like to thank Mathanja Berger for her truly excellent work in copyediting this volume. There is no doubt that the text would have contained many more violations of linguistic order without her meticulous attention to detail.

Contents

Introduction to the Philosophy of Creation Order, with Special Emphasis on the Philosophy of Herman Dooyeweerd.................. 1
Gerrit Glas and Jeroen de Ridder

Part I The Concept of Creation Order

Natural Law, Metaphysics, and the Creator 33
Eleonore Stump

Is the Idea of Creation Order Still Fruitful? 51
Danie Strauss

Creation Order in the Light of Redemption (1): Natural Science and Theology 67
Henk G. Geertsema

Part II Creation Order, Emergence, and the Sciences

Christianity and Mathematics..................................... 97
Danie Strauss

Properties, Propensities, and Challenges: Emergence in and from the Natural World 119
Marinus Dirk Stafleu

Nuancing Emergentist Claims: Lessons from Physics................ 135
Arnold E. Sikkema

Order and Emergence in Biological Evolution....................... 151
Denis Alexander

Is There a Created Order for Cosmic Evolution in the Philosophy of Herman Dooyeweerd? 171
Jitse M. van der Meer

Creation Order and the Sciences of the Person 203
Gerrit Glas

Beyond Emergence: Learning from Dooyeweerdian Anthropology?... 231
Lydia Jaeger

Part III Creation Order and the Philosophy of Religion

For the Love of Wisdom: Scripture, Philosophy, and the Relativisation of Order 257
Nicholas Ansell

A Contribution to the Concept of Creation Order from a Lutheran Perspective 289
Hans Schaeffer

Out of the Ashes: A Case Study of Dietrich Bonhoeffer's Theology and the Orders of Creation 301
Annette Mosher

Sergei Bulgakov's Sophiology as the Integration of Sociology, Philosophy, and Theology 317
Josephien van Kessel

Erratum E1

Index 337

Introduction to the Philosophy of Creation Order, with Special Emphasis on the Philosophy of Herman Dooyeweerd

Gerrit Glas and Jeroen de Ridder

Abstract In this introductory chapter, we provide some background to the main theme of these two volumes, to wit, creation order. We start with a quick historical sketch of how the traditional connection between the experienced orderliness of the world around us and the will of a divine Creator came under pressure as a result of various philosophical and scientific developments. We also show how scholars from Christian and other religious traditions responded in different ways to these developments. Next, we offer a brief overview of some key themes from the work of the Dutch philosopher Herman Dooyeweerd. We believe that his thought offers insights that can profitably be used to advance the contemporary discussion on creation order, as is evidenced by a number of contributions to these two volumes. The chapter closes with a brief overview of what can be expected in the chapters that follow.

Keywords Law of nature · Creation order · Herman Dooyeweerd · Reformational philosophy · Modal aspects · Christian philosophy

Context

This is the first of two volumes based on presentations that were given at the international conference "The Future of Creation Order," organized by the (Dutch) Association for Reformational Philosophy in collaboration with the Department of Philosophy at Vrije Universiteit Amsterdam, at the occasion of the 75th anniversary celebration of the association, in August 2011. The purpose of the conference was to delve deeper into the current health of the philosophical concept of (creation) order and of such related concepts as law, structure, necessity, change, emergence, and principle.

G. Glas (✉) · J. de Ridder
Department of Philosophy, Vrije Universiteit Amsterdam, Amsterdam, The Netherlands
e-mail: g.glas@vu.nl; g.j.de.ridder@vu.nl

© Springer International Publishing AG 2017
G. Glas, J. de Ridder (eds.), *The Future of Creation Order*, New Approaches to the Scientific Study of Religion 3, https://doi.org/10.1007/978-3-319-70881-2_1

In this introductory chapter, we will first provide a historical and intellectual update with respect to the theme of the conference, sketching how developments in the sciences and humanities led to the demise of the idea of a pregiven created order and presenting an overview of how scholars from Christian and other religious traditions have responded to this challenge.

We will then continue with a brief introduction to the systematic philosophy of Herman Dooyeweerd, one of the founders of what has become known as *neo-Calvinist* or *reformational philosophy*. This introduction enables readers to understand authors who develop and expound their ideas in discussion with this philosophy.

The chapter closes with an overview of the content of the book. The first part of this volume focuses on the philosophical discussion about creation order. The second part delves into the relevance of the creation order concept for the sciences and pays special attention to the notion of emergence. The third part of the volume investigates the possible role of theology and philosophy of religion with respect to the understanding of creation order.

Order, Orderliness, and Law

People of all times have experienced the natural world as expressing an overwhelming beauty, coherence, and order. In the great monotheistic traditions, this beauty, coherence, and order have been related to the will or nature of a divine Creator and this relation has been variously described with metaphors derived from the world of art, manufacturing, and agriculture. For a long time, philosophers found the notion of a divinely imposed order unproblematic. Natural philosophy transformed these lay conceptions into metaphysical concepts such as demiurge, first mover, highest being, and origin. This highest being was seen as the origin of *natural law*, which was reflected in the social and moral order of the world.

Things changed with the application of mathematical methods to issues that previously were dealt with in natural philosophy. The metaphysical order of natural philosophy became the object of scientific inquiry. Speculation was replaced by mathematics and mechanics, and the metaphor of the clock expressed the newly gained insight into the way the world was ruled (Harrison 1998).

The concept of law of nature emerged in this period. It not only reflected confidence in humanity's capacity to unravel the mysteries of the universe, but it also reinforced trust in God as the One who wills and maintains these laws (Harrison 2008). The Deist interpretation of the universe combined both: it maintained belief in God as sovereign monarch and it fostered trust in science as the way to read the book of nature.

Nowadays, scientists mostly ignore the theological background of the notion *law of nature*. As the Deist universe lost plausibility, the notion of a superior divine intelligence fell into disrepute and a deep and long-standing confusion about the status of laws of nature began. For one thing, laws denote both the way things are and how they, in some sense, ought to be. The gradual disappearance of a Creator who is willing these laws weakened the support for the idea that laws of nature are willed and that they, therefore, hold with necessity (Clayton 2008). This is the background of the dispute between the so-called *regularity view* and *necessity view* of laws of nature (Armstrong 1983).

Evolutionary theory complicated the debate even more. The development of this theory mimicked developments in physics by replacing natural history, just like physics had replaced natural philosophy. The first edition of Darwin's *Origin of Species* drew an analogy between the law of gravity and the law of development. The law of development, he argued, would change our view of the organic world in much the same way as the law of gravity changed our view of the physical world. The laws of evolutionary biology soon came to be seen as deeply historical and contingent, challenging not only the theistic but also the deterministic view of laws of nature. Laws appeared to be nothing more than regularities or patterns holding for a certain period of time.

Reformational Philosophy

Ever since the emergence 75 years ago of the branch of Christian philosophy that has come to be called *reformational* (or *neo-Calvinist*) *philosophy*, the concepts of order and law (or principle, structure) have been at its heart. In fact, it is fair to say that this tradition can be characterized as a philosophy of creation order.

Among its central concerns is a long-standing debate over the nature of *law*: its origin, its status as boundary between God and creation, its validity, and its relationship to history and human agency. Firmly rejecting both scholastic metaphysics and Deism, reformational philosophers have maintained the notion of law as *holding for reality*, thereby preserving a variant of the necessity view of laws. However, questions have arisen about the nature of such laws. For instance, can the philosophical concept of law be equated with the concept of law in physics and biology? Or is it in fact a concept at the level of worldviews; that is, an articulation of a fundamental assumption about the origin, nature, and destiny of the world? Does *law* mean more than just "orderliness"? More recently, the issue has been raised as to whether laws of nature have always existed or, rather, "emerge" in the process of the disclosure of reality (Klapwijk 2008). There have been discussions on the universality of laws, on their possible susceptibility to change, and on the difference between law and the law-side of reality (see below). Developments in the life sciences have challenged the distinction between creation and temporal becoming—a distinction that has contributed to the acceptance of evolutionary theory by some of the major figures in the movement.

Enlightening as these discussions have been, it is also widely acknowledged that they are far from finished. Some critics of reformational philosophy, for example, still tend to interpret its view of law and of order as a variant of Deism, Platonism, or Aristotelianism (for a discussion, see Wolters 1985; Henderson 1994, chap. 5;

Echeverria 2011). Others have reproached it for offering a fundamentally static, essentialist, and/or monarchical conception of order (see Chaplin 2011, 51–54; Strauss 2009, 201–204). Theologians have questioned the implicit assumption that it is possible to gain access to creation order independent of the cross and of mediation by the church (see Douma 1976). How cogent and compelling have reformational theories of law and order been if, 75 years later, they are still confronted with criticisms such as these (for earlier discussions, see especially Van der Hoeven [1981, 1986])?

The Current Debate in the Sciences and Humanities

In addition to what has been said about the theological roots of the notion *law of nature* and the ambivalences surrounding it, there are at least three major theoretical developments that have helped shape the current debate on the nature of law and (creation) order and especially its application to man and society: (1) evolutionary theory, (2) postmodern and constructivist social theory, and (3) philosophy. Although in the remainder of this book much emphasis will be put on the anthropological, social, and moral dimensions of creation order, it is good to keep in mind that the very idea of law itself cannot be seen apart from its theological reminiscences.

1. *Evolutionary theory* has exerted an enormous influence on the way we conceive of ourselves and the world. If the living world is the product of accident and chance, then order can be, at best, the product of a process of development, but never its presupposition. Order, in other words, is neither pregiven nor unchangeable. If the evolutionary account of order is true, then it appears that not only the living world but, more specifically, the existence of the human race itself is utterly contingent. Evolutionary theory, especially in its ultra-Darwinist renderings, has eroded the distinctness of humankind and the intuition that we are "at home in the universe."
2. The *social sciences* have contributed via other routes to the demise of order and lawfulness. Historicism and postmodernism have been especially influential and dominant, propounding the view that order is constructed and should be seen as the product of human interaction and interpretation. Society, therefore, will inevitably be "plural." This plurality initially denoted a factual condition but gradually evolved into a directive norm—i.e., "pluralism." On such a view, there can be no universally valid rules and norms. Such rules and norms are simply reports of particular standpoints, whether personal or shared with the like-minded, but never a reflection of a moral order transcending human (inter)subjectivity.
3. *Philosophy*, finally, has also been of critical importance in the dramatic change of perceptions of order in our time. Philosophy was the birthplace of historicism and postmodernism. Together with Marxism, Nietzschean perspectivism, existentialism, and critical theory, these traditions were responsible for the downfall of the classical philosophical conception of order. Philosophers drew subversive

conclusions from the life sciences and announced the fundamental contingency of human existence.

Today's intellectual climate seems to be shifting and may bring us to the verge of a significant reorientation. Even the discipline of economics—probably the most persistently modernist among the social sciences—is in disarray after the near-catastrophic financial crisis of 2008 and later. Standard models of economic equilibrium have been discredited in an unprecedented way. Fundamental reflection on the presuppositions of economic order is called for. It seems we have been confronted with boundaries that cannot be violated without damaging repercussions—not least in the possibly irreversible destruction of a sustainable natural environment.

Signs of a reorientation are also noticeable in other areas. Cultural anthropologists are looking for models that go beyond standard constructivist approaches. Social and political scientists are seeking concepts that may deepen or even replace the usual language of a limitless pluralism. Even the discipline of public administration shows renewed interest in such notions as *reasonableness* and *fairness*. Psychiatrists are trying to find a way beyond the uneasy dichotomy between naturalist (biomedical) and constructivist approaches to disease. In short, there are signs that scholars in at least some fields are trying to unearth order again by looking for a middle ground between antiquated, static conceptions of order and newer approaches which give priority to contingency and plurality.

Christian Philosophical Responses

Christian philosophers have responded in different ways to the downfall of order.

Hermeneutically oriented Christian philosophers have mainly concentrated on the issue of pluralism and have argued for a position that draws a distinction between plurality and pluralism. Recognition of plurality does not imply full-blown relativism in ethics, they have suggested. The issue of relativism itself needs to be scrutinized because it seems to presuppose some form of absolute and/or objective truth. Charles Taylor has developed an approach to plurality by relating it to the even more fundamental notion of recognition.

Christian philosophers inspired by *Anglo-American* (analytical) philosophy have concentrated on logic, language, knowledge, and the semantics of possible worlds. Their conceptual armory has helped lay bare foundationalist tendencies in both naturalist and (creative) antirealist accounts. As a response to both naturalism and creative antirealism, Alvin Plantinga (2000) has developed the idea of *proper function*—with the idea of (divine) design in the background. Nicholas Wolterstorff (1995, 2008) has cleared the ground for a philosophical understanding of the claim that God speaks and, recently, for a theistic conception of justice and human rights (albeit one set against a notion of "justice as right order"). All this work has been enormously important for the preservation of philosophical ground for notions such as *law*, *design,* and *intrinsic quality*. However, much work still needs to be done in relating this philosophical work to discussions in the life sciences and social sciences.

Neo-Thomist philosophers have made significant advances in reformulating classical notions of law, substance, and natural order. Moreland and Rae (2000), for instance, have argued for a substantialist account of human nature and for a concept of the person that is firmly rooted in a substance view of the soul. They see both contemporary philosophy of mind (especially the approach known as non-eliminative physicalism) and postmodern philosophy as threats to the classical Christian doctrine of humanity, and argue for the continuing importance of the idea of natural law. This argument is echoed in the recent encyclical *Caritas in Veritate*, which urges that truly compassionate responses to human need must remain rooted in stable moral truth.

Christian thinkers in the tradition of *radical orthodoxy* distance themselves from modern notions of order conceived as absolute, rational, and universal order. John Milbank, for instance, argues for an approach that does not focus on an independent creation order as point of departure for ethics and political theory but rather calls for a "counter-ontology" and a "counter-ethics" narratively constructed by the church, as an *altera civitas*, and based on her specific experience of participation in divine, creative love (Milbank 2006; Milbank et al. 1999). Such a counter-ontology would, they propose, help us discern again the deeper intentions of classical Christian notions of metaphysical order, which have been obscured in Christian capitulations to modernity.

Christian thinkers inclined toward *process philosophy* and/or *chaos theory* have also distanced themselves from classical concepts of natural law and creation order, and tended to a view in which God is seen as creative counterpart in the process of the development of creation from its inception. According to Ian Barbour (1997), we should give up the monarchical view of an almighty God who rules the world through unchangeable laws. We must instead be open to a *panentheist* view in which God participates in reality. This participation should be "located" in the openness and indefiniteness of creation, in the receptive side of both the living and the non-living worlds. In this approach, order is temporal and, therefore, inevitably in constant change.

Finally, some Christian philosophers in the tradition of *reformational philosophy* have been ready to give up the idea that biological species are rooted in an originally given creation order, yet without distancing themselves from the idea of creation order as such. They have not been unanimous, however, in the way they characterize the laws by which such order is constituted. Are laws best understood as philosophical concepts referring to the boundary between God and creation, as knowable structures such as physical laws, or as terms referring to the theological notion of divine providence? And should the notion of order be conceptualized in terms of dynamic principles that are waiting for (various forms of) disclosure, or as a fixed, pregiven order? With respect to societal and moral order, reformational philosophers have often proposed that a knowable framework exists of *structural principles* or *normative structures* for social institutions. However, there is a variety of interpretations of these principles and structures, both with respect to their nature and to the range of their possible implementations (or *positivizations*).

To summarize, the impetus for the conference and, therefore, for the contributions in these two volumes comes from two sides: on the one hand, rapid developments in the natural and social sciences, the humanities, and particularly philosophy; on the other hand, the challenge presented by the wide diversity of views in Christian philosophical circles. Both evolutionary theory and social philosophy challenge the idea of the pregivenness of norms, laws, and structures. They suggest that there are no such norms, laws, and structures and that what we know under this heading are products of either the process of natural selection or human construction and subjective interpretation. Christian philosophers have responded to the collapse of order in a variety of ways. There are strands of Christian philosophy that still argue for the idea of a pregiven, if not fixed, world order. Other scholars, however, see this idea of a stable creation order and/or natural law as redundant and in need of thorough rethinking. These two volumes give a sketch of the landscape and are an attempt to answer the question whether there is still room for affirmation of pregiven norms—or what one could call *ontic normativity*—while also acknowledging the particularity and "situatedness" of our articulation of those norms.

Herman Dooyeweerd: A Short Introduction

Since many contributions in these volumes refer to and make use of terminology that was originally developed in Dooyeweerd's systematic work, we will give a brief introduction to some of these terms and the fundamental ideas behind them.

Herman Dooyeweerd (1894–1977) was one of the founders of what has become known as *reformational philosophy*. He developed his philosophy in the interbellum, in an intellectual climate that was characterized by uncertainty, deep divides between philosophical traditions, and the presentiment of the decline of Western culture.[1] A former president of the Royal Dutch Academy of Arts and Sciences once called him "the most original philosopher the Netherlands has ever produced, even Spinoza not excepted" (Langemeijer, quoted in Kalsbeek [1970, 10]). Dooyeweerd studied law and philosophy at Vrije Universiteit Amsterdam. He had a brief career in public administration before he became secretary of the Abraham Kuyper Foundation, a precursor of the scientific institute of the Anti-Revolutionary Party, one of the Christian political parties in the Netherlands before World War II. Dooyeweerd worked there between 1922 and 1926. From 1926 until his retirement in 1965 he was professor of philosophy and history of law at Vrije Universiteit Amsterdam.

---

[1] Dooyeweerd himself also refers to the popular and highly influential book *The Decline of the West* (*Der Untergang des Abendlandes*) by the German thinker Oswald Spengler, of which the first volume appeared in 1918.

He wrote extensively and had an almost encyclopedic knowledge not only of philosophy and the history of philosophy (in particular neo-Kantianism and ancient philosophy), but also of the sciences of his time, especially mathematics, physics, biology, law, and social and political sciences. Most of his ideas in systematic philosophy were developed and refined in the interaction with the sciences.

To understand Dooyeweerd's philosophy it is useful to keep in mind that it is built up around two main themes: (1) the distinction between what he called different *modal aspects* (or ways of functioning); and (2) the idea that all human activity is rooted in what he called the *heart* and which denotes a kind of concentration point within our existence where we are known deepest by others and respond to our ultimate concerns.

In the following sections we will discuss first the theory of modal aspects, then the theory of entities or *individuality structures*, followed by the concepts of law and cosmic order, next the idea of the heart and, finally, what Dooyeweerd saw as the fundamental flaw of Western philosophy and science: its absolutization of the theoretical attitude of thought.

Modal Aspects

Dooyeweerd says in an interview that the idea of modal diversity came to him in a flash during a walk in the dunes somewhere around 1921 when he was overwhelmed by the astonishing diversity and the incredible coherence in the way things exist and are functioning (Van Dunné et al. 1977, 37). Everything exists in many different ways—ways that are both distinguishable and interconnected in our ordinary experience of the world. A flower, for example, exists in a spatial, a physical, and a biotic way: it occupies a certain space, it has physical properties (e.g., mass), and it functions as a biotic entity, because it grows, blossoms, and reproduces. A flower may also function in other spheres or aspects—for instance, in the economic or the aesthetic aspect. It then functions as an economic or aesthetic object, respectively. All these different ways of functioning are woven together in a seemingly self-evident and natural way in our everyday experience. We do not even notice the manifold differences.

However, we can become aware of them in certain contexts in which a particular feature stands out, or when we step back and reflect on the differences. For example, a flower seller will be more aware of certain physical properties of plants and flowers, such as their weight, because he has to handle them. Consumers on the other hand may be more interested in how flowers function in the aesthetic and economic aspects. Similarly, a physician will explicitly take notice of specific features of a wound (e.g., color, temperature, size) whereas the patient will be inclined to focus on the pain the lesion causes. Dooyeweerd calls the most general ways of existing (or functioning) *aspects*, or *modal aspects*—other terms he employs for the same notion are *functions*, *modal functions*, and (more technically) *law-spheres*. The term *modal* does not refer to logic (modal logic) but to the Latin term *modus* which

Table 1 Modal aspects and meaning nuclei

Modal aspects	Meaning nucleus of each aspect
Numerical (or quantitative) aspect	Discrete quantity
Spatial aspect	Continuous extension
Kinematic aspect	Uniform movement/flow, constancy
Physical aspect	Energy
Biotic aspect	Life
Psychic (or sensitive) aspect	Feeling, sensitivity
Logical aspect	Analysis
Historical (or cultural-historical) aspect	Formative power/control
Lingual (or sign) aspect	Signification, articulation
Social aspect	Social life
Economic aspect	Frugality
Aesthetic aspect	Allusivity, imaginativity
Jural aspect	Retribution
Moral aspect	Love, sincerity, honesty, integrity
Pistic (or certitudinal/fiduciary) aspect	Trust, confidence, reliability

means "way of existing or functioning." Dooyeweerd discerns 14 (and later, 15) of these modes of functioning (see Table 1).

Everything functions in a number of *spheres*; not in one sphere at a time, diachronically, but in all spheres synchronically, in an orderly way and in close conjunction. A sphere is not a layer—an ontic cross section, so to say—within an entity. It is a way of functioning or existing, not a part of a substance.

Things function in these spheres in basically two ways; namely, as subject or as object. Flowers exist (or function) within a numerical, a geometrical, a kinematic, a physical, and a biotic sphere. That is, they exist numerically, spatially, physically, and biotically; which means that, in their existing, flowers manifest discreteness (numerical sphere), spatial continuity (geometrical sphere), persistence/constancy (kinematic sphere), qualities such as mass and energy (physical sphere), and generation and tendency to self-maintenance (biotic sphere). In these five spheres, flowers function as *subject*—that is, they manifest these qualities themselves, *actively*. Flowers also function in other spheres as *object*; for instance, as object of scientific analysis (logical sphere) or as object with aesthetic qualities (aesthetic sphere). In all these other spheres flowers function as objects, *passively*, in their interaction with human beings (or animals and plants). Object-functions are, in other words, latent as long as their qualities are not disclosed (or opened up) by other subjects.

It is safe to say that modal aspects refer to kinds of properties rather than properties per se. The term *property* is usually understood as referring to an instantiation of a more general category. When I say, "This car is black," then black is a property of the car. It is an instantiation in this particular car of the general category of blackness. Analytic philosophers often refer to property instantiations as tropes and to properties as universals. Modes, or modal aspects, refer neither to property instantiations, nor to properties, but to the general categories or families of properties—

more specifically, to kinds of properties. The distinctness of these kinds has something to do with the distinctness, or *sovereignty*, of laws and/or principles, according to Dooyeweerd. More precisely, the distinctness of kinds of properties is a reflection of the irreducible distinctness in the way laws and principles determine and delimit what exists and occurs.

The idea of distinctness picks up a theme from the thought of Abraham Kuyper, the nineteenth-century theologian, philosopher, statesman, and prolific author who probably inspired Dooyeweerd most. Kuyper had developed the idea of *sphere sovereignty* in order to understand how different social spheres can overlap but retain relative independence at the same time. For example, the activities of the church and of the state overlap, but are at the same time *sovereign in their own sphere*, because both obey to their own normative principles. These normative principles are ultimately not man-made but intrinsic to our social existence and "given"—although their specific implementations are, of course, influenced by culture and local circumstances. Dooyeweerd applied this sociological principle of sphere sovereignty to all kinds of laws. The cosmos we inhabit manifests an order with a manifold of laws. Each of these laws belongs to a particular type and these types represent modal spheres.

Dooyeweerd devoted the entire second volume of his magnum opus *A New Critique of Theoretical Thought* (1953–1958) to the analysis of and distinction between these modal aspects. He distinguished 15 modal aspects (see Table 1), each with its own typical character, or meaning. As mentioned earlier, modal aspects are not layers or components but rather modes of existence—they concern the *how* and not the *what*. In the next section we will explore what Dooyeweerd has to say about the *what*—i.e., about entities and their structure.

In his systematic philosophy, the modal analysis precedes and is more fundamental than the analysis of entities. Dooyeweerd is, of course, aware that most scientists are primarily occupied with entities; i.e., with things, part-whole relationships, and with the interactions and relations between particular types of things. He nevertheless maintains that science starts by selecting a particular modal point of view. It is only after having gone through "the gate of modal analysis" that the scientist will study the relationships between and within things. To be sure, this modal point of view does not have to be similar to the modal aspect that qualifies the thing to be studied. Physics, for instance, became a science not by adopting a physical point of view but by applying mathematical principles to physical phenomena.

The Difference Between Modes and Entities

The failure to recognize the relevance of the distinction between modes and entities is the cause of much trouble in the sciences, according to Dooyeweerd. Mental phenomena, for instance, are often conceived as expressions of a mental part (substance, layer, or component) within the organism. In other cases, they are seen as products that are causally brought about by some mechanism in the brain. This mechanism is usually considered to be non-mental.

Both ways of conceptualizing mental phenomena are problematic in the Dooyeweerdian view. The implicit assumption of the first position is that if there are mental properties and functions, they can only exist if there also exists some mental "stuff" that serves as a bearer of these properties and functions. This view leads inevitably to a form of mind–body dualism, which most scientists and philosophers today find unattractive (as did Dooyeweerd). From a Dooyeweerdian perspective, the argument is based on a non-sequitur—namely, a confusion between the modal and the entitary point of view. Mental properties and functions are not just mental: the phenomena they are referring to are always also biological, social, and moral, to mention but a few of the most obvious other candidate spheres. The term *mental* is in itself slightly confusing because it refers to so many kinds of psychological activities and experiences. But all these psychological kinds of functioning presuppose entities (e.g., activities, processes, actions) that realize them. And these entities also have other qualifications: they involve the working of certain brain circuits (biotic sphere), they presuppose molecular and metabolic processes in these circuits (physical sphere), and they have a developmental history with social, cultural, and moral characteristics.

In short, mental phenomena are entities that function in all modal spheres. To see them as immaterial expressions of an immaterial part within the organism is mixing the modal point of view (their modal qualification) with the entitary point of view; in other words, the psychological *aspect* of the thing (i.e., thought or feeling) is held to be a proof of the existence of a psychic thing (*entity*) in us. Dooyeweerd would, for this reason, be very hesitant to speak of levels or layers as ontic realities. Such layers often do not exist in reality and are, in fact, the product of the ontologizing of a modal point of view. The fact that we can distinguish biological, psychological, social, and moral aspects in our functioning as human beings does not warrant the conclusion that we are composites built up of biological, psychological, social, and moral components.

The implicit assumption of the second and more popular position (i.e., viewing mental phenomena as products of an underlying mechanism or brain process) is also based on the confusion between the modal and the entitary point of view, but in a different way. Here, the point is that mental phenomena are first isolated and then conceptualized as products of the preceding operation of another entity or component in the organism, usually the brain, which is conceived as a biotic entity. Then, the transition from the biotic to the mental becomes problematic, because how can mental phenomena be the output of biotic processes in the brain? The difficulty we have in imagining this is the result of the preceding conceptual separation between production process and product. This separation runs parallel with a tendency to reify both process and product. The picture of the brain as an organ that produces mental phenomena is as old as the sciences of psychology and psychiatry and offers a clear example of this reification tendency. It is problematic because it construes causal relationships between processes that logically and factually imply one another. The phenomena that are produced cannot be separated from the producing process. Mental processes, for example, are in many respects embodied; their existence can never be seen apart from a context of embodiment and embeddedness. The production metaphor is therefore wrong: it suggests that mental processes and brain processes are separated in time and that they belong, logically and factually, to a different order.

Contemporary adherents of this second approach reject mind–body dualism and therefore have difficulty with the idea of the mental as more or less independent output of the brain. Their rejection of the mental as substance (a rejection with which Dooyeweerd would concur) leads to a reduction of the mental to either an illusion or an epiphenomenon of something else, i.e., processes in the brain. To put it differently: if mental phenomena are not conceived as intrinsically connected with brain processes and if mind–body dualism is not an option, then mental phenomena can only be seen as either illusions (a position which is known as *eliminative physicalism*) or epiphenomena of material processes in the brain (a position known as *non-reductive physicalism*). In the first case, there are no mental phenomena—they only exist in the mind of the perceiver. In the latter case, mental phenomena are only epistemic realities and not ontic—again, they do not really exist.

Dooyeweerd would reject the very presuppositions of this line of reasoning; most notably the idea that modal distinctness should be taken as proof of entitary distinctness. This in turn leads to unjustified substantialization (or reification) of both mental activities and (brain) processes. It is true that neurobiology is an important gateway to the study of the brain. But brain functioning is always embedded in the functioning of the nervous system as a whole, and the nervous system can only function in its interlacements with the body. The body in turn functions in its interlacements with all other aspects of who we are—persons who are interacting with their environment. This relatedness and these interactions are not secondary but constitutive for what it is to be a brain and to function as a brain.

One cautionary remark needs to be added, however, which saves the general point but allows us to qualify this account. In the case of humans—and in fact all living beings—the functioning in some modal aspect might still be the result of the functioning of a part of the entity. Whether or not this is the case is a contingent, empirical matter. It seems relevant and adequate to observe that there are parts in the human body (such as livers, spleens, intestines, and certain metabolic processes) that function relatively independently of other parts. Dysfunctioning of these relatively autonomous parts and processes leads to biological and other symptoms that are more or less immediate expressions of that part's function. Other processes lack such relative independence and are, as it were, absorbed within the functioning of the whole. In the initial phases of collaboration between parts they may retain their relative independence. Later on in the process—for instance, under the influence of certain environmental constraints—this independence may be given up for the benefit of the system as a whole: the whole becomes not only more than its parts, but the parts are also no longer identifiable as parts because they are absorbed within the system. The brain might be a candidate for this absorbing type of part-whole relationships. Whether or not this is the case is an empirical matter, however, if one reasons along Dooyeweerdian lines.

Dooyeweerd on Laws, Order, and Transcendental Ideas

Let us return to the relationship between modal aspects and laws. The distinctness of kinds of properties, we said, is a reflection of the irreducible distinctness in the way laws determine what exists and occurs. What does this mean?

Dooyeweerd is a kind of realist with respect to laws and, consequently, order. He adheres to the transcendental view on laws according to which laws exist as conditions; this means that without laws, the things for which they hold would not exist. For something to be *transcendental* is for it to be a necessary presupposition of something else. This necessity is not only epistemic, as in the Kantian, idealistic version of transcendental philosophy, but also ontic (or cosmic—Dooyeweerd's own preferred term). Without these transcendental conditions not only logical thinking would be impossible, but also our everyday experience, and even existence itself. The existence of flowers, for instance, requires—that is, necessarily presupposes—laws or lawful principles in the spatial, kinematic, physical, and biotic spheres. In order to do justice to the pretheoretical intuition that the biotic aspect is fundamentally distinct from the spatial, kinematic, and physical aspects of the flower, we need a concept of modal distinctness that does not reduce the distinctions to cultural and/or subjective expressions (as in constructivistic epistemologies). In addition, the concept should reflect the insight that the distinctive features of the respective modal aspects cannot be objectified nor conceptually grasped. In other words, the proper meaning of *being biotic*—that is, of *biotic functioning*—can neither be reduced theoretically to notions derived from another modal aspect, nor restlessly scientifically defined. This meaning transcends the logical conception of it, so to speak.

An analogy with beauty might be helpful here. Beauty as a real-life phenomenon entails more than what is meant with the theoretical, or logical, concept of beauty. What it is for a piece of music or poem to be beautiful cannot be fully scientifically defined in the form of a precise concept of beauty, because scientific concepts can only capture the logically graspable aspects of things. Concepts are logical artifacts; the concept of beauty is a derivative of the original meaning of beauty. What beauty is, is difficult to express, even in ordinary language. Its deeper and original meaning comes to expression in the coherence of the aesthetic aspect with other aspects of reality. Beauty has to be experienced, for instance (psychic aspect)—it emanates as an enigmatic quality of our experience, which is itself based in our bodily existence (biotic aspect). The experience of beauty, in turn, is not a solipsistic event—it can be shared with other people (social aspect). It is undeniably there, and we can approach it in different ways: by undergoing the experience, by practicing the relevant form of art, by studying a piece of art, and by learning from what others say about their experiences. But, in doing all of this, we do not grasp a fixed conceptual structure behind or within the phenomenon itself. Rather, we intuit that there is something special and distinct in the phenomenon of beauty that we can approach by paying attention to the richness of the phenomenon, without ever being able to completely grasp and/or define it.

Let us now, for the sake of the argument, presume that beauty is the essential feature of the aesthetical aspect. Dooyeweerd would then contend that just as the essential feature (or *meaning kernel*, as he would say) of the aesthetic aspect cannot be grasped conceptually, theoretically, or scientifically, so, too, cannot the biotic aspect's essential feature be defined conceptually, theoretically, or scientifically. And something similar holds for all the other aspects: they have essential features that transcend theoretical conceptualization and should be thought of as presupposed (i.e., as always being already there) rather than as results of theoretical reflection. This is what Dooyeweerd has in mind when he speaks about the irreducible nature of the modal aspects. Table 1 provides the entire list of the modal spheres' essential qualities.

So far, we have connected Dooyeweerd's conception of law with his ideas about the modal aspects, especially their logical irreducibility. We can now see why this connection can be made. Rather than being a logical order, the lawful order turned out to be transcendental (i.e., necessarily presupposed). The first qualification of the laws by which the order exists is modal. Laws hold, in other words, first of all by determining *how* things exist (and only later by determining *what* exists). Transcendental conditions are special in the sense that they refer to realities of which we have a pretheoretical intuition but to which we have no immediate epistemic access. Our pretheoretical intuitions can be deepened and explicated in the form of theoretical intuitions. These theoretical intuitions are what Dooyeweerd—and Kant and the neo-Kantians before him—calls *transcendental ideas*; i.e., intuitions without which we cannot have the experience or knowledge we appear to have. Behind this looms a distinction between *concept* and *idea* which stems from the Kantian tradition. Concepts can be grasped by pure—that is, theoretical/scientific—reason; they can be defined accurately. Transcendental ideas cannot be accurately defined; they are a kind of intuitions that give a clue about the diversity (i.e., the irrevocable modal distinctness) of reality.

Before proceeding, we should first clarify that what scientists call a law of nature—or principle, lawful regularity, or lawful pattern—is not the same as the ideas of law and order on which Dooyeweerd is focusing. Scientific laws are interpretations, or approximations, of a lawful order; they are not the order itself. The terms *law* and *order* are therefore, strictly speaking, boundary concepts for theoretical thinking. Scientists and philosophers have to presuppose the existence of a lawful order in order to make sense of the regularities and causally relevant relations they discover in their sciences—relations which appear to manifest a fundamental distinctness in the ways in which things function. The presupposition that such an order exists entitles and enables them to discern the more mundane laws of their respective branches of science, Dooyeweerd suggests. The laws that the sciences discover are thus interpretations, fallible attempts to grasp the order of reality.

Let us now take one final step with respect to this notion of transcendental order. We have strong pretheoretical intuitions about the fundamental diversity of reality—a diversity which has its origin in a cosmic order that transcends our conceptual abilities and of which we have no empirical proof. In a similar fashion Dooyeweerd also speaks of other important (transcendental) features of our knowledge, experience, and existence: the *coherence*, *unity*, and *origin*, respectively, of the cosmic order we inhabit. The fundamental diversity in our experience of reality is a reflection of the fundamental distinctness of laws and of how they hold.

At the same time, however, there is also a fundamental connectedness between the things in the world and between us and the world. This experience of connectedness is a reflection of what is theoretically expressed with the term *coherence* as transcendental idea. Coherence and diversity belong together: they are two sides of the same coin and indicate the first transcendental idea. The experience of meaning—the fact that things refer to one another, together with the suggestion that their connectedness reflects a deeper wholeness and unity—is the pretheoretical precursor of the second transcendental idea, to wit, the idea of *unity* or *wholeness* (totality) of meaning. This sense of unity and wholeness requires in turn a notion of origin of meaning, according to Dooyeweerd. This is expressed in the third transcendental idea, which is the idea of an *origin* of meaning. Diversity (in conjunction with coherence), unity, and origin are the three most fundamental ideas, or transcendental presuppositions, of our experience of reality—and of reality itself, as Dooyeweerd would add.

Do We Need This Framework?

Do we really need these difficult and unpopular transcendental conceptions of law and of an underlying lawful order? Does Dooyeweerd have arguments for this? Would not a more pragmatic notion of law suffice, such as laws as models for the lawful regularities we encounter in our objects of study, or laws in the sense of the regularities, constants, and definitions scientists are working with in their everyday practices?

This is very much the topic of this book, and different authors will respond differently to this question. We will give an example of a possible response to show how Dooyeweerdian intuitions can be brought into contact with current discussions. This response says that maybe we do not need this florid notion of law in the everyday practice of laboratory research or while digesting large amounts of data, for instance, in genetic or epidemiological research. But as soon as the artificiality of the laboratory and other experimental conditions is left behind, the old and well-known questions return. How does it all fit together? What does this theory say about the object under study in its larger context? Are our simplifying models valid and if so, to what extent? Thus, coherence and wholeness will inevitably, if perhaps sometimes implicitly, be on the table, not only in science itself, but also in its applications. References to these notions emerge when scientists tell their own story of what they have been doing, or when they are educating the public. Pointing out what we know about the brain, the origin of the universe, or our genes, will inevitably bring us to a point where implicit assumptions can no longer be kept implicit.

For Dooyeweerd there are also philosophical reasons for this difficult transcendental route. The argument for cosmic order and for the modal distinctness of laws is indirect and in fact a philosophical conjecture, as he would say—a conjecture that is defensible given the unattractiveness of some of the alternatives. Dooyeweerd suggests that if we try to account for what we know without presupposing the ideas of order and law, this will sooner or later lead to inconsistency.

More precisely, without the presupposition of a (transcendental) cosmic order, our philosophical position would either become nominalistic or rationalistic. Nominalists traditionally believe that laws only exist in our minds. Rationalists think that the order of reality is intelligible and that laws can be accessed and known by (theoretical) reason. Both positions are unattractive for Dooyeweerd. He rejects nominalism because it removes the point of contact between the sciences and everyday experience. For Dooyeweerd, pretheoretical intuitions are important, not because they are always true, but because they give a clue as to the different forms of distinctness and coherence at the level of the cosmic order—a distinctness and sense of coherence that are lost as soon as the scientific attitude is adopted. From a more practical perspective, nominalism is also unattractive because of its sharp contrast with the commonsense realism of most scientists. Their hypotheses are conjectures about what they think really exists. Thus, nominalism with respect to laws would introduce a contradiction in the heart of the empirical sciences.

Dooyeweerd's rejection of rationalism is based on his objections against classical realism (as espoused by Aristotle and the scholastics). According to classical realism, we can grasp the order of reality with our intellectual faculties—the capabilities of our intellect, in other words, correspond to the intelligibility of reality. Dooyeweerd rejects this position because ultimately it absolutizes theoretical reason or, more precisely, the theoretical attitude of thought. For Dooyeweerd theoretical reasoning is always a derivative of our everyday understanding of the world. What is lost in the theoretical attitude is the "indissoluble interrelation" among the modal aspects which present themselves as completely interwoven within our everyday experience (Dooyeweerd 1953–1958, 1:3). This everyday understanding is of course less precise and more subject to error than scientific knowing, but it is characterized by a holism and sense of diversity that fades away as soon as the scientific attitude is adopted. To express this in yet another way: the experience of coherence and diversity is so fundamental that we cannot go behind it—not by reasoning or experience, not even by a scientific reconstruction of the phenomenological properties of our experience.

We have no doubt that Dooyeweerd's nomenclature will raise questions, if not eyebrows. Dooyeweerd's position seems to thrive on what some would call a certain *mysterianism*; i.e., on pretheoretical intuitions that can be made explicit only partially in the form of transcendental ideas. Let us just say that for the reader with a primary scientific or theological interest it is much more important to understand the idea that something is lost when the scientific attitude is adopted—most fundamentally, our pretheoretical sense of diversity, coherence, and wholeness—than to understand the transcendental framework in which Dooyeweerd captures and expresses these ideas. The discussion about this framework is for specialists, and it has no immediate bearing on the discussion about absolutization and reification.

Nowadays, the scientific image of the world has merged with our pretheoretical understandings in many respects. It is very much in line with Dooyeweerd's philosophy to see this as a challenge—that is, to see it as the inevitable consequence of our culture's differentiation, not as something which is in itself wrong or deplorable. It is important to recognize where scientific images are conflated with pretheoretical understanding, to investigate what this means and implies, and to evaluate such conflations in terms of their benefits and disadvantages.

Dooyeweerd on the Heart

Let us finally turn to Dooyeweerd's other core idea which concerns our functioning as humans, both individually and collectively. It is Dooyeweerd's deep conviction that all human functioning is ultimately rooted in what he calls the "heart." The heart is the concentration point of our existence. The term refers to the idea that humans are driven by fundamental concerns, motivations, commitments, and convictions. These are not only individual psychological realities, but they also have an existential and moral/spiritual core and are typically aimed at what is beyond the horizon of our knowledge and experience. The heart itself is not something that can be studied by empirical means. Rather, it is presupposed and it expresses itself in all human functioning, most clearly and explicitly in a person's worldview.

Dooyeweerd's Christian, neo-Calvinist inspiration is definitely important in these ideas of rootedness in the heart and of striving beyond the horizon of our knowledge and experience. However, Dooyeweerd goes to great lengths to support his claim that the idea of the heart as origin of an existential/religious dynamic refers to a reality that is structurally given, and is not limited to a philosophical translation of insights taken from a specific religious tradition. All human beings have an inclination to transcendence, so to speak, independent of their ethnicity or religion. This inclination is ingrained and structurally given. Philosophy cannot answer the question as to what (or who) the existential dynamic is aimed at. It can only argue that the deepest and most central human commitments are aiming at an ultimate meaning of which the source cannot be grasped or explained or immediately experienced. Dooyeweerd calls this source "origin" or "origin of meaning"; it lies beyond the horizon of experience and reflection.

The difficulty with conceptualizing the notion of the heart is that it cannot be equated with functioning in one of the modal aspects nor identified with a part of our biopsychosocial existence. It is not the same as aesthetic feeling, or moral sensitivity, or religious openness, though all of these may represent what is in our hearts. We could call the functioning of the heart a *dynamic*; a dynamic that is itself not bound to one modal aspect and resonates with a person's character, morality, ethos, and worldview. If the modal aspects are plotted on the y axis, then the activity of the heart could be plotted on the x axis. Followers of Dooyeweerd have described this as a relation between structure and direction. *Structure* refers then to modal functioning (*y* axis) and *direction* to the existential, moral, and/or religious dynamic within the person, group, or culture (*x* axis).

Dooyeweerd on Absolutization

With this in mind, we can easily understand at what point things go wrong in philosophy and the sciences, according to Dooyeweerd—namely, when philosophers or scientists ignore the transcendence of the notion of order and the transcendental nature of the ideas of diversity/coherence, unity, and origin. Ignoring this transcendence inevitably leads to the identification of some aspect of reality with what these transcendental notions stand for. This identification means that something within reality (e.g., elementary particles and the laws they are subjected to) is held to be the ultimate foundation (origin), the most unifying element (unity), and/or ultimate binding principle (coherence) of reality. Dooyeweerd calls such unjustified identification of an aspect of reality with these principles of coherence, unity, and origin *absolutization* (from the Latin *absolvere*, "to loosen" or "to set apart"). A part or aspect is set apart and treated as if it were self-sustaining (as substance), had a meaning by itself, and were the ultimate source (origin) of the world.

These absolutizations are characteristic of what he calls *immanence philosophies*—i.e., philosophies which take one element or aspect of the world as the basic "material" of or most fundamental explanatory principle for all that exists. The isms in the sciences are good examples of such absolutizations: physicalism, biologism, psychologism, and so on. Other examples, less bound to one modal aspect, are related to major themes in cosmology, epistemology, or anthropology: the absolutization of individual freedom, for instance, or the exclusive reliance on reason and scientific thinking, or utopian ideas about the malleability of the social and cultural world.

It is one of the main thrusts of Dooyeweerd's philosophy to unmask these absolutizations. It is his philosophical bet that the absolutization of an aspect or part of reality always leads to inner contradictions in one's overall conception of reality. This is because absolutization in itself already distorts reality. Distortions resulting from absolutization will inevitably lead to tensions in one's overall picture of reality and, therefore, to what Dooyeweerd calls *inner antinomies*. One-sided emphasis on scientific or technological control will, for instance, cause problems in one's concept of freedom. One-sided emphasis on human freedom is incompatible with the idea that we are in many ways determined by our biology and culture.

Something similar holds for the analysis of the Western culture. Long before Horkheimer and Adorno published their landmark study *Dialectic of the Enlightenment* (1947), Dooyeweerd had already pointed out that there is a fundamental tension between technocratic control and individual freedom—or, between "the ideal of science" and the "ideal of personhood" (i.e., being a free person). This tension, or polarity, is irresolvable without a fundamental critique of the presumed autonomy of scientific reason. Dooyeweerd speaks of an irresolvable dialectic in the ground motive of our culture. The root of this dialectic is a dogmatic adherence to the idea of autonomy of (scientific) reason. A genuinely critical philosophy will adopt a reflective attitude toward its own biases, including the bias of a one-sided scientistic view on reality.

Dooyeweerd engaged in extensive dialogue with his fellow philosophers—neo-Kantians, positivists, phenomenologists, and philosophically minded scientists alike—to show the inner antinomies in their thinking. His most important target was the absolutization of theoretical thought itself, especially in the neo-Kantian tradition. It is the presumed autonomy of theoretical reasoning, he thought, that lay at the heart of the crisis in the philosophy and culture of his days. Philosophers and scientists who accept and proclaim the idea of autonomy of theoretical reason often think that their work is critical, independent, and objective. Dooyeweerd attempts to show that, instead, their work is not critical enough. These thinkers forget that all human activity, theoretical thinking included, is always rooted in a broader conception of reality and, in fact, in life itself. They do not see that knowledge of an abstracted part of reality cannot function as the source of meaning, principle of unity, and explanation of diversity. By implicitly assuming that it can, these scholars inadvertently turn legitimate reduction into illegitimate reductionism.

Overview of the Book

As mentioned in the first section of this chapter, this volume is divided in three parts. Part I is devoted to the concept of creation order as such. Part II connects the notion of creation order with work in the special sciences and, especially, with the notion of emergence. The focus is on mathematics, physics, biology, and psychology/psychiatry. Part III investigates how the idea of creation order is conceptualized in three different theological approaches.

Eleonore Stump begins Part I with a chapter on natural law, metaphysics, and the Creator. After an exposition of the *secularist scientific picture* and Aquinas' metaphysics of natural law, she focuses on the topic of reductionism. Drawing on the work of Dupré and Hendry, she argues that there are molecules that have biotic/chemical properties which cannot be derived from the properties of the molecules' physical constituents. She endorses a form of *substance causation*; i.e., the idea that substances can have causal powers with effects on their constituent parts, in virtue of their form or organization. This idea of wholes endowed with causal capacities is also applicable in other fields, such as developmental neuroscience. One example is *joint attention*, which is the phenomenon that mother and child learn to attune to each other very early in infancy, thereby sharing each other's psychological engagement with the world. In cases such as this, the components or parts may determine *how* wholes function, but *what* the whole does is a function of the causal power had by the whole in virtue of the form or configuration of the whole. Stump closes by arguing that rejecting ontological reductionism as such does not establish the truth of theism and the theistic interpretation of (natural) law. However, reductionism does not fit well with theism. And if reductionism is false and if our ontology can include all kinds of things that can initiate causal chains, from water molecules to persons, then the motive force driving towards atheism seems considerably diminished.

The next chapter, authored by Dooyeweerd scholar *Danie Strauss*, sketches the background of Herman Dooyeweerd's conception of creation order. This conception is firmly rooted in the neo-Calvinist conception of a God whose existence and laws are beyond created reality and who manifests himself in his words and in creation—in the holding of an incredibly complex meshwork of laws and principles. Dooyeweerd rejects both the substantialization and the functionalization of these laws and principles. Modern science has adopted a functional view on laws: laws are patterns or relations that help us understand how things function; they are the expression of a certain (scientific) way of looking at things, not the expression of how things really are or are meant to be. Functionalism is historically connected with nominalism, in Strauss' historical reconstruction. Strauss sketches the enormous influence of nominalism (and its merger with rationalism) on scientific thinking and the modern worldview. He suggests that even Dooyeweerd has fallen prey to the temptations of nominalism, given his inclination to deny the universality of factual reality and his tendency to conflate the lawfulness of reality with the holding of laws. Strauss maintains that it is crucial to recognize that universality is an ontic reality, and not just a way of looking at things. Strauss does not discuss the subject, but it could be added here that Dooyeweerd would probably disagree with Stump's defense of substance causation. His systematic framework implies that causation can only be founded upon the holding of laws and lawful principles and not upon the existence of substances (wholes) per se.

Henk G. Geertsema connects the notion of creation order to its future: a promised new creation at the end of history. *Creation order* does not refer to the laws we discover by scientific research and theoretical models, but first and foremost to the concrete order that we live in, experience, and understand. We become acquainted with this order when we learn to walk, to speak, and to relate to others. Philosophy and science are concerned with certain aspects that we can analyze theoretically, but these aspects are not in themselves the order of creation. They are abstract elements of the full order of creation. The eschatological perspective sheds new light on a number of issues: on the continuity between now and then, between the current and the new creation, on the impossibility of adopting a God's eye point of view, on the importance of hope and faithful expectation, and also on the question as to whether and how a single event—to wit, Christ's death and resurrection—can have an impact with a universal meaning. The new creation is a fulfilled creation, in which all structures will be "structures of answering" within our relationship with God and other creatures. With this Geertsema builds on earlier work in which he depicts our existence as a *responding* existence—as determined by the call to respond to God's promise-command. This basic structure will not change in the new creation.

Part II starts with another chapter by *Danie Strauss*. He claims that Christian philosophy with its non-reductionist ontology has a meaningful contribution to make with respect to mathematics. His discussion focuses on the concept of infinity, especially the distinction between the *successive* (or potential) infinite and the *at once* (or actual) infinite. Key to the notion of the successive infinite is the idea of a sequence of rational numbers that converge to a certain limit (1/2, 2/3, 3/4, 4/5, and so on, converging to 1). This approach to the notion of limit (and infinity) is based on the geometrization of mathematical relations. The at once (or actual) infinite is based on the concept of a purely arithmetical continuum of points. The continuum is seen as an infinite totality of non-space-occupying, purely arithmetical points.

The history of mathematics shows that both approaches run into difficulties. Strauss argues that in order to avoid the one-sidedness of arithmeticism—which overemphasizes number—and geometricism—which overemphasizes spatial continuity—mathematics should acknowledge both the uniqueness of and mutual coherence between number and space. This is possible with the systematic philosophy of Herman Dooyeweerd, which construes the *numerical mode* as determined by distinctness (discreteness) and (order of) succession, and *continuity* as a spatial concept entailing both simultaneity (an order at once) and the notion of wholeness, or totality. The idea of an infinite sequence of numbers points to both succession and to wholeness/simultaneity. It is, in other words, based on a spatial deepening of the primitive numerical meaning of *infinity* (of succession) toward the idea of an infinite totality. Based on these Dooyeweerdian notions, Strauss criticizes attempts to ground the idea of the infinite in Christian theology—i.e., in the infinity (omnipresence, eternal existence) of a God who transgresses our conceptual understanding in every respect. For Strauss, it should be the other way around. Mathematics should not derive its basic concepts from theo-ontological speculations; rather, theology should inform itself about basic concepts within the sciences and then formulate its own boundary concepts. This position is in very sharp contrast to some of the views that will be discussed in the theological chapters in Part III of this volume.

Next, *Marinus Dirk Stafleu* explores how to make sense of the notion of emergence within the context of an (adapted) Dooyeweerdian systematic philosophical framework. He begins by making a distinction between emergence within, emergence of, and emergence from the physical world. The latter type of emergence is, obviously, the most difficult to explain. Stafleu defends the view that laws for the different modal spheres are God-given and preexistent. But this does not imply that the emergence of life out of the physical world can be explained by the presence of the biotic and higher modal spheres. Stafleu develops a view in which analogical anticipations within the physical sphere toward the biotic and higher spheres give rise to—still physical—propensities that under certain very special circumstances may lead to the emergence of entities that also have biotic features. Over the course of history, DNA molecules (physical sphere) began to gain (self-)replicating properties that turned out to become a condition for genetic relationships. On this view, the order of creation is, and remains, the same on the law-side of the created world, in spite of the impressive developments and catastrophes that have taken place during the astrophysical and biological evolution of the universe. Such developments occur at the subject-side of reality, which is the side that is characterized by subject-subject and subject-object relations. By drawing a distinction between a philosophical conception of an order of laws and law-spheres, on the one hand, and a scientific approach to developments at the subject-side of reality, on the other hand, Stafleu can maintain the classical notion of creation order—at least, one particular version of it—and do justice to developments in the sciences, especially physics and biology. It appears again—as in the previous chapter on mathematics—that the Dooyeweerdian framework offers helpful suggestions for a better understanding of basic concepts in the sciences, and also for the translation of scientific findings to broader audiences.

Arnold E. Sikkema continues this discussion by exploring emergentist claims in the context of physics. This chapter is clearly about emergence *within* a particular field of science. After having pointed out that the notion of emergence is not at all clear in itself, he discusses a variety of examples of emergence within physics: the forming of crystalline structures, the coming into existence of correlated electron systems, the Rayleigh-Bénard convection cells, and others. He outlines how reformational philosophical concepts such as idionomy, encapsis, and anticipation can help make sense of these physical phenomena. Sikkema connects the concept of idionomy to the notions of underivability, unpredictability, and (even) contingency. He associates encapsis with the phenomenon of synchronic emergence, whereas he likens the idea of anticipation to the state of certain molecules which are predisposed, or prepared, to evolve into new, emergent states.

Denis Alexander's contribution shifts the attention toward biology. He discusses two large topics; to wit, *progress* and *purpose*. He describes the long history of evolutionary thinking and its adumbration of the idea of evolution as inherently progressive—an idea that has proven to be persistent and that recurs even in the work of Richard Dawkins. In the light of this history, the idea of evolution as a random, blind, contingent process, with humanity as an utterly unlikely, cosmic accident, is a fairly recent development. Alexander thinks that theology does not commit Christians to any particular theory about progress in evolutionary biology, except in the rather weak sense that God fulfills his intentions and purposes through evolutionary processes. God can bring about these intentions and purposes even in a contingent universe. But what does this mean with respect to purpose? Here the tensions seem more apparent, at first sight. Purpose seems self-evident to the Christian; it is an important element in all mainstream Christian theology. But to the atheist, it is not self-evident at all: without a God, and looking through the window of biology alone, there is nothing that forces the atheist to adopt a narrative of ultimate purpose. So, on the one hand, there is no evolutionary theory that allows us to derive a theology of purpose from it. But, on the other hand, there is also no variant of evolutionary theory that necessitates us to accept the idea of a universe without a plan or purpose.

The discussion seems undecided, at least at the most fundamental level. Alexander nevertheless suggests that our current understanding of biology offers a number of clues that make it likely that there are more law-like patterns, uniform principles, and converging trends in the evolutionary process than are compatible with the idea of a totally random, algorithmic process of natural selection. He mentions seven of these clues which give the impression that evolution occurs in a way that is more organized and constrained than mainstream biology has traditionally suggested. This organization and these constraints even lead to a certain degree of predictability of evolutionary processes. The idea of a highly organized and constrained evolutionary history is consistent with the theological claim that there is a God who has intentions and purposes for the world in general and for us in particular.

Emergence is the main topic of *Jitse M. van der Meer*'s extensive chapter. The field under study is mainly biology, especially the emergence of life out of the physical world. Van der Meer deviates considerably from the position that Stafleu develops in his chapter. According to van der Meer, emergence is a real thing—it is, therefore, crucial to conceive of it as a causal process. We should look for the source of causality in the material world and not in a preexisting order of laws and lawful principles. Laws are equated with lawfulness and the lawful functioning of objects and processes. In the course of evolution new structures with their own novel lawfulness emerge bottom-up, as it were. It makes sense to relate the new orderings to philosophical frameworks such as the Dooyeweerdian doctrine of modal aspects with their different kinds of laws. But philosophy should not stand in the way of empirical science by ruling out the emergence of new structures without the help of preexistent idionomic principles. We may speak about lawful structures that emerge, but only *post hoc*, so to say. These structures are neither transcendentally nor religiously foreshadowed (e.g., in an idea of creation order containing the seeds of what will finally emerge) in the empirical world. God created the world, according to Scripture, and it makes sense to speak of divine decrees. But this worldview language should not be mixed with philosophical or scientific language. The term *creation order* may still be used—not in the sense of a pregiven order, but as an order that gradually unfolds during the process of evolution.

There is no place in van der Meer's account for the idea that laws hold. This explains why causality becomes so crucial for the understanding of emergence, and also why the law–subject distinction fades away. Laws in the Dooyeweerdian sense are *abstract objects*, according to van der Meer, and abstract objects have no causal power. Attributing more to these abstract objects than their existence in the scientist's mind leads to essentialism. Dooyeweerd's theory of cosmic time order is therefore a form of essentialism.

Whatever one might think of this position, van der Meer is right in asserting that it is deeply problematic when philosophers—on whatever grounds—deny a priori that new structures (and even species) can emerge from existing ones. True, Dooyeweerd's own interpretation of type laws seems to exclude such emergence. However, the chapters by Stafleu and Glas suggest that one does not need to give up the Dooyeweerdian framework to do justice to emergent phenomena. Van der Meer argues for a broader concept of causality, a new conception of type laws (allowing for causal isolation of parts as a condition for the emergence of new structures), and a rejection of the difference between naïve experience and theory.

Gerrit Glas' chapter continues the discussion of emergence against the background of the sciences of the person—most notably, neuroscience and psychology. Three questions are central in Glas' contribution: (1) Does it make a difference for the sciences of the person to maintain a strong notion of law (*strong* in the sense that laws are considered as preexistent and necessary)? (2) Can the apparent tension between the creation order view and evolutionary accounts of lawfulness and order be diminished by employing the concept of emergence? (3) Can the concept of emergence be made compatible with a strong concept of law? Glas' answer to the first two questions is yes, and his answer to the third is a conditional yes. He argues that, given the slipperiness of the concept of emergence, it is best to take *emergence* in a primarily heuristic and paradigmatic sense; i.e., as a boundary concept at the background of a broad research program. With this he draws a line between emergentism as a heuristic paradigm and emergentism as an implicit ontology which emphasizes lawfulness as "caused" by bottom-up processes (a position that Glas rejects). Something similar holds for the strong view on laws. This view, too, can best be considered a philosophical working hypothesis that may turn out to need adaptations in the form of auxiliary hypotheses. Especially the necessity claim might need adaptations from a worldview perspective. Nevertheless, Glas defends a position in which the notion of holding is retained and taken in a realist sense—in other words, as belonging to reality. There is lawfulness of and within reality which reflects the holding of an order that is in some way related to the intentions of a Creator. This position maintains the law–subject distinction; it allows emergence, including the emergence of new orderings; and it makes firms distinctions between science, philosophy, and worldview.

The chapter by *Lydia Jaeger* offers a fine commentary on the chapters on emergence, especially Glas' chapter on neuroscience and psychology. It also gives a look behind the screens of theology and philosophy of religion with respect to creation order, providing a smooth transition to the third part of the book, which treats the theology of creation order. Jaeger draws on Dooyeweerdian insights to criticize two types of emergentism that are also discussed in Glas' chapter: a non-reductive physicalist variant (with Jaegwon Kim and Philip Clayton as proponents) and a dynamical systems theory variant (with Francisco Varela, Michel Bitbol, and Evan Thompson as representatives). Her criticism on Kim and Clayton runs parallel with Glas' critique, but she also discusses the religious, especially Buddhist, background of the dynamical systems theory of people such as Varela, Thompson, and Bitbol.

The contrast guiding her investigation is one between emptiness and substance. Jaeger focuses on Thompson's account of dynamic co-emergence. His refusal to identify a base level from which new properties emerge—as in physicalist accounts of emergence—and the general difficulty of identifying a substrate (i.e., the ground from which dynamical co-emergence is emerging) are direct consequences of the fundamental role of the Buddhist notion of emptiness. Everything refers circularly to everything, with an empty hub in the middle, which implies that our thinking, theoretical or otherwise, cannot gain a firm conceptual foothold.

This is different in the creation view, which postulates the existence of a Creator and of products of his hands. A substance view seems most appropriate to account for the relative independence and the reality of these products, on the provision that this substantiality is completed with a notion of personhood, which is guaranteed by the biblical notion of man as image of God. *Personhood* seems a spiritual notion on Jaeger's account, in the sense that it can be seen as an expression of the divine *Logos*, which permeates created reality and serves as life-saving and life-bringing connection with God as origin. Recognizing Dooyeweerd's motives for doing so, Jaeger nevertheless rejects his criticism of the metaphysics of substances and of the so-called *logos* speculation. To be sure, substances are not things-in-themselves and the *logos* speculation has Platonic and rationalistic features. But these imperfections can be remedied. We are not the first to try this. As an example, consider the balance between nature and personhood, which is so important for anthropology. This balance was already a widely discussed and crucial subject for the church fathers in their painstaking search for the right formulation of the existence of a God whose divine being is one in nature (substance) and, at the same time, consists of three persons—Father, Son, and Holy Spirit (personhood).

The first chapter in Part III by *Nicholas Ansell* offers a biblical-theological evaluation of creation order thinking from the perspective of the wisdom literature in the Old Testament. According to creation order thinking, right living requires our lives to be aligned to the God-given structure of existence. Life is thus conceived of as a *going with*, rather than a *going against*, what some have called "the grain of the cosmos." This conception is widely believed to be grounded in, and supported by, the wisdom literature of the Old Testament. However, building on the work of Roland Murphy and others, Ansell argues against this view. What Scripture means by *wisdom* is best interpreted not as conformity to a (hidden) normative order, but as a way that consists of the life-giving interplay between God's blessing and creation's participation in that blessing. Based on a careful reading of Proverbs, especially chapter 30, Ansell concludes that a creation order reading tends to obscure certain facets of meaning and experience—facets that point to a mystery-affirming appreciation of creation. Proverbs 30:19 asks us, after all, not to turn our gaze to the "fixed order" of the stars (Jer. 31:35), but to the singular, unrepeatable path of an eagle and the subtle and supple way of the serpent. Mystery involves much more than hidden order. By being attuned to the original blessing with which the biblical narrative begins, Ansell concludes, the mystery-affirming appreciation of creation does not lead to an *anti*-nomian eradication of order; rather, it leads to its *ante*-nomian relativization. This is another way of saying that blessing, historically and systematically, precedes order.

Hans Schaeffer strikes a similar note in his contribution on the concept of creation order from a Lutheran perspective. According to Luther, God's work in creation can be divided in three estates or "hierarchies": *ecclesia*, *oeconomia*, and *politia*. Church (*ecclesia*) is the primal relationship between human beings as creatures to God the Creator. Economy (*oeconomia*) denotes everything which in current society is differentiated as marriage, family, economy, education, and science. The third is the state (*politia*), which is protecting us from chaos and shapes human life by laws and regulations. These estates or hierarchies do not refer to pre-established fixed orders, but should be seen as God's address to us by which he upholds (institutes) his relationship to us and to the world. The role of this doctrine of estates is mainly heuristic, pointing at what theologian Bernd Wannenwetsch calls "life-forms" or "life-giving forms." God's words open spheres of human life in which humans, living in a sinful context, are called to respond to his call. These spheres are the means that God provides for the sanctification of our lives. It should be noted that the description of this call looks very similar to Geertsema's reinterpretation of Dooyeweerd's law-subject relation as a relation between God's promise-command and human responding.

In the second part of the chapter Schaeffer discusses Lutheran criticisms on classical neo-Calvinist approaches to creation and creation order. The argument largely parallels well-known (older) criticisms within reformational philosophical circles. The Reformed view on creation order is seen as depending on the idea of a fixed order; as an exclusively backward-looking instead of redemption-oriented, eschatological doctrine; as leading to essentialism; and as being in support of moral conservatism and the ruling class. The Lutheran perspective is then depicted as more flexible and as leaving more room for hamartiological, redemptive, and eschatological perspectives on human existence. At the end of the chapter Schaeffer mentions some topics for further discussion, such as how the notion of vocation is related to a conceptual framework that distinguishes between normatively distinct spheres of human responding; how Dooyeweerd's transcendental critique can be brought into contact with Lutheran theology; and how the notion of hope can inform discussions about the future of creation order.

In the work of Dietrich Bonhoeffer, we encounter an even more radical critique on creation order theology. *Annette Mosher* describes how Bonhoeffer, already before 1933, began to criticize theologians such as Althaus and Hirsch, who defended a *volk* theology with a vocabulary derived from the doctrine of creation order. It is in the history and spirit of what is called the *volk* (a mixture of ethnicity, nationality, and character, together with a sense of being entitled to a certain historical role) that creation ordinances become manifest, according to these theologians. The human person becomes a person as a result of his/her relation to the *volk*. Bonhoeffer left no room for such *volk* theology outside Christ. Bonhoeffer accepted Luther's two kingdom theology along with the creation order theologians, but gave it a radically different theological interpretation. Community with God exists only through Christ and Christ is present only in his church-community. This is another version of the idea that it is only possible to gain access to the understanding of creation order through the cross and the church.

Creation order theology looks toward the beginning, but Bonhoeffer argues that we can never know the beginning—it is an infinite question that cannot be answered. Trying to find the ordinances of the beginning is looking for the old things of the world instead of the new world that the church finds in Christ. It is enthroning reason in the place of God. Bonhoeffer argues that our focus should be on the reality of Christ in the present, which is in the middle between the past and the future. There is no revelation in history or in nature—only in Christ. These ideas are further elaborated in Bonhoeffer's *Ethics*, where the order of preservation is transformed into a doctrine of four mandates. The mandates are work, marriage, government, and the church. They are not so much spheres of life on their own as they are gifts and duties that should be directed towards Christ. Work is serving and glorifying Christ by participating in the world. Marriage exemplifies the life-bringing union between Christ and the church. The government is given to sustain what exists; not as a final goal and end in itself, but as a way to preserve the functioning of all that is reality in

Christ. True community, finally, is not a nationalistic division, but is found in the church which embraces and envelops all of humanity.

The final chapter by *Josephien van Kessel* is devoted to the Sophiology of Sergei Bulgakov. In the work of this Eastern Orthodox thinker and priest we return to a version of the *logos* speculation discussed in Lydia Jaeger's chapter. *Sophia* is a difficult concept to understand. It refers, of course, to wisdom—God's wisdom—but also to God's love and providence. At the same time, wisdom forms the hidden order of creation. This order is not the expression of divine wisdom, nor is wisdom in the human mind a reflection of the hidden order of creation. The order is Sophia and Sophia is order—not as a scholastic substance or a cosmic blueprint in God's mind, but as an embodiment of contrasts, as an in-between; between transcendence and immanence, between the divine and the human world, and between the one and the manifold. A Dooyeweerdian thinker would call it a boundary concept, and Bulgakov in fact uses such terms as *boundary* and *border* to indicate Sophia's nature. But Sophia also represents the connection and the "between" between immanence and transcendence. This "between" is an antinomy that helps us escape from what Bulgakov calls *immanentism*. Sophia is not an abstraction but a reality, a living reality or living being that directs our spiritual attention beyond immanent concerns. As such, Sophia is marked by countenance and personhood. It is a kind of hypostasis; not as a separate substance next to God as Father, Son, and Holy Spirit, but as God's nature, beyond time and space, and yet also as the root of the world and of human existence—and therefore fully immanent.

There is a cosmological and a theological approach to Sophia. Human knowledge, and especially science, is limited in the sense that it is rationalistically and unilaterally oriented toward this world. Bulgakov's quest for a religious revival aims at reaching deeper—existentially as well as ontologically—than traditional, rationalistic Western philosophies. It is true that Bulgakov is inspired by Hegel's dialectic of stages of consciousness in religion, art, and philosophy. But his ultimate aims are broader and more religious when he describes a hierarchy of possibilities to experience and to express the absolute through successive levels of myth making. Bulgakov's thoughts are difficult to grasp and somewhat remote from neo-Calvinist philosophy, yet there are more than superficial similarities; for instance, between Sophia and Dooyeweerd's prism metaphor as an expression of the relation between the supra-temporal origin of meaning and the many temporal manifestations of this origin; and between Sophia and the primacy of blessing over rationally understandable order, as described in Ansell's account.

Conclusion

This is not the place to draw firm conclusions. There are, however, a few strands of thought which deserve to be mentioned and which offer hints as to where we are and how the discussion is moving forward.

Within reformational philosophical circles there are divergent interpretations of cosmic order: from classic defenses of the distinction between law and subject (Strauss) to the view that laws are just abstract objects (van der Meer). Nevertheless, the reformational philosophers who contribute to this volume tend to stress the distinction between worldview, philosophy, and science—even more than Dooyeweerd. As a result there is a tendency

- to put less emphasis on the idea of pregivenness and the necessity of cosmic order;
- to highlight the association between order and the trustworthiness, depth, beauty, and wisdom of God and his concern with the work of his hands (Geertsema);
- to locate the concept of creation order in the sphere of worldviews and not (or not primarily) in the sphere of philosophy or science (Geertsema; Glas);
- to create considerably more conceptual distance between scientific formulations of laws of nature, on the one hand, and the philosophical (transcendental) notion of law, on the other hand, and to see the philosophical notion of law primarily as something we need in order to make sense of the notion of the holding of laws in science and in everyday life (Stafleu; Sikkema); and
- to recognize a variety of ways in which laws hold, which leads, among other things, to an appreciation of the importance of dispositional approaches to lawfulness (van der Meer; Jaeger; Glas).

Another strand of thought focuses on the concept of emergence and its possible role in counteracting reductionism. *Creation order* is often seen as an important concept in this context, also by those contributors who do not primarily associate their philosophies with reformational philosophy (Stump; Jaeger). However, there appear to be several notions of emergence (Stafleu; van der Meer; Jaeger) and no consensus about the relation between law, causality, and emergence. A possible way forward may be to think of *emergence* as a heuristic, philosophical boundary concept within the sciences; i.e., as a research program, rather than a panacea against reductionism (Glas).

Alexander's contribution merits special mention in this context, because he does not seem to need the concept of emergence. This is, Alexander suggests, because there are sufficient arguments within biology itself to counter atheism and reductionism. This point is especially interesting in light of a suggestion made earlier in this chapter, namely, that evolutionary theory is based on the assumption that evolutionary—and cosmic—processes are fundamentally contingent. Alexander's account suggests that contingency is an empirical rather than a philosophical presumption, and that there are empirical grounds for thinking that the evolutionary process has a direction.

In the theological chapters one can discern a strong tendency to reject the idea of creation order as a fixed and necessary order, existing from the beginning of the universe. The reasons for this rejection are that

- creation order turns our attention in the wrong direction—i.e., toward the beginning instead of toward the end (Geertsema; Ansell; Schaeffer; Mosher);

- it stresses static order rather than development and dynamism, which should be connected with the theological notions of sin and the need for salvation (Schaeffer; Mosher); and
- it is strongly associated with rationalism, essentialism, moral conservatism, and false (or even idolatrous) ideologies (Ansell).

It should be obvious that the creation order view that most theologians object to is not the view that the majority of reformational philosophers defend. We hear in the theological accounts a strong rejection of rationalistic approaches to creation order and a desire to do justice to the reality and transforming power of God's grace and blessing (Ansell; Van Kessel)—a desire, also, to connect with the very fact of our createdness by connecting with the life and work of Jesus Christ. These desires are obviously legitimate and—probably—not discussed enough in reformational philosophical circles. They indicate the need for further discussion between theologians and philosophers on the intriguing subject of creation order and its future.

References

Armstrong, David M. 1983. *What Is a Law of Nature?* Cambridge: Cambridge University Press.
Barbour, Ian G. 1997. *Religion and Science. Historical and Contemporary Issues*. Revised and expanded edition of *Religion in an Age of Science*. New York: HarperCollins.
Chaplin, Jonathan. 2011. *Herman Dooyeweerd: Christian Philosopher of State and Civil Society*. Notre Dame: University of Notre Dame Press.
Clayton, Philip. 2008. Contemporary Philosophical Concepts of Laws of Nature: The Quest for Broad Explanatory Consonance. In *Creation: Law and Probability*, ed. Fraser Watts, 37–58. Aldershot: Fortress.
Dooyeweerd, Herman. 1953–1958. *A New Critique of Theoretical Thought*. 4 vols. Amsterdam/Philadelphia/Paris: Presbyterian and Reformed Publishing Company.
Douma, Jochem. 1976. *Kritische aantekeningen bij de wijsbegeerte der wetsidee*. Groningen: De Vuurbaak.
Echeverria, Eduardo. 2011. Review Essay: The Philosophical Foundations of Bavinck and Dooyeweerd. *Journal of Markets & Morality* 14 (2): 463–483.
Harrison, Peter. 1998. *The Bible, Protestantism, and the Rise of Natural Science*. Cambridge: Cambridge University Press.
———. 2008. The Development of the Concept of Laws of Nature. In *Creation: Law and Probability*, ed. Fraser Watts, 13–36. Aldershot: Fortress.
Henderson, Roger D. 1994. *Illuminating Law: The Construction of Herman Dooyeweerd's Philosophy 1918–1928*. Amsterdam: Buijten & Schipperheijn.
Horkheimer, Max, and Theodor W. Adorno. 1947. *Dialektik der Aufklärung*, exp. ed. Amsterdam: Querido.
Kalsbeek, Leendert. 1970. *De wijsbegeerte der wetsidee. Proeve van een christelijke filosofie*. Amsterdam: Buijten & Schipperheijn.
Klapwijk, Jacob. 2008. *Purpose in the Living World? Creation and Emergent Evolution*. Translated and edited by Harry Cook. Cambridge: Cambridge University Press.
Milbank, John. 2006. *Theology and Social Theory*, 2nd ed. Oxford: Blackwell.
Milbank, John, Catherine Pickstock, and Graham Ward, eds. 1999. *Radical Orthodoxy: A New Theology*. London: Routledge.

Moreland, James Porter, and Scott B. Rae. 2000. *Body and Soul. Human Nature and the Crisis in Ethics*. Downers Grove: InterVarsity Press.

Plantinga, Alvin. 2000. *Warranted Christian Belief*. Oxford: Oxford University Press.

Spengler, Oswald. 1918–1923. *Der Untergang des Abendlandes. Umrisse einer Morphologie der Weltgeschichte*. 2 vols. Munich: Oskar Beck.

Strauss, Danie. 2009. *Philosophy: Discipline of the Disciplines*. Grand Rapids: Paideia Press.

Van der Hoeven, Johan. 1981. Wetten en feiten. De "wijsbegeerte der wetsidee" temidden van hedendaagse bezinning op dit thema. In *Wetenschap, wijsheid, filosoferen. Opstellen aangeboden aan Hendrik van Riessen bij zijn afscheid als hoogleraar in de wijsbegeerte aan de Vrije Universiteit te Amsterdam*, ed. P. Blokhuis, B. Kee, J.H. Santema, and E. Schuurman, 99–122. Assen: Van Gorcum.

———. 1986. Na 50 jaar: Philosophia Reformata—Philosophia Reformanda. *Philosophia Reformata* 51: 5–28.

Van Dunné, Jan Meinardus, Peter Boeles, and Arend Jan Heerma van Voss. 1977. *Acht civilisten in burger*. Zwolle: Tjeenk Willink.

Wolters, Albert M. 1985. The Intellectual Milieu of Herman Dooyeweerd. In *The Legacy of Herman Dooyeweerd: Reflections on Critical Philosophy in the Christian Tradition*, ed. C.T. McIntire, 1–19. Lanham: University Press of America.

Wolterstorff, Nicholas. 1995. *Divine Discourse: Philosophical Reflections on the Claim that God Speaks*. Cambridge: Cambridge University Press.

———. 2008. *Justice: Rights and Wrongs*. Princeton: Princeton University Press.

Part I
The Concept of Creation Order

Natural Law, Metaphysics, and the Creator

Eleonore Stump

Abstract In this paper, I contrast the notion of natural law on a secularist scientific picture with the notion of natural law in the thought of Thomas Aquinas. I show the way in which the highly various metaphysics of the two worldviews give rise to such divergent notions. In this connection, I look at contemporary arguments against reductionism in the sciences and in recent metaphysics. I argue that this new antireductionist approach sits more easily with the Thomistic worldview than with the secularist scientific view.

Keywords Reductionism · Thomas Aquinas · Natural law · Metaphysics · Emergence · Joint attention

Natural Law as the Laws of Physics

Trying to summarize the view of the world given by the secularist appropriation of science now common in Western culture, Simon Blackburn describes things this way:

> The cosmos is some fifteen billion years old, almost unimaginably huge, and governed by natural laws that will compel its extinction in some billions more years, although long before that the Earth and the solar system will have been destroyed by the heat death of the sun. Human beings occupy an infinitesimally small fraction of space and time, on the edge of one galaxy among a hundred thousand million or so galaxies. We evolved only because of a number of cosmic accidents.... Nature shows us no particular favors: we get parasites and diseases and we die, and we are not all that nice to each other. True, we are moderately clever, but our efforts to use our intelligence ... quite often backfire.... That, more or less, is the scientific picture of the world. (Blackburn 2002, 29)

I will call a view such as this 'the secularist scientific picture' (SSP, for short), to distinguish it from a mere summary of contemporary scientific data. It remains a

---

E. Stump (✉)
Saint Louis University, St. Louis, MO, USA
e-mail: eleonore.stump@slu.edu

widely held picture of the world, even though, as I will show in what follows, research in various areas is making inroads against some parts of this view.

On SSP, as I will understand it for purposes of this paper, the natural laws Blackburn refers to are typically taken to be the laws of physics. Some versions of SSP have included theoretical reductionism, although this view is now less in favor. On theoretical reductionism, all sciences reduce to physics, and all other scientific laws are reducible to the natural laws of physics. A more common version of SSP includes ontological reductionism, which still has widespread adherence.[1] On ontological reductionism, all things in the world are thought to be reducible to the fundamental units of matter postulated by physics and governed by the natural laws of physics. Even if the *laws* of all other sciences cannot be reduced to the laws of physics, all *things* can be reduced to the elementary particles postulated by the laws of the ultimately correct theory of physics; and everything that happens is the result of the law-governed causal interactions of these elementary particles.

One important presupposition commonly held by adherents of SSP is a metaphysical rather than a scientific principle, namely, that constitution is identity. On this presupposition, for anything made of parts, that thing is identical to the parts that are its constituents. There is nothing to a whole other than the sum of its parts. And, of course, the same holds for each of the parts. Each part is also nothing more than the sum of *its* parts, and so on down to the most fundamental level. Ultimately, everything is identical to the most fundamental parts that constitute it. On SSP, these are the elementary particles governed by the natural laws of physics. Theories that accept the principle that constitution is identity incorporate an ontological reductionism. As Robin Findlay Hendry puts it, "the reductionist slogan is that x is reducible to y just in case x is 'nothing but' its reduction base, y" (Hendry 2010, 209).[2]

The appeal of reductionism was greatly enhanced by scientific developments in the twentieth century, especially in molecular biology and genetics. Describing the growth in adherence to reductionism in consequence of these scientific developments, Cynthia and Graham Macdonald say,

> The use of chemical theory in all these developments [in biology was] ... crucial, suggesting that biology was reducible to chemistry and thereby to physics, given that the reducibility of chemistry to physics was thought to have been demonstrated by the physical explanation of chemical bonding. The major trend in all of this scientific work was to explain processes at the macro-level by discovering more of the detail of microprocesses. Reductionism looked to be an eminently suitable research strategy. (Macdonald and Macdonald 2010, 3)

Reductionism is often thought to rest on another metaphysical claim as well, namely, the claim that there is causal closure at the level of physics. On this claim, apart from quantum indeterminacy, there is a complete causal story to be told about everything that happens; and that complete causal story takes place at the level of the elementary particles described by physics. On the view of natural laws in SSP,

---

[1] Adherents of ontological reductionism also include some theists, as I will explain below in the discussion of the position held by Peter van Inwagen.

[2] For a fuller discussion of these issues, see also Hendry (2012).

then, any causality found at the macro-level is just a function of the causality at the micro-level of physics. Because there is causal closure at the lowest level, the causal interactions among the fundamental particles of a thing are not open to interference by anything which is not itself at the most fundamental level and governed by the natural laws operating on that level. And everything that happens at any higher level, from the chemical to the psychological, happens as it does just because of the causal interactions among the fundamental physical particles involved.

So, for example, any act of a human being is a function of events at the level of bodily organs and tissues; these are a function of events at the level of cells; these are a function of events at the level of molecules; these are a function of events at the level of atoms—and so on down to the lowest level, at which there are the causal interactions among the elementary particles postulated by physics and governed by the natural laws of physics. The causal interactions of things at this lowest level thus account for everything else that happens, including those things human beings do.

Or, to put the point of this example in a more provocative way, on SSP love and fidelity, creativity, the very achievements of science, and any other thing that makes human life admirable or desirable is itself just the result of the causal interactions of elementary particles in accordance with the natural laws of physics. *Every* human act is determined by the causal interaction of elementary particles governed by these natural laws; and even the belief that SSP is correct is so determined. SSP has so strong a hold on some contemporary philosophers that they see no alternative to holding that all mental states are causally determined by the physical states of the brain, which are in turn causally determined by causal interactions at the lowest level.

To philosophers in the grip of SSP, libertarian free will can seem impossible. Since neural states are part of a causal chain that is determined by causal interactions at the level of the microphysical and there is causal closure at that lowest level, not only states of the will but in fact all mental states, considered as mental, seem causally inert. Cynthia and Graham Macdonald summarize this position this way:

> The physicalist is thought to be committed to the "basic" or fundamental character of the physical, and an expression of this is contained in the assumption that the physical domain is causally closed: any event that has a cause has a complete (sufficient) physical cause. The thought that a mental event (or a mental property) could cause an effect without relying on, or working through, physical events (or properties) was rightly deemed inimical to physicalism.... If the higher-level property's causal power is constituted by the contribution from the lower-level–realizing properties, then it is difficult to see how its causal power could fail to be exhausted by that contribution—it seems that it will contribute nothing of its own to the effects it is said to cause. (Macdonald and Macdonald 2010, 8 and 12)

It is not surprising, then, that as regards freedom of the will, philosophers who accept something like SSP tend also to accept compatibilism, the theory that the will of a human person can be both free and also causally determined. Compatibilism appears to be a sort of corollary to the scientific picture that embraces reductionism and causal closure at the microphysical level. If all macrophenomena are reducible to microstructural phenomena and if there is a complete causal story to be told at the micro-level, then whatever control or freedom we have as macroscopic agents has to be not only compatible with but in fact just is a function of the complete causal story at the micro-level.

For many people, me included, the implications of SSP seem highly counterintuitive. Is every macroscopic thing reducible to the sum of its elementary particles? Is everything that happens really completely determined by causal interactions at the microphysical level? Could it really be the case that the mental states of a person are causally inert as far as his own actions are concerned? Could an act of will really be both free and yet also causally determined?

Natural Law in the Thought of Thomas Aquinas

It is instructive to reflect on SSP by contrasting it with the very different view of the world held by the medieval philosopher Thomas Aquinas. Aquinas talks of natural law, too; but, as is well known, the notion of natural law in the thought of Aquinas is nothing like the notion of natural law in SSP.[3] With respect to the notion of natural law in Aquinas's thought, human persons and human agency are not rendered marginal or even invisible, as they seem to be in SSP. On the contrary, they are at the center of the discussion.

Aquinas's notion of natural law has been the subject of extensive discussion,[4] and different characterizations of it have been given. Sometimes natural law is described as if, for Aquinas, it were a matter of innate and incorruptible knowledge of moral truths.[5] Sometimes it is characterized more as a set of moral principles, or a set of some especially fundamental sort of moral truths.[6] And sometimes it is categorized as a matter of metaphysics, as something metaphysical that grounds morality.[7] Aquinas's characterization of natural law is complicated enough to provide some justification for all these different descriptions.

When Aquinas explains his notion of natural law in his *Summa Theologiae* (*ST*), he says that the *natural* law is a participation on the part of a human person in the *eternal* law in the mind of God (*ST* IaIIae.91.2).[8] And, when he explains the *eternal* law, he says that it is the ordering of all created things as that ordering is determined in the mind and will of the Creator (*ST* IaIIae.91.1). Clearly, this ordering will

---

[3] For more discussion of Aquinas's notion of natural law and its place within Aquinas's metaethics and normative ethics, see the chapter on goodness in Stump (2003).

[4] For one recent and sophisticated treatment of the topic, see Murphy (2001).

[5] Ralph McInerny explains it this way: "Natural law is reason's natural grasp of certain common principles which should direct our acts" (McInerny 1992, 110).

[6] McInerny himself says, "natural law is a dictate of reason"; and a little later he remarks that there is a way in which "natural law is a claim that there are moral absolutes" (McInerny 1997, 46 and 47).

[7] For example, in a discussion of the relation of rights and law, John Finnis says, "if I have a natural—as we would say, human—right I have it by virtue of natural law" (Finnis 1998, 135).

[8] The translations in this paper are mine; but, in translating *ST*, I am often guided by the translation of the Fathers of the English Dominican Province. Their choice of English terms for Latin technical vocabulary has become standard by now, and their ability to render Aquinas's scholastic Latin into intelligible English is enviable.

include what contemporary philosophers think of as natural laws as well as the moral truths that are the focus for Aquinas. For a created person to participate in the eternal law of God is for that person to have a mind and will which reflect their origin in the Creator: the natural law in created human persons is an analogue of the eternal law in the Creator.

The ordering of creation in the eternal law includes all the organization of the created world; but it is the moral ordering in the eternal law that is at the heart of the natural law. A human person's intellect has enough of the natural light of reason to be able to discern what is good and what is evil, and his will has some natural inclination to follow reason's light (*ST* IaIIae.91.2). The natural law in a created person is therefore a participation in the Creator in two ways: first by way of knowledge about good and evil in the intellect and secondly by way of an inward principle in the will that moves to action in accordance with the deliverances of the intellect.[9]

So, for example, Aquinas says,

> all acts of virtue pertain to the natural law ... for everything to which a human being is inclined in accordance with his nature pertains to the natural law. Now everything is naturally inclined to an operation appropriate to it in accordance with its form.... And so since the rational soul is the proper form of a human being, there is in every human being a natural inclination to act in accordance with reason. But this is to act in accordance with virtue. (*ST* IaIIae.94.3)

And he goes on to explain,

> if we are talking about virtuous acts in themselves, that is, insofar as they are considered in their proper species, then in this way not all virtuous acts belong to the natural law. For many things are done virtuously to which nature at first does not incline; rather human beings come to find them by the investigation of reason, as useful for living well. (Ibid.)

In fact, Aquinas maintains that the eternal law as it is in the mind of the Creator can be made known to created persons not only by the natural light of reason[10] but also by revelation. Just as human beings are made in the image of God, so the free agency of created persons is an image of the free agency of the Creator. God can exercise his free agency by choosing to share his mind with his creatures; and so God can reveal to human beings parts of the eternal law that might not be available to them by the natural light of reason alone. Human persons therefore have access to the natural law both because of the natural light of reason and also because of the Creator's willingness to reveal some part of the eternal law to them (*ST* IaIIae.19.4 ad 3).

Aquinas describes law in general as an ordinance of reason for the common good which is made by a person who has the care of the community and which is promulgated. A question therefore arises for Aquinas whether the natural law is also promulgated. In reply, he says that the natural law is promulgated just in virtue of the

---

[9] See, for example, *ST* IaIIae.93.6. There Aquinas remarks that both ways are diminished in the wicked because their knowledge of the good and their inclination to it are imperfect.

[10] See, for example, *ST* IaIIae.19.4 ad 3. There Aquinas remarks that a created person can know and will in a general way what God wills because a created person can know that God wills what is good. And so, Aquinas says, "whoever wills something under some description (*ratio*) of the good has a will conformed to the divine will as far as the description of what is willed [is concerned]."

Creator's instilling it into a created person's intellect as a matter of natural knowledge.[11] In the will, however, what the Creator instills is not a habit of knowledge but rather an innate inclination for the good, as perceived by the intellect.

Even so, the will is master of itself and free. Contrary to the compatibilist account of free will typically taken to be implied by SSP, for Aquinas the will is free in the strong sense that nothing, not even the intellect, acts on the will with efficient causation.[12] Although the general precepts derived from the divinely implanted habit of knowledge in the intellect cannot be completely wiped out even in evil people, secondary precepts derived from these general precepts can be blotted out; and even the application of the most general precepts to particular actions can be hindered by the effects of moral evil on a person's intellect (*ST* IaIIae.94.6).

Aquinas makes an analogous point as regards the will's natural inclination to act in accordance with the good. He argues that even the natural inclination to the good can be undermined by moral evil. In the wicked, not only is the natural knowledge of the good corrupted by the passions and morally evil habits, but also "the natural inclination to virtue is corrupted by habits of vice" (*ST* IaIIae.93.6).

So one way to understand Aquinas's account of natural law is as a gift of the Creator to the human persons he has created. It consists in a pair of habits, one in the will and one in the intellect, which is given to human beings either by means of the natural light of reason or through the Creator's revelation of his own mind to his creatures. Although, apart from revelation, these gifts are implanted innately, they are so far in the control of the creature that a person's exercise of his free will in evil acts can corrupt them. Nothing about God's rendering the natural law innate in human persons takes away from them their free agency.

Just as many people find the implications of SSP counterintuitive, so, for many people, the implications of Aquinas's account of natural law, grounded as it is in his metaphysics and theology, seem counterintuitive too. Can everything in the world really be traced back to an omnipotent, omniscient, perfectly good Creator? Could it really be the case that a human person has the causal powers of intellect and will which reflect the eternal law in the mind of the Creator? Or, to put the question in a less theological way, could the action of something at the macro-level, such as a human being, exercise causality, from the top down, as it were, without being itself determined at the micro-level? Could it really be, for example, that human beings have libertarian free will?

---

[11] See, for example, *ST* IaIIae.90.4 ad 1. But the natural knowledge in question consists in very general moral precepts, the precepts of the natural law, such as that the good is to be done and the bad is to be avoided (*ST* IaIIae.94.2).

[12] I have argued the case for this claim in the chapter on freedom in Stump (2003).

Double Vision

Any attempt to hold in one view the very different notions of natural law in SSP and in the outlook of Aquinas can induce vertigo. How is one to understand the differences in worldview between the two, and how could one even begin to adjudicate their competing claims, or the competing accusations of being counterintuitive?

It will be profitable to begin by considering their highly differing foundational metaphysics.

As has often been remarked, one notable difference between the notion of natural law in SSP and the Thomistic notion of natural law is that, for Aquinas but not for SSP, natural law is the law of a law-giver, whose mind is the source of the law and whose relation to and care for other persons lead him to promulgate the law. On the view of natural law in SSP, the whole notion of law is only metaphorical or analogous. A natural law of physics understood as SSP sees it is a generalization describing the nature of the world at the microphysical level, or its general statements expressing relations between universals, or something else along these lines. It is not promulgated; it is not prescriptive, in the way that a law promulgated by a law-giver is; and it is not the result of an act of intellect and will on the part of a law-giver.

This dissimilarity is correlated with a much greater difference as regards the ultimate foundation of reality, of course. On SSP, the ultimate foundation of reality consists in those elementary particles described by the ultimately correct version of contemporary physics and their causal interactions governed by the natural laws of that physics. There is ontological reduction; everything that there is is reducible to the elementary particles composing it. Persons are no exception to this claim. Persons too are reducible to the elementary particles that constitute them. At the ultimate foundation of all reality, therefore, is only the nonpersonal.

What is challenging for SSP therefore is the construction of the personal out of the impersonal. The mental states of persons, their free agency, their relations with each other all have to be understood somehow as built out of the physically determined interaction of the nonpersonal. On Aquinas's view, things are in a sense exactly the other way around. That is because, for Aquinas, the ultimate foundation of reality is God the Creator, who exists in the three persons of the Trinity.[13] Furthermore, on the doctrine of the Trinity, none of the persons of the Trinity is reducible to anything nonpersonal. That is, it is not the case that the persons of the Trinity are reducible to the Godhead or to being qua being or to anything else at all. On the Thomistic worldview, the ultimate foundation of reality is therefore precisely persons.

It would not be hard, I think, to trace the notable differences between SSP and Aquinas's worldview, as implied by their varying notions of natural law, back to the great dissimilarity in their metaphysical views regarding the ultimate foundation of

---

[13] On the traditional, orthodox doctrine of the Trinity, the word 'person' is used in a technical sense. But, even on that technical sense, each person of the Trinity has mind and will. And so it is also true that each member of the Trinity counts as personal in our ordinary sense of being characterized by mind and will.

reality. But, given this radical difference between SSP and the Thomistic worldview as regards such foundational matters, is it so much as possible to reason about their competing claims?

Of course, people who are very much in the grip of SSP might suppose that there is no point in trying to do so. For them, an evaluative comparison of the two differing pictures of the world is not worth the effort; the Thomistic view of the ultimate foundation of reality is no longer a live option. No doubt, those committed to a Thomistic worldview return the compliment as regards SSP, which is not a live option for them either.

Nonetheless, even in the face of this great divide, I want to see what can be done by way of an evaluative comparison; and I want to do so without addressing the question of the existence of God. Even if the recent history of philosophy did not make us pessimistic about the prospects for success when it comes to arguing over the existence of God, it is clear that it would not be profitable in a short paper to tackle a disagreement of this magnitude head-on. It is, however, possible to evaluate these two differing worldviews with regard to one much smaller metaphysical issue. This is the issue of reductionism.

The brief sketch of Aquinas's views given above makes clear that Aquinas's metaphysics is incompatible with reductionism, unlike SSP, which is committed to it.[14] As I explained at the outset, reductionism has come under increasing attack in recent years, in science as well as in philosophy. In what follows, I will sketch a little of this attack and argue that the rejection of reductionism it supports is right. And then I will not argue but only suggest that such a rejection of reductionism is more at home in a worldview such as that of Aquinas, which sees the ultimate foundation of reality as personal.

Reductionism

In virtue of supposing that everything is reducible to the elementary particles composing it, reductionism holds that ultimately all macro-level things and events are a function only of things and events at the microstructural level. That is one reason why reductionism typically includes a commitment to causal closure at the microphysical level.[15] One way to understand reductionism, then, is that it ignores or

---

[14] For a defense of the claim that Aquinas's metaphysics rejects reductionism, see Stump (2003, chap. 1).

[15] For a helpful discussion of the general problem of reductionism relevant to the issues considered here, see Garfinkel (1993). Garfinkel argues against theoretical reductionism by trying to show that reductive microexplanations are often not sufficient to explain the macrophenomena they are intended to explain and reduce. He says, "A macrostate, a higher level state of the organization of a thing, or a state of the social relations between one thing and another can have a particular realization which, in some sense, 'is' that state in this case. But the explanation of the higher order state will not proceed via the microexplanation of the microstate which it happens to 'be'. Instead, the explanation will seek its own level" (449). Aquinas would agree, and Aquinas's account of the

discounts the importance of levels of organization or configuration, or form, as Aquinas would put it, and the causal efficacy of things in virtue of their organization or form.

This feature of reductionism also helps explain why it has come under special attack in philosophy of biology (see, for example, Garfinkel [1993] and Kitcher [1993]). Biological function is frequently a feature of the way in which the microstructural components of a thing are organized, rather than of the intrinsic properties of the microcomponents themselves. Proteins, for example, tend to be biologically active only when folded in certain ways, so that their function depends on their three-dimensional configuration. But this is a feature of the organization of the protein molecule as a whole and cannot be reduced to properties of the elementary particles that make up the atoms of the molecule. In fact, for large proteins, even an omniscient knowledge of the properties of the elementary particles of the atoms that comprise the protein[16] may not be enough to account for the configuration of the folded protein and the causal powers consequent on that configuration,[17] because the activity of enzymes is required to catalyze the folding of the components of the molecule in order for the protein to have its biologically active configuration.[18]

---

relation of matter and form in material objects helps explain Garfinkel's point. A biological system has a form as well as material components, so that the system is not identical to the components alone; and some of the properties of the system are a consequence of the form of the system as a whole. Consequently, neither ontological reductionism nor theoretical reductionism is acceptable. Garfinkel himself recognizes the aptness of the historical distinction between matter and form for his argument against theoretical reductionism. He says, "the independence of levels of explanation ... can be found in Aristotle's remark that in explanation it is the form and not the matter that counts" (149). See also Kitcher (1993).

[16] There is some room for ambiguity and confusion here, because one can think of the microstructure of the system or the properties of the parts in different ways. In particular, one can think of the properties of the parts either as (i) the properties of the parts taken *singillatim*, that is, the properties had by the molecule's constituent elementary particles, taken individually, or as (ii) the properties the parts in fact have when they are organized into the whole, that is, the properties the constituent elementary particles have in the configuration which the molecule has in its final, biologically active form. I am taking 'properties of the parts' in sense (i) here. In sense (i), it is true to say, as biochemists do, that the folded shape of a protein cannot always be derived from even perfect knowledge of the biochemical properties of the components of the protein, including their causal interactions (since it might be the case that the protein achieves that folded shape only with the help of enzymes, for example). It would not be true to say this in sense (ii). If we take 'properties of the parts' in sense (ii), then we smuggle the configuration, or the form of the whole, into the properties of the parts of the whole. In sense (ii), it would be very surprising if there were features of the whole system that were not predictable on the basis of or determined by the causal interactions of the parts of the whole, since the features of the system are a function of the configuration of the whole and that configuration is in effect being counted among the properties of the parts.

[17] For a connection between reductionism and predictability, see Crane (2010, 28). Crane says, "Sometimes it is said that emergent properties are those properties of a thing whose instantiation cannot be predicted from knowledge of the thing's parts.... Properly understood, the idea of predictability contains the key to emergence."

[18] See, for example, Richards (1991). According to Richards, for relatively small proteins, folding is a function of the properties and causal potentialities among the constituents of the protein; but "some large proteins have recently been shown to need folding help from other proteins known as chaperonins" (54).

In his magisterial attack on reductionism from the perspective of philosophy of biology, John Dupré takes the examples in his arguments against reductionism from ecology and population genetics, rather than molecular biology (Dupré 1993; see especially chaps. 4, 5, and 6). He summarizes his rejection of reductionism this way:

> [On ontological reductionist views,] events at the macrolevel, except insofar as they are understood as aggregates of events at the microlevel—that is, as reducible to the microlevel at least in principle—are causally inert.... [But] there are genuinely causal entities at many different levels of organization. And this is enough to show that causal completeness at one particular level [the microlevel] is wholly incredible. (Dupré 1993, 101)

In philosophy of science, as Dupré's work illustrates, arguments against reductionism have frequently been directed against the possibility of reducing biological things and events to things and events at the level of physics; but analogous arguments can be used to undermine even the project of reducing the things and events described by chemistry to those at the level of physics.[19]

So, for example, consider the chirality or handedness of a molecule. The same constituents of a molecule chemically bonded in the same way can form different chiral analogues or enantiomers, depending on whether those very same similarly bonded constituents are in a left-handed or a right-handed form. Enantiomers of the same molecule can behave very differently. It turns out, for example, that different enantiomers of organophosphates, which are a mainstay of insecticides, have radically different toxicities for freshwater invertebrates. Testing for the ecological safety of one enantiomer alone can give very misleading results about the safety of an insecticide (see, for example, Liu et al. [2005]). Each enantiomer has its own form or configuration, and the causal power of the enantiomer is a function of the configuration of the whole, not of the particular intrinsic properties of the constituents of the molecule.

In a detailed defense of the claim that things at the level of chemistry cannot be reduced to things at the level of physics, Hendry takes as one of his examples ethanol and methoxymethane. Hendry says,

> These are distinct substances, though each contains carbon, oxygen, and hydrogen in the molar ratios 2:1:6. Clearly, the distinctness of ethanol and methoxymethane as chemical substances must lie in their different molecular structures, that is, the arrangement of atoms in space. (Hendry 2010, 214)

Discussing another of his examples, hydrogen chloride, Hendry says,

> if the acidic behavior of hydrogen chloride is conferred by its asymmetry, and the asymmetry is not conferred by the molecule's physical basis according to physical laws, then ontological reduction fails because the acidity is a causal power which is not conferred by the physical interactions among its parts. (Ibid., 215)

And he goes on to remark,

---

[19] For a detailed discussion of ontological reduction and ontological emergence, especially as regards chemistry, see Stump (2012).

> the explanation of why molecules exhibit the lower symmetries they do would appear to be holistic, explaining the molecule's broken symmetry on the basis of its being a subsystem of a supersystem (molecule plus environment). This supersystem has the power to break the symmetry of the states of *its* subsystems without acquiring that power from its subsystems in any obvious way. That looks like downwards causation. (Ibid., 215–216)

In addition to work such as this in the philosophy of biology and chemistry, there have also been attacks on reductionism in recent work in philosophy of mind and metaphysics. This work has attempted to undermine the credibility of the claim that there is causal closure at the microphysical level.

So, for example, Alexander Bird has argued that a substance is what it is in virtue of having the causal powers it does; it could not be the substance it is and have different causal powers. For philosophers such as Bird, causal powers are vested in substances, not in events[20]; all causation is substance causation, not event causation. The particular configuration of a thing, its intrinsic properties and internal organization, is another way of picking out what Aquinas would see as the form of the substance. For Bird, as for Aquinas, the organization or form of a substance gives that substance certain specific capacities for acting. A substance can act to exercise the causal power it has in virtue of its form or configuration. For this reason, a causal chain can be initiated by any substance at any level of organization. Consequently, there can be top-down causation, as well as bottom-up causation.[21]

One way to think about such recent antireductionist moves in philosophy is to see them as adopting a neo-Aristotelian metaphysics of a Thomistic sort. For Aquinas, a thing's configuration or organization, its form, is also among the constituents of things; and the function of a thing is consequent on the form of the whole. Dupré himself is aware of the Aristotelian character of his position. In making his case for the rejection of both reductionism and the causal closure view, he says that there is no reason for "attaching preeminent metaphysical importance to what things are made of…. Why should we emphasize matter so strongly to the exclusion of form?" (Dupré 1993, 92–93).

On philosophical views such as these, a thing is not just the sum of its parts, ontological reductionism fails, and there is not causal closure at the microphysical level. The component parts of a whole can sometimes determine *how* the whole does what it does. But *what* the whole does is a function of the causal power had by the whole in virtue of the form or configuration of the whole.

---

[20] There are differing ways of understanding this claim. For one interpretation which allows a role to events, taken in a certain way, see O'Connor and Churchill (2010, 45).

[21] For my own attempt to explain top-down causation, especially in light of Thomistic metaphysics, see Stump (2012).

An Example Drawn from Neuroscience and Psychology

Some contemporary philosophers, such as E.J. Lowe, for example, suppose that similar lessons apply with regard to the mind and to mental states.[22] Like Bird and others, Lowe argues that genuine causal powers belong only to substances and that substances have the causal powers they do in virtue of the properties (or the form) of the whole. On his view, a mental act is an exercise of the causal power had by a human being in virtue of the complex organization had by human beings.

In fact, recent discoveries in neuroscience and developmental psychology suggest that we should go even further in this direction. These discoveries suggest that some cognitive capacities arise as a consequence of a system that comes into existence only when two people are acting in concert. Research on some of the deficits of autism has helped to make this clear.

Autism in all its degrees is marked by a severe impairment in what some psychologists and philosophers call 'mindreading' or 'social cognition.' We are now beginning to understand that mind reading or social cognition is foundational to an infant's ability to learn a language or to develop human cognitive abilities in other areas as well. For an infant to develop normally as regards mind reading, the infant's neural system has to be employed within the active functioning of a larger system composed of at least two persons, the infant and a primary caregiver. The system provides for shared attention or joint attention between a child and its caregiver.

It is not easy to give an analysis of joint attention. One developmental psychologist, Peter Hobson, says: "Joint attention … occurs when an individual … is psychologically engaged with someone else's … psychological engagement with the world" (Hobson 2005, 188). Researchers are now inclined to think that a there is a foundational deficit in autism which has to do with what they call "dyadic shared attention." As one scientist says, "this is the most direct sharing of attention and the most powerful experience of others' attention that one can have" (Reddy 2005, 85). Many lines of recent research are converging to suggest that autism is most fundamentally an impairment in the capacity for dyadic joint attention.[23] Trying to summarize his own understanding of the role that the lack of shared attention plays in the development of autism, Hobson says that autism arises "because of a disruption in the system of child-in-relation-to-others" (Hobson 2004, 183). By way of explanation, he says,

> my experience [as a researcher] of autism has convinced me that such a system [of child-in-relation-to-others] not only exists, but also takes charge of the intellectual growth of the

---

[22] Lowe's view differs from Dupré's in that Lowe hopes to make his view consistent even with the acceptance of causal closure principles.

[23] One group of researchers in this area says, "Early research findings focusing on the joint attention impairment [of autistic children] initially emphasized a specific impairment in triadic interactions rather than dyadic interactions.... Recently, however, the tide has begun to turn. Several studies show group differences in dyadic interaction between children with autism and those with other developmental delays.... The research shows that certain measures of dyadic interaction predict diagnosis of autism several years later." See Leekam (2005, 207).

infant. Central to mental development is a psychological system that is greater and more powerful than the sum of its parts. The parts are the caregiver and her infant; the system is what happens when they act and feel in concert. The combined operation of infant-in-relation-to-caregiver is a motive force in development, and it achieves wonderful things. When it does not exist, and the motive force is lacking, the whole of mental development is terribly compromised. At the extreme, autism results. (Hobson 2004, 183)

On Hobson's views, then, autism results from impairment in a complex system involving two human beings, an infant and its primary caregiver. Commitment to ontological reductionism and causal closure at the level of the microphysical cannot account for the role of the jointness in attention critical for normal infant development. On the contrary, as the phrase indicates, joint or shared attention cannot be taken even just as a function of the properties of one human being considered as a whole, to say nothing of the lowest-level components of a human being. Rather, it is a function of a system comprising at least two human beings acting in concert. Bonded into a unity, whose organization makes possible novel causal powers, each human being in the unity of the pair acquires the new power for mind reading that the joint attention makes possible. A system of at least this much complexity is needed to make possible the shared attention and mind reading that in turn enables typical infant development.

The Implications of the Rejection of Reductionism

SSP supposes that all macrophenomena are reducible to micro-level phenomena and that there is a complete causal story to be told at the micro-level. The converging lines of research in the sciences and several areas of philosophy, however, make a good case that ontological reductionism is to be rejected. And if ontological reductionism is rejected, then it is not true that everything is reducible to the elementary particles composing it or that everything is determined by causal interactions at the level of the microphysical. And it is therefore also not true that things at the macro-level are causally inert. Rather, causal power is associated with things at any level of organization in consequence of the configuration or form of those things.

For SSP, whatever control or freedom human beings have as agents at the macro-level has to be not only compatible with but in fact dependent on the complete causal story at the micro-level. But if reductionism is rejected, as the new work in philosophy and the sciences suggests it should be, then there can be causal efficacy at various levels of organization, including at the level of human agents. A person's intellect and will can exercise real causal efficacy, from the top down, in the way Aquinas supposes they do.

Dupré puts the point this way:

> There is no reason why changes at one level may not be explained in terms of causal processes at a higher, that is, more complex, level. In the case of human action, the physical changes involved in and resulting from a particular action may perfectly well be explained in terms of the capacity of the agent to perform an action of that kind. (Dupré 1993, 216–217)

He can take this position, because having rejected reductionism he is free to hold that

> humans have all kinds of causal capacities that nothing else in our world has.... There is no good reason for projecting these uniquely human capacities in a reductionist style onto inanimate bits of matter. Nor is there anything ultimately mysterious about particular causal capacities' being exhibited uniquely by certain very complex entities. (Dupré 1993, 216)

On this view, there is nothing mysterious about assigning such causal power to human beings. On the contrary, compatibilism looks like an unnecessary concession, an attempt to preserve what we commonly believe about our control over our actions in the face of a mistaken commitment to reductionism. With reductionism rejected in favor of a metaphysics that allows causal power vested in substances at any level, the causality exercised in libertarian free will is one more case of a kind of top-down causation that any complex system, even molecules, can manifest.

In a metaphysical system of this antireductionist sort, the place of persons is not imperiled. In fact, even a human pair bonded in love, as a parent and child are, can be a sort of whole, with causal power vested in their bondedness.

The Moral of the Story

If ontological reductionism is rejected, as the work I have canvassed argues it should be, then with respect to this one issue the Thomistic worldview is more veridical and more worthy of acceptance than SSP is. By itself, of course, this conclusion certainly does not decide the issue as regards the central disagreement between SSP and the Thomistic view. It cannot adjudicate the issue regarding the ultimate foundation of reality. And so, as far as the evidence canvassed in this paper is concerned, the central disagreement between SSP and the Thomistic view remains an open issue. Clearly, it is possible to reject reductionism and accept atheism.

For that matter, it is possible to reject atheism and accept reductionism. As I have described it, SSP is a secular view that combines contemporary scientific theories with certain metaphysical claims. But it is possible to have an analogue to SSP in which a reductionist scientific view of the world is combined with a commitment to religious belief, even religious belief of an orthodox Christian sort. That is, SSP can have a theistic analogue which includes most of the scientific and metaphysical worldview of SSP but marries it to belief in an immaterial Creator.

So, for example, consider Peter van Inwagen's explanation of God's providence. Trying to explain God's actions in the created world, van Inwagen says that God acts by issuing decrees about elementary particles and their causal powers: "[God's] action consists in His ... issuing a decree of the form 'Let *that* [particle] now exist and have such-and-such causal powers'" (van Inwagen 1995, 49). For van Inwagen, apart from miracles,[24] God's actions in the world consist just in creating and sustaining

---

[24] On van Inwagen's view, miracles are a matter of God's supplying "a few particles with causal powers different from their normal powers" (van Inwagen 1995, 45).

elementary particles and their causal powers. This, van Inwagen says, "is the entire extent of God's causal relations with the created world" (van Inwagen 1995, 44).

For most people conversant with religious discourse in the Judaeo-Christian tradition, this religious analogue to SSP will seem a very odd mix. Could it really be possible, as van Inwagen is apparently claiming, that decrees concerning the existence and causal powers of particles exhaust the rich panoply of divine interaction with human persons that Judaism and Christianity have traditionally ascribed to God? Most orthodox interpreters of Judaism and Christianity have assumed that, apart from doing miracles, God also intervenes in the lives of the person he has created by engaging in personal relationships with them. On this view, God not only issues decrees (about particles or anything else). God also cajoles, threatens, instructs, illumines, demands, comforts, and asks questions. At the heart of all these activities is the direct (but nonmiraculous) interaction between God and human persons of the mind-reading sort.

Even if, per improbabile, all this and more could be reduced to decrees about particles, the reduction would have lost the direct personal connection that in both Judaism and Christianity has been the most important element in the relations between God and human persons. For van Inwagen, God interacts directly with particles and thereby, indirectly, with things composed of particles, including human beings. Most religious believers have supposed that God interacts directly, at least sometimes, with human persons. In fact, in the book of Job, God describes to Job God's interacting directly, playfully and lovingly, even with nonhuman animals, such as the ostrich, for example. And it makes a certain kind of sense that God would act in this way. What is the point of God's creating things such as ostriches, or people, if he does not want to interact directly with *them*—or, for that matter, play with them?

For these reasons, reductionism does not fit well with theism. I am not claiming that it is incompatible with theism. The point is only that there is something awkward or forced or otherwise implausible about reductionism in a theistic worldview. It is not natural there, one might say. On a worldview that takes persons to be the ultimate foundation of reality, reductionism to the level of elementary particles is not really at home.

By the same token, it seems to me that the rejection of reductionism is harder to square with a worldview in which the ultimate foundation of reality is impersonal. Here too the issue is not the compatibility of the two positions. I am not claiming that the rejection of reductionism is incompatible with a worldview on which the impersonal is foundational. The point is rather this. The rejection of reductionism leaves room for the place ordinary intuition accords persons in the world. And this in turn seems to deflate some of the impetus towards atheism. If all there really is in the world is elementary particles and their causally determined interactions, then adding in to this picture the existence of a personal God seems jarring, or embarrassing, or in some other way dissonant. But if reductionism is rejected and our ontology can include all kinds of things that can initiate causal chains, from water molecules to persons, then the motive force driving towards atheism seems considerably diminished.

Or if someone supposes that this is too much to say, then this at any rate seems to me right. The metaphysics that gives persons and their top-down causal agency the place ordinary intuition assigns them is more readily intelligible on a worldview that sees persons as the ultimate foundation of reality. Figuring out how to make it cohere with the picture Blackburn paints, even if we subtract reductionism from that picture, strikes me as much harder to do.[25]

References

Blackburn, Simon. 2002. An Unbeautiful Mind. *New Republic*, 5 and 12 Aug 2002.
Crane, Tim. 2010. Cosmic Hermeneutics vs. Emergence: The Challenge of the Explanatory Gap. In *Emergence in Mind*, ed. Cynthia Macdonald and Graham Macdonald, 22–34. Oxford: Oxford University Press.
Dupré, John. 1993. *The Disorder of Things: Metaphysical Foundations of the Disunity of Science*. Cambridge, MA: Harvard University Press.
Finnis, John. 1998. *Aquinas: Moral, Political, and Legal Theory*. Oxford: Oxford University Press.
Garfinkel, Alan. 1993. Reductionism. In *The Philosophy of Science*, ed. Richard Boyd, Philip Gasper, and J.D. Trout, 443–459. Cambridge, MA: MIT Press.
Hendry, Robin Findlay. 2010. Emergence vs. Reduction in Chemistry. In *Emergence in Mind*, ed. Cynthia Macdonald and Graham Macdonald, 205–221. Oxford: Oxford University Press.
———. 2012. *The Metaphysics of Chemistry*. Oxford: Oxford University Press.
Hobson, Peter. 2004. *The Cradle of Thought: Exploring the Origins of Thinking*. Oxford: Oxford University Press.
———. 2005. What Puts the Jointness into Joint Attention? In *Joint Attention: Communication and Other Minds; Issues in Philosophy and Psychology*, ed. Eilan Naomi et al., 185–204. Oxford: Clarendon Press.
Kitcher, Philip. 1993. 1953 and All That: A Tale of Two Sciences. In *The Philosophy of Science*, ed. Richard Boyd, Philip Gasper, and J.D. Trout, 553–570. Cambridge, MA: MIT Press.
Leekam, Sue. 2005. Why Do Children with Autism Have a Joint Attention Impairment? In *Joint Attention: Communication and Other Minds; Issues in Philosophy and Psychology*, ed. Naomi Eilan et al., 205–229. Oxford: Clarendon Press.
Liu, Weiping, Jianying Gan, Daniel Schlenk, and William A. Jury. 2005. Enantioselectivity in Environmental Safety of Current Chiral Insecticides. *Proceedings of the National Academy of Sciences of the United States of America* 102 (3): 701–706. Published online 4 Jan 2005. https://doi.org/10.1073/pnas.0408847102.
Macdonald, Cynthia, and Graham Macdonald, eds. 2010. *Emergence in Mind*. Oxford: Oxford University Press.
McInerny, Ralph. 1992. *Aquinas on Human Action: A Theory of Practice*. Washington, DC: Catholic University of America Press.
———. 1997. *Ethica Thomistica: The Moral Philosophy of Thomas Aquinas*, rev. ed. Washington, DC: Catholic University of America Press.
Murphy, Mark. 2001. *Natural Law and Practical Rationality*. Cambridge Studies in Philosophy and Law. Cambridge: Cambridge University Press.
O'Connor, Timothy, and John Ross Churchill. 2010. Is Non-reductive Physicalism Viable Within a Causal Powers Metaphysic? In *Emergence in Mind*, ed. Cynthia Macdonald and Graham Macdonald, 43–60. Oxford: Oxford University Press.

---

[25] I am grateful to Jeroen de Ridder for his helpful comments on an earlier draft of this paper.

Reddy, Vasudevi. 2005. Before the 'Third Element': Understanding Attention to Self. In *Joint Attention: Communication and Other Minds; Issues in Philosophy and Psychology*, ed. Naomi Eilan et al., 85–109. Oxford: Clarendon Press.
Richards, Frederic M. 1991. The Protein Folding Problem. *Scientific American* 264 (Jan): 54–63.
Stump, Eleonore. 2003. *Aquinas*. New York/London: Routledge.
———. 2012. Emergence, Causal Powers, and Aristotelianism in Metaphysics. In *Powers and Capacities in Philosophy: The New Aristotelianism*, ed. Ruth Groff and John Greco, 48–68. New York/London: Routledge.
van Inwagen, Peter. 1995. *God, Knowledge, and Mystery. Essays in Philosophical Theology*. Ithaca/London: Cornell University Press.

Is the Idea of Creation Order Still Fruitful?

Danie Strauss

Abstract Dooyeweerd noted that the idea of cosmic order is present throughout the history of philosophy. The legacy of Plato and Aristotle was uprooted by modern nominalism which challenged the Greek-medieval realistic metaphysics by eliminating what Christianity saw as the God-given *order for* (law for) creatures and the *orderliness of* creatures. Denying universality outside the human mind eliminated any God-given order for and orderliness of creatures. This created a vacuum quickly filled by nominalism, for now human understanding took over the role of law-giver (Kant). Historicism and the linguistic turn pursued the road to an unbridled irrationalism and relativism. All of this adds up to a systematic elimination of the idea of a creational order. Clearly these diverging developments bring to expression the abyss between the spirit of modern humanism and reformational Christianity. Alternatively, reformational philosophy explores the idea of a creational order by turning away from an epistemic point of departure towards an ontic perspective, making possible a new approach towards the various dimensions of reality. This new approach is designated as the transcendental-empirical method. It advances a new way of articulating the foundational role of a creational order or a cosmic law-order. This is illustrated by the provision of a definition of a natural law and of norming principles. The argument concludes by pointing out that the future of the idea of a creational order depends on a proper understanding of the constancy and universality of such an order—embedded in a non-reductionist ontology.

Keywords Cosmic order · Law-conformity · Cosmonomic idea · Continuous flux · Conceptual rationalism · Historicism · Ground motive · Ground idea · Ontic laws and principles · Natural laws · Norming principles · Positivizations

D. Strauss (✉)
North-West University, Potchefstroom, South Africa
e-mail: dfms@cknet.co.za

How Did the Idea of Cosmic Order Permeate the History of Philosophy?

Dooyeweerd's decision to designate his philosophy as the philosophy of the law-idea (*wetsidee*; philosophy of the cosmonomic idea) was informed by the fact that during the history of philosophy various prominent philosophers explicitly connected their philosophical reflection to the idea of an encompassing cosmic order. Different philosophical systems, from Greek antiquity and medieval philosophy up to modern philosophy, "expressly orientated philosophic thought to the Idea of a divine world-order, which was qualified as lex naturalis, lex aeterna, harmonia praestabilita, etc." (Dooyeweerd 1997, 1:93–94). But let us first consider the question how the idea of cosmic order did permeate the history of philosophy.

From the outset Dooyeweerd was concerned with the world and life views and ultimate commitments surfacing in philosophical ideas of cosmic order. He noticed that Western philosophy originated in Greek antiquity and was transformed during the medieval period by attempting to obtain a synthesis between Greek philosophy and biblical Christianity. Since the Renaissance, modern humanism transformed the Greek-medieval legacy as well as Christendom into the new humanistic motive of a free and autonomous human personality, which in turn gave birth to an encompassing natural science ideal. The tension was immediately manifest, for if reality is entirely explained in terms of a deterministic understanding of cause and effect, then the assumed human freedom is also reduced to a cause among the causes and an effect among the effects. In his *Roots of Western Culture*, Dooyeweerd (2012, see 149–217) shows how this humanistic personality ideal generated the modern natural science ideal.

Dooyeweerd designated the basic dualism present in Greek culture as that between two original and opposing principles (*Archai*), namely, *matter* and *form*. Within Greek philosophy itself these mutually excluding principles of origin are also designated by the opposition between the *one* and the *many*, between what is *static* and *dynamic*, and between what is *constant* and *changing*. Dooyeweerd provides us with an extensive analysis of how these two poles of the basic motive of Greek philosophy commenced by giving primacy to the *matter pole* (the pole of the formless stream of life) but eventually switched to giving priority to the *form pole* (the motive of form, measure, and harmony). Since both these principles are persistently present in the thought of the Greek philosophers, the only option was to give primacy to the one or the other. Heraclitus, for example, accentuates changefulness and at the same time accepts the world law (*logos*) as an untransgressable measure (i.e., as something constant): "The sun will not transgress his measures: were he to do so, the Erinyes, abettors of Justice, would overtake him" (Comperz 1964, 73).

In the thought of Plato, the problem of change is approached on the basis of what is enduring. If the world of becoming (of sense perception) is in a state of continuous flux, nothing could be known. Therefore, Plato postulated supra-sensory ontic forms (*eidè*) to account for knowledge (see *Cratylus* 439c–440a). He had learned from Heraclitus that all things accessible to sensory perception are in an ever-fluctuating state. It is therefore impossible to know these things. This conclusion rests on the presupposition that everything is changing. But in that case, what Plato considers to be the essential being of things (their static *eidos*) should also be constantly changing. However, this is unacceptable to Plato. His desire is to acknowledge that the so-called essence of things could not also be subject to continuous change.

In his dialogue *Phaedo*, Plato characterizes the domain of the static *eidè* as being divine, immortal, conceivable, simple, indissoluble, constant, and "self-identical" (80b1–6). This shows that Plato stumbled upon what Dooyeweerd designates as God's law for creatures. The tension present within the Greek motive of matter and form is strikingly highlighted in the absence of an *eidos* (ontic form) for matter (the formless—see Dooyeweerd [1997, 2:9]).

Aristotle did not acknowledge Plato's transcendent realm of *eidè*. He introduced a so-called secondary substance, i.e., the universal substantial forms inherent to things. This universal form (*formula*) is not subject to becoming. The "being of house is not generated, but only the being of *this* house" (cf. *Metaphysics* 1039b23; cf. *De Anima* 412b16ff.). The *houseness* of a house, its *being-a-house* is the universal way in which any particular (individual) house shows that it conforms to the conditions (law) for being a house.

The key terms employed in this context are that of the idea of *orderliness, lawfulness*, or *law-conformity*. Only what is subject to an *order for* or to a *law for* can display the feature of being *law-conformative* or *orderly*. This entails that the terms *law* and *order* may be used as synonyms. In general, Dooyeweerd holds that the law for or the order for creatures is delimiting and determining their existence. We shall return to this point below.

The synthesis between the Greek basic motive of matter and form and the biblical motive of creation, fall into sin, and redemption resulted in a new basic motive, namely, that of nature and grace. The form-matter split now appears both within the nature pole and the grace pole. Thomas Aquinas' law-idea accepts the dual teleological order of Aristotle, accommodated within the new nature-grace divide. As the encompassing community within the natural domain, the state merely serves as the natural foundation (matter) for the church as overarching superstructure, as the supernatural institute of grace (the form). The state carries human beings to their highest natural aim in life, namely, goodness (within the *societas perfecta*), whereas the church elevates them to their supra-temporal perfection, to wit, eternal bliss (the church institute as *Corpus Christianum*).[1]

The cosmonomic idea of the realistic metaphysics of the medieval era accommodated Plato's ontic forms as ideas in God's mind (*universalia ante rem*), whereas the universal substantial forms of Aristotle became universals inhering in individual things (*universalia in re*).

---

[1] The persistence of this view is still found in the famous papal encyclical "Quadragesimo anno" (15 May 1931), which explicitly states: "Surely the church does not only have the task to bring the human person merely to a transient and deficient happiness, for it must carry a person to eternal bliss" (cf. Schnatz 1973, 403).

Denying the Order for and Orderliness of Creation

The challenge to this reigning medieval cosmonomic idea came from late scholastic nominalism. This movement denied both the *universalia ante rem* (before creation in God's mind) and the *universalia in re* (inherent in things). We noted that this distinction pertains to God's law for creatures and the orderliness of creatures. Nominalism rejects universality outside the human mind, both in respect of the ideas in God's mind and regarding the universal substantial forms inherent in things. In doing this it at once eliminates the God-given order for (law for) creatures and the orderliness of creatures. But when creation is stripped from God's determining and delimiting law as well as of its law-conformity, then reality collapses into a chaotic multiplicity of (structureless) individual entities.

This created a vacuum that was soon filled by an overestimation of human understanding. Von Weizsäcker describes this transition very well:

> This state of affairs is characteristic of modernity. It is not the world in which I find myself that guarantees my existence. This guarantee is not lost, for when I recover the world then it is as the object of my self-assured thinking, that is to say, as an object which I can manipulate. (Von Weizsäcker 2002, 130–131)

It should not be surprising that this new inclination soon gave rise to the ideal of *logical creation*. At the same time, the classical realistic view that truth is the correspondence of thought and being (reality—*adequatio intellectus et rei*) was challenged. For nominalism truth only involves compatibility of concepts. Ernst Cassirer explains this by saying that truth is not inherent to things since it belongs to the names and their comparison as it occurs in statements (see Hobbes, *De Corpore* 1.3.7 and 8; quoted in Cassirer [1971, 56]).

Since Descartes, modern nominalism has considered (number and) all universals as mere modes of thought. Galileo envisaged a thought experiment—regarding a body in motion—from which he deduced the *law of inertia* (see Galileo [1683] 1973). It inspired Immanuel Kant to elaborate this new motive of logical creation to its extreme rationalistic consequences.[2]

Kant takes the thought experiment of Galileo a step further because he wants to understand how it is possible to formulate such a thought experiment, deduce from it a law of nature, and then apply it to things in nature. His solution carries the rationalistic element present in nominalism to its ultimate consequences because he

---

[2] In general, Dooyeweerd defines *rationalism* as an absolutization of the law-side of reality and *irrationalism* as the absolutization of the (individual) factual side of reality. However, if one accepts universality on the factual side of reality and in addition accepts that conceptual knowledge always embraces what is universal (either the universal law for or the universality of what is lawful), it is clear that one should rather say that rationalism absolutizes conceptual knowledge and that irrationalism absolutizes idea-knowledge (that is, concept-transcending knowledge). Interestingly, in his contribution to the *Festschrift* of Van Til, Dooyeweerd actually defines rationalism correctly: "Rationalism as absolutization of conceptual thought" (Dooyeweerd 1971, 83). Nominalism is rationalistic and irrationalistic at once, for it acknowleges universality within the human mind and what is purely individual outside the human mind.

elevates human understanding (and the thought categories) to be the a priori formal law-giver of nature.[3]

The supreme reign of conceptual rationalism was soon undermined by the element of relativity carried to its extreme by historicism at the beginning of the nineteenth century. The preceding eighteenth century is the era of conceptual rationalism, followed by romanticism and post-Kantian freedom-idealism. Enlightenment was both rationalistic and individualistic. Romanticism initially switched to an irrationalistic individualism but owing to its anarchistic consequences it moved on to a full-blown irrationalistic universalism. Throughout this development humanism remained faithful to its inherent ideal of autonomy. Rationalism turned the subject (*autos*) into a mere reflex of the universal law (compare Kant's categorical imperative), whereas irrationalism explored the opposite direction by giving primacy to subjectivity—with law merely reflecting what is unique and individual: whether *autos* or *nomos*, the end result is still *auto-nomy*.

The influence of historicism increasingly questioned the supposed universally valid construction of reality by human reason, though at once it turned into a victim of the relativistic consequences of its own orientation. In combination with the linguistic turn by the end of the nineteenth century and the beginning of the twentieth century, this process inspired the idea that the lifeworld of a person is the product of one's own making and that social reality is also solely a human construction.[4] All of this unfolded against the background of the linguistic turn which switched from the cause-effect relation—the stronghold of the classical humanistic science ideal—to meaning and interpretation (see Appleby et al. 1996, 1). The emphasis on language reinforced the relativism of historicism because every interpretation calls forth a (slightly) different interpretation. For this reason, postmodernism embodies the combining force of historicism and the linguistic turn which increased the threat of relativism. The initial claim that everything is subject to historical change is taken further in the claim that everything is interpretation.

In spite of diverse attempts to relativize human endeavours, also found in the developments within the neo-Kantian Baden school of thought (including Windelband, Rickert, and Weber), all these attempts continually got stuck in conditions for logicality, historicity, and linguisticality, which are all universal features holding for being human.

---

[3] "Understanding creates its laws (*a priori*) not out of nature, but prescribes them to nature" (Kant [1783] 1969, 2, 320, sect. 36). See also Holz (1975, 345–358).

[4] The titles of the following books underscore this development: *The Social Construction of Reality: A Treatise in the Sociology of Knowledge* (Berger and Luckmann 1969); and *Der sinnhafte Aufbau der sozialen Welt* (The meaningful construction of the social world—Schutz 1974).

How Does the Idea of Creation Order Relate to the Idea of a Cosmic Order and Law-Order in Reformational Philosophy?

Dooyeweerd notes that Abraham Kuyper pointed out that the great movement of the Reformation could not continue to be restricted to the reformation of the church because its "biblical point of departure touched the religious root of the whole of temporal life and had to assert its validity in all of its sectors." Kuyper "began to speak of 'Calvinism' as an all-embracing world view which was clearly distinguishable from both Roman Catholicism and Humanism" (Dooyeweerd 2013, 1). Yet later on Dooyeweerd rejected the term *Calvinistic* and simply preferred to speak about *Christian philosophy* without any further qualification (Dooyeweerd 1997, 1:524). One of the lasting expressions eventually employed to designate this philosophical trend is *reformational philosophy*.

God Is Not Subject to Laws, But He Is Also Not Arbitrary

In their reformational philosophy, Dooyeweerd and Vollenhoven distinguish between God and creation—with *law* as the boundary connecting God and creation. Vollenhoven points out that the expression *Deus legibus solutus* is derived from Calvin who opposed aristocratic nominalism which accepted the absoluteness of monarchical power (Vollenhoven 1933, 295–296).[5] Dooyeweerd mentions that Duns Scotus at least wanted to tie the absolute despotic power of God (*potestas Dei absoluta*) to the first table of the law, as an expression of "God's holy and good Being." However, Ockham's appeal to God's despotic arbitrariness led to the subsequent process of secularization of modern humanism (Dooyeweerd 1997, 1:186ff.).

The way in which Calvin sidestepped the unacceptable position of nominalism concerns his acknowledgement of the elevation of God above his law for creation. Although God is not subject to his creational law, in his providential care he is faithful to his law—without becoming lawless or arbitrary. The classical formulation is *Deus legibus solutus est, sed non exlex* (God is not subject to laws, but he is also not arbitrary). Bohatec adds that for Calvin God's almighty will is arranged by equity and by what is just (see Bohatec 1940, 1–28).

In the first volume of his *A New Critique of Theoretical Thought*, Dooyeweerd explains his assessment of Occam's position under the heading "The nominalistic conception of the *potestas Dei absoluta* entirely contrary to its own intention places God's Creative Will under the boundary-line of the lex" (Dooyeweerd 1997,

---

[5] In Vollenhoven's "Notes" (*Aantekeningen*) to this work he mentions places in the writings of Calvin where this formulation is found: "The places are De aeterna-praedestinatione, 1552 (Corpus reformatorum 36, colomn 361) and Commentarius in Mosis libros V, 1563 (Corpus ref. 52, colomn 49 and 131)" (Vollenhoven 1933, 27; see n. 480).

1:186–188). Occam allows for the possibility that God just as well could have sanctioned an "egoistic" ethics and even holds that the first table of the Decalogue is "a mere product of divine arbitrariness" (ibid., 187). However, Dooyeweerd correctly points out that "arbitrariness" presupposes a normative standard violated by anti-normative (arbitrary) actions. Ascribing arbitrariness to God places God's will under the borderline of the law (*lex*).

When the Vrije Universiteit was established in 1880 it was a result of drivenness by the Christian life and world view. Kuyper aimed at a university free from the interference of both church and state. Western culture currently experiences a situation in which the Christian life and world view is still threatened by non-Christian orientations—mainly physicalistic and evolutionistic in nature. Kuyper's assessment in 1892 still holds today: "The theory of evolution is the 'formulary of unity' ... which currently unites all priests of modern science in their secularized temple" (quoted in Zwaan 1977, 40).[6]

The Encompassing Ontic Scope of the Creation Order: Its Four Dimensions

When Kuyper discusses the "ordinances of God" (such as in Kuyper 1959, 56) he knows that the term *law* does not merely intend the "Ten Commandments; not even the Mosaic law, nor the moral or ceremonial law." Instead, "what must come into view is that whole concatenation of laws, in every creaturely thing, by which everything exists that God created on, or above, or under the earth" (quoted in Veenhof 1939, 30).

This perspective highlights the idea that there is a cosmic order or creation order. But it was Dooyeweerd who elaborated these seminal insights into a more articulate understanding of the creation order (cosmic order). He does that by distinguishing between four dimensions within created reality: (i) the central religious dimension, subject to the central command of love; (ii) the dimension of cosmic time; (iii) the dimension of modal aspects; and (iv) the dimension of individuality structures.[7]

Ground Motive and Ground Idea

Within each one of these cosmic dimensions a strict correlation between law (law-side) and factual side (sometimes also designated as subject-side) prevails. Furthermore, on the factual side one finds subject-subject relations and subject-object relations (except for the numerical aspect in which solely subject-subject

---

[6] "De evolutietheorie is het 'formulier van eenigheid,' dat op dit oogenblik alle priesters der moderne wetenschap in hun geseculariseerden tempel vereenigt."

[7] See the introductory chapter to this volume where Gerrit Glas and Jeroen de Ridder articulate a brief understanding of these basic distinctions within reformational philosophy.

relations are present). According to Dooyeweerd, theoretical thinking finds its religious starting point in the central ground motive operative in the root of humankind—these (above-mentioned) ground motives are therefore communal in character (the form and matter motive; the biblical motive of creation, fall into sin, and redemption through Christ in the communion of the Holy Spirit; the scholastic motive of nature and grace; and the modern humanistic motive of nature and freedom—see Dooyeweerd [2012]).

Whereas the ground motive directing human life from its core is supra-theoretic, every scholarly (scientific) act proceeds on the basis of a triunity of (transcendental) ideas, concerning the diversity and coherence, the radical unity, and the ultimate origin of the universe. The term *transcendental* here intends to capture those conditions (law-order) which make our experience of reality possible. Taken together Dooyeweerd refers to them as the transcendental ground idea of philosophy. The transcendental ground idea, as directed by the biblical basic motive, represents what Dooyeweerd initially intended with the Dutch title of his magnum opus, *Dewijsbegeerte der wetsidee*. The triunity of transcendental ideas, the transcendental ground idea of philosophy, cannot be deduced from its ground motive.

Ismic Orientations Embody Alternative Transcendental Ground Ideas

Since the history of philosophy and the special sciences are constantly burdened by multiple one-sided monistic orientations, such as the mechanistic main tendency of classical physics, subsequent materialistic physicalism, biologistic vitalism, psychologism, logicism, historicism, and so on—it should be kept in mind that each one of them depends upon a specific transcendental idea regarding the diversity of modal aspects. Every *ismic* stance did discern a genuine structural element of God's creation order, however much it may be distorted by its reification or absolutization. Before a scholar criticizes a particular *ismic* orientation, the following question ought to be answered: What feature of the creation order was noticed in such an orientation? Clearly, without detecting the numerical aspect first it would be impossible to develop a distorting arithmeticistic perspective on it.[8] The same applies to the other just-mentioned monistic orientations. All of them proceed from a particular transcendental idea regarding the unity and diversity within reality.

Moreover, if the notion of a creation order is equivalent to the idea of a cosmic law-order, then a crucial issue would be to develop a more precise idea of what a *law* in an ontic sense is—keeping in mind that ontic laws are not merely *epistemic* in nature, such as found in the nominalistic conviction of Descartes who asserts that

---

[8] Kurt Gödel was very "fond of an observation that he attributes to Bernays": that a "flower has five petals is as much part of objective reality as that its color is red" (quoted in Wang 1988, 202). To account for "mathematical objects" Gödel introduces the idea of "semiperceptions" which may represent "an aspect of objective reality" (ibid., 304).

"number and all universals are mere modes of thought" (*Principles of Philosophy* 1.58) or in Kant's above-mentioned view of human understanding as the formal law-giver of nature.

Articulating the Idea of Ontic Laws and Ontic Principles: The Transcendental-Empirical Method

Dooyeweerd had a sound understanding of the destructive effects of modern nominalism, particularly regarding its rejection of ontic universality. Unfortunately, he did not escape from its aftereffects because he accepts the nominalistic legacy by denying the universality of factual reality and—as a consequence—by identifying law and lawfulness.

From the perspective of the history of philosophy, Dooyeweerd's reformational philosophy is unique in the way it identifies and distinguishes between the dimension of cosmic time, the dimension of modal aspects, and the dimension of concrete (multi-aspectual) entities (designated by him as individuality structures).

The idea of a creation order entails that these dimensions of reality are lying at the basis of our experience of reality—in the transcendental sense of making it possible. Investigating what makes our experience possible is actually directed at God's law-order. Another way to address this state of affairs is to depict it as a *transcendental-empirical* method of research.

The physicist Stafleu and the jural scholar Hommes prefer to designate Dooyeweerd's new approach as employing this transcendental-empirical method. One of the reasons for the powerful impact of this method is certainly related to the intermodal criterion of truth which it employs. A mere logical contradiction remains enclosed within the sphere of logicality, because logic can only assert that two contradictory statements cannot both be true at the same time, as already observed by Kant. But it cannot tell which one is true. It is the principle of sufficient ground (reason) that directs thought beyond the confines of logic. Yet it is the irreducibility of the various aspects of reality that serves as the foundation for the more-than-logical ontic principle of truth, the principle of the excluded antinomy (*principium exclusae antinomiae*). Confusing two spatial figures, such as a square and a circle, is *intra*-modal because it concerns realities within one aspect only.[9] But confusing two distinct aspects, such as the aspects of space and movement in Zeno's paradoxes, gives rise to an *inter*-modal clash of laws, a genuine *antinomy* (*anti* = against; *nomos* = law).

The transcendental-empirical method is a straightforward exploration of the idea of a creational order or a cosmic law-order. Accepting this method entails a rejection of the biblicistic idea that the creation order is revealed in the Scriptures. It explains

---

[9] Cassirer mentions the example of a "round square" (Cassirer 1910, 16), but Kant already first introduced it in the following form: "a square circle is round/is not round" ("ein viereckiger Zirkel ist rund/ist nicht rund"; see Kant [1783] 1969, 341, sect. 52b).

Dooyeweerd's opposition to every conception of "a scriptural philosophy that looks for support in specific Bible texts for intrinsically philosophical and in general scholarly problems and theories. It actually merely boils down to 'positing a few privileged issues' about which the Bible would give explicit statements, while for the rest, where such special texts are not found, one at leisure can continue to fit into a mode of thinking driven by intrinsically un-biblical motives" (Dooyeweerd 1950, 3–4).[10]

The transcendental-empirical method has an inherent dynamic openness:

1. By virtue of the radical depth of sin, all human insights (also those discovered by applying the transcendental-empirical method) remain *provisional, fallible*, and *open to improvement*.
2. The wealth of human experience, embedded in God's creational order, is constantly deepened through scientific and technical advances, implying that every investigation may be altered and even refuted in the light of discovering new states of affairs.

Moreover, the idea of a creation order also entails a non-reductionist approach, rejecting the absolutization or deification of anything within creation. At the same time a non-reductionist ontology in principle takes serious the principle of the excluded antinomy.

The Idea of a Creational Order: Modal Laws and Type Laws

Introducing this distinction should be seen as the reaction of reformational philosophy to the above-mentioned Greek-medieval legacy of the *substance* concept and the modern inclination to overemphasize *functional* relations. The significant contribution of reformational philosophy is that every modal functional law holds for whatever there is, and whatever there is either has a subject function or an object function within all modes or aspects of reality—as explained in Glas and De Ridder's introduction.

The modal universality of modal aspects is unspecified. When directing our theoretical attention at the modal aspects or functions of reality we are not classifying entities according to the kinds or *types* to which they belong. The mere distinction between economic and uneconomic, for example, is not specified in any typical way. Diverse societal entities (such as universities, nuclear families, states, and firms) have to observe economic principles (act frugally and avoid what is excessive). Both a state and a business can waste their money (and thus act uneconomically) and both are called to function under the guidance of economic considerations of frugality.

---

[10] Kock underscores the same point: "We believe, however, with the philosophy of the cosmonomic idea, that religion (as pre-scientific root-dynamics) provides only orientation and direction to thought and that by it no single scientific problem is brought to a solution" (Kock 1973, 11; see 12).

Although we may differ about what norm-conformative or antinormative behaviour may be, such a difference of opinion presupposes the acceptance of the ontic reality of contraries like these. They are found in all post-psychical aspects where they analogically reflect the coherence between the logical and the other normative aspects of reality.

Whereas modal laws hold universally for all possible classes of entities, type laws only hold for a limited class of entities. Therefore the universality of type laws is specified. The law for being an atom or being a state holds universally for all atoms and all states, respectively. But this universality is specified because not everything in the universe is an atom or a state.

Kant already understood this distinction, for he distinguishes between (a priori) formal laws (thought categories—Kant's account of modal universality) and "empirical laws of nature" (individuality structures)—where the former (a priori universal laws of nature) do not require special perceptions while the latter do presuppose particular observations (the equivalent of what leads positivism to its emphasis on experimentation—on the road to the discovery of type laws/individuality structures).

Dooyeweerd's philosophy made a significant contribution to the analysis of the concept of a law and a principle. Let us now briefly explain some of the intricacies involved in such an analysis. The focus will first be on (natural) laws and then on (norming) principles.

Natural Laws

When it is asserted that modal laws hold universally (everywhere), the spatial aspect is prominent. Without an awareness of space we would be unable to understand the universality displayed by laws. Likewise, in spite of all the spectacular changes that occurred in the history of the universe, physics still has to acknowledge universal constants. The remarkable fact is that they are spread over the first four aspects as the basic units of measurement: number (mass/kilogramme), space (length/meter), kinematical (duration/velocity/second), and physical (charge/quantum/energy).

Natural laws not only strike us in their universality and constancy (persistence) but also in their operation, in the effect they have on events, and in the fact that they are valid in the sense of being in force. What is normally seen as the necessity of laws therefore relates to the universality of their effect and to their enduring being in force.

Clearly describing a law requires an awareness of number (it is *one* law), of space (its *universality*), that it is persistent (*constancy*), and that it holds (it is being *in force*). Without highlighting the aspects from which these terms are derived, the paleontologist Schindewolf contemplates the fact that geologists suggest that natural laws are subject to change and therefore not constant. His reaction is categorical:

> If one defines the laws of nature as rules according to which processes always take place in the same way everywhere, there can naturally be no question of mutability and development over time. It is only our formulation of the laws of nature that is mutable. As soon as we learn from experience that the concept of a law is not universally applicable because it is somehow contingent upon time, then the law should be excluded from the formulation. (Schindewolf 1993, 5)

From this brief analysis it is also clear that within the creation order no single law "stands on its own." For example, the physical aspect of reality presupposes the aspects of number, space, and movement, each with their own unique laws. In fact, the meaning of the physical aspect is constituted by its interconnections with non-physical aspects. For example, the first main law of thermodynamics, traditionally designated as the law of *energy conservation*, should actually be called the law of *energy constancy*, which displays a kinematic retrocation (backward-pointing analogy) to the original meaning of the kinematic aspect of uniform (constant) motion (cf. Strauss 2011c).

Summarising an extensive analysis may provide us with a complex definition of a law of nature. The compound or complex basic concept of a natural law may therefore be formulated as follows:

As a unique, distinct, and universally valid order for what is factually correlated with and subjected to it, a natural law constantly holds (either in an unspecified way as in the case of modal laws or in a specified way as in the case of type laws) within the domain in which it conditions what is subjected to it.

Norming Principles

An analysis of the nature of cultural norms or principles has to follow the same path, because it is equally complex in the sense that it comprises terms derived from multiple modal aspects. We have noted that normativity always reflects the logical principle of non-contradiction. Furthermore, principles are not valid per se such as natural laws, because principles require human intervention to be made valid, to be enforced. Diverging philosophical traditions designate giving shape to pre-positive principles by positivizing them. Such positivizations ought to take unique historical circumstances into consideration. Conservatism sticks to positivized principles and is therefore reluctant to alter positivizations while a revolutionary approach attempts to break all connections with the past. Reformation by contrast observes both an element of continuity and an element of discontinuity in history. Within the science of law the legacy of natural law accepts above positive law another jural order supposedly holding for all times and places in its positivized form. Legal positivism, in turn, does not want to accept constant principles.

Insofar as norming principles are universal and constant they are not yet positivized—and insofar as they are positivized they are no longer universal in an unspecified sense of the word. Let us now look at an equally complex definition of a principle.

A principle is a universal and constant point of departure that can only be made valid through the actions of a competent organ (person or institution) in possession of an accountable (responsible) free will enabling a normative or antinormative application of the principle concerned relative to the challenge of a proper interpretation of the unique historical circumstances in which it has to take place (see Strauss 2011a, b).

Note that this formulation implicitly employs the gateway of various modal aspects—which underscores that the term *principle* is a complex or compound fundamental scientific concept—in distinction from the elementary basic concepts in the various disciplines that only appeal to a single particular analogy within the structure of an aspect of reality. The nature of modal analogies, viewed within the context of a distinction between law-/norm-side and factual side, opens up the possibility of a philosophical analysis of modal principles. Every analogy on the law-side of a normative aspect provides us with a fundamental modal principle.

Within the natural aspects of creation a similar coherence on the law-side could be discerned. The meaning of the physical aspect, for example, is constituted by its interconnections with non-physical aspects. The first main law of thermodynamics, traditionally designated as the law of *energy conservation*, should actually be called the law of *energy constancy*, which is (on the law-side) a kinematic retrocipation (backward-pointing analogy) to the original meaning of the kinematic aspect of uniform (rectilinear/constant) motion.

Regarding the philosophical foundations of the special sciences, this method first of all requires an analysis of the analogical basic concepts of a discipline—also known as its elementary basic concepts. Examples of elementary basic concepts within the science of law are found in expressions such as *jural causality* (physical analogy of cause and effect within the jural aspect), *legal order* (arithmetical analogy), *legal constancy* and *dynamics* (kinematic and physical analogies taken in their mutual coherence), *legal differentiation* and *integration* (biotic analogy), *jural accountability* (logical-analytical analogy), *juridical expression* and *interpretation* (analogy of the sign mode), and *juridical economy* (avoiding what is excessive—economic analogy).

These concepts cut across all the subdivisions of law because they partake in the modal universality of the jural aspect. Once the elementary basic concepts have been analyzed, a discipline should contemplate its *complex* or *compound basic concepts*, that is, those *totality concepts* erected by incorporating multiple elementary basic concepts at once. Concepts such as legal subject, legal object, legal personality, and legal normativity are examples of compound basic concepts within the science of law. Only at this point could the transition be made to typical concepts, involving also the societal entities found within a differentiated society.

It should be kept in mind that Dooyeweerd first tested his new systematic philosophical distinctions within his own field of speciality, the discipline of law, before he made public his general philosophical distinctions and insights. Elaborating this immense task also embodies his positive appreciation of the meaning of the creation order for an inner reformation of the various academic disciplines.

Without an Order of Creation There Could Be No Future

The general theme of the 2011 conference in Amsterdam was captured in the following question: "Is the idea of creation order still fruitful?" The key point in what we have discussed thus far is that without a creational order there is no future for the universe. This assessment entails that constancy is indeed appreciated as the condition for change and that we therefore have to avoid the postmodern emphasis on change at the cost of constancy. It may result in an emphasis on the *eschaton* at the cost of the *proton*. Olthuis correctly remarks:

> The current eschatological orientation in theology which tends to seek even the beginning in the end will need revision. The Bible begins with Genesis and Genesis begins with creation. The Scriptures see the Gospel as the link connecting creation and consummation. And this link between past and future is revealed as the Word which connects the end with the beginning, the consummation with the creation. "I am the Alpha and the Omega, the first and the last, the beginning and the end" (Rev. 22:12). A proper vision of the consummation requires a proper appreciation of the beginning. Without this understanding, the fulfilment lacks substantial content and tends to evaporate into pious words about hope. A non-robust view of creation emasculates the gospel, for it is the creation which is brought to fulfilment in Jesus Christ even as it began in him. (Olthuis 1989, 32–33; see also Strauss 2009, 197)

The distinction between *order for* (*law for*) and *orderliness of* (i.e., *law-conformity*) implicitly underscores the insight that without an order of creation there could be no future. Julian Huxley provides us with an example of confusing this distinction. In 1959, at the occasion of commemorating the appearance of Darwin's *Origin* in 1859, he said: "This is one of the first public occasions on which it has been frankly faced that all aspects of reality are subject to evolution, from atoms and stars to fish and flowers, from fish and flowers to human societies and values—indeed, that all reality is a single process of evolution" (Huxley 1960, 249).

Note the discrepancy between the claim that "all aspects of reality are subject to evolution" and the statement that "all reality is a single process of evolution": the first one elevates evolution to an all-encompassing law to which all aspects of reality are subjected and the second one reduces all laws to what "all reality is," namely, "a single process of evolution"! Of course, the latter statement is contradicted by the assumption that random mutation and natural selection have to be constants which determine the ongoing process of evolution, for if these elements of constancy are not accepted, no evolutionary change would be possible. In other words, if "all reality is a single process of evolution" it does not allow for the determining and delimiting "law-role" of the combination of random mutation and natural selection.

Concluding Remark

The long-standing insight that constancy lies at the basis of change contains the key to a positive interpretation of the future of the idea of a creational order. We argued that the Greek-medieval legacy stumbled upon the difference between law for (Plato's ideas) and lawfulness of (Aristotle's universal substantial forms). However, the influence of modern nominalism undermined the idea of universality outside the human mind by stripping creation of its order for and orderliness of—thus depreciating factual reality into a state of chaotic structurelessness. The vacancy thus created was filled by elevating the human being to become a law unto itself—up to the extreme rationalistic position assumed by Kant in his claim that human understanding is the a priori formal law-giver of nature. The idea of a creational order reverts from an epistemic to an ontic perspective, which makes possible an alternative approach to the various dimensions of reality, known as the transcendental-empirical method. This method enables a new way to articulate the idea of a creational order or a law-order, illustrated by the provision of a definition of a natural law and norming principles. The future of the idea of a creational order depends on a proper understanding of the constancy and universality of such an order. Whenever an attempt is made to reduce the rich diversity of aspects and entities within creation, theoretical antinomies surface. The alternative option is to pursue a non-reductionist ontology.

References

Appleby, Joyce Oldham, Elizabeth Covington, David Hoyt, Michael Latham, and Allison Sneider. 1996. *Knowledge and Postmodernism in Historical Perspective*. New York: Routledge.
Berger, Peter L., and Thomas Luckmann. 1969. *The Social Construction of Reality: A Treatise in the Sociology of Knowledge*. London: Allen Lane.
Bohatec, Josef. 1940. Autorität und Freiheit in der Gedankenwelt Calvins. *Philosophia Reformata* 5 (1): 1–28.
Cassirer, Ernst. 1910. *Substanzbegriff und Funktionsbegriff*. Darmstadt: Wissenschaftliche Buchgesellschaft.
———. 1971. *Das Erkenntnisproblem in der Philosophie und Wissenschaft der neueren Zeit*. Vol. 2. 3rd ed. Darmstadt: Wissenschaftliche Buchgesellschaft.
Comperz, Theodor. 1964. *A History of Ancient Philosophy*, 7th imp. London: William Clowes & Sons Ltd.
Dooyeweerd, Herman. 1950. De strijd om het schriftuurlijk karakter van de wijsbegeerte der wetsidee. *Mededelingen van de Vereniging voor Calvinistische Wijsbegeerte,* July 1950: 3–6.
———. 1971. Cornelius Van Til and the Transcendental Critique of Theoretical Thought. In *Jerusalem and Athens: Critical Discussions on the Philosophy and Apologetics of Cornelius Van Til*, ed. E.R. Geehan, 74–89. Nutley: Presbyterian and Reformed Publishing Company.
———. 1997. *A New Critique of Theoretical Thought,* The Collected Works of Herman Dooyeweerd, Series A. Vol. 1–4. Lewiston: Edwin Mellen Press.
———. 2012. *Roots of Western Culture: Pagan, Secular, and Christian Options*, The Collected Works of Herman Dooyeweerd, Series B. Vol. 15. Grand Rapids: Paideia Press.

———. 2013. *Christian Philosophy and the Meaning of History*, The Collected Works of Herman Dooyeweerd, Series B. Vol. 13. Grand Rapids: Paideia Press.
Galileo, Galilei. (1683) 1973. *Unterredungen und mathematische Demonstration über zwei neue Wissenszweige, die Mechanik und die Fallgesetze betreffend*. Darmstadt: Wissenschaftliche Buchgesellschaft.
Holz, Friedrich. 1975. Die Bedeutung der Methode Galileis für die Entwicklung der Transzendentalphilosophie Kants. *Philosophia Naturalis* 15 (3): 344–358.
Huxley, Sir Julian. 1960. The Evolutionary Vision. In *Evolution After Darwin: The University of Chicago Centennial, Issues in Evolution*, ed. Sol Tax and Charles Callender, vol. 3, 249–261. Chicago: University of Chicago Press. https://archive.org/stream/evolutionafterda03taxs/evolutionafterda03taxs_djvu.txt. Accessed 10 April 2017.
Kant, Immanuel. (1783) 1969. *Prolegomena einer jeden künftigen Metaphysik die als Wissenschaft wird auftreten können*. Hamburg: Felix Meiner.
Kock, Pieter de Bruyn. 1973. *Christelike wysbegeerte: Standpunte en probleme*. Bloemfontein: VCHO.
Kuyper, Abraham. 1959. *Het Calvinisme*, 3rd imp. Kampen: Kok.
Olthuis, James. 1989. The Word of God and Creation. *Tydskrif vir Christelike Wetenskap* 25 (1st & 2nd quarter): 25–37.
Schindewolf, Otto H. 1993. *Basic Questions in Paleontology: Geologic Time, Organic Evolution, and Biological Systematics*. Chicago: University of Chicago Press.
Schnatz, Helmut, ed. 1973. *Päpstliche Verlautbarungen zu Staat und Gesellschaft, Originaldokumente mit deutscher Übersetzung*. Darmstadt: Wissenschaftliche Buchgesellschaft.
Schutz, Alfred. 1974. *Der sinnhafte Aufbau der sozialen Welt*. Frankfurt am Main: Suhrkamp. First published 1932.
Strauss, Danie F.M. 2009. *Philosophy: Discipline of the Disciplines*. Grand Rapids: Paideia Press.
———. 2011a. Normativity I – The Dialectical Legacy. *South African Journal of Philosophy* 30 (2): 207–218.
———. 2011b. Normativity II – An Integral Approach. *South African Journal of Philosophy* 30 (3): 105–128.
———. 2011c. Relasiebegrippe in die lig van die vraag of Einstein in die eerste plek 'n "relatiwiteitsteorie" ontwikkel het. *Suid-Afrikaanse Tydskrif vir Natuurwetenskap en Tegnologie* 30 (1), art. #28, 8 pages. https://doi.org/10.4102/satnt.v30i1.28.
Veenhof, Cornelis. 1939. *In Kuyper's lijn*. Oosterbaan: Goes.
Vollenhoven, Dirk H.Th. 1933. *Het Calvinisme en de Reformatie van de wijsbegeerte*. Amsterdam: H.J. Paris.
von Weizsäcker, Carl F. 2002. *Große Physiker, Von Aristoteles bis Werner Heisenberg*. Munich: Deutscher Taschenbuch Verlag.
Wang, Hao. 1988. *Reflections on Gödel*. Cambridge: MIT Press.
Zwaan, J. 1977. On Evolution and Evolutionism. *Beweging* 41 (June): 40.

Creation Order in the Light of Redemption (1): Natural Science and Theology

Henk G. Geertsema

Abstract This chapter discusses the order of creation in relation to new creation. The first part is about science, eschatology, and the natural world. It argues that theoretical reflection should choose its starting point in our integral experience, not in scientific knowledge. I propose to relate the continuity between creation and new creation to what is called in Dooyeweerdian philosophy *modal aspects* and *typical structures*. Discontinuity can be connected with the specific laws and concrete properties that characterize these aspects and structures. The second part is about resurrection and personal identity. I reject substance dualism and its derivatives in which continuity between this life and the next is connected with mental and discontinuity with material properties. As an alternative, I propose the distinction between personal and structural identity. Personal identity can only be expressed in such personal terms as *I* and *you*. Structural identity covers both material and mental properties. I argue that there will be continuity and discontinuity in both respects. This may even apply to the intermediate state between death and resurrection, when we are "with Christ." I suggest the possibility that we will have both material and mental properties in that state. As a general point it is argued that creation order should primarily be understood as the order that we live and experience, not as the abstract order described in scientific laws.

Keywords Creation order · Creation–new creation · Continuity–discontinuity · Time · Dooyeweerdian philosophy · Substance dualism · Second-person perspective · Personal identity

Introduction

Is there a future for creation order? The idea of creation order is meant to characterize the basic nature of the world we live in as coming from God and, therefore, full of meaning. Of course, not everything of the order of the world goes back to the

H. G. Geertsema (✉)
Vrije Universiteit Amsterdam, Amsterdam, The Netherlands
e-mail: hgg@bos.nl

creation or, better, to God the Creator. There is human formation; there is distortion; but according to the idea of creation order, the basic constituents of our world and its order have their foundation in God's act of creation. As such, I contend, the idea is part of the Christian faith.

But the idea raises also serious questions. How are we to distinguish, in relation to the order (and disorder) we find around us, between what comes from God and creation, what from human formation, and what from distortion by evil? And how do we relate the results of science to the God-given order of creation? Actually, the issue goes even deeper: What is the reality of the order we observe? To what extent is it, as such, a human construct? Is it possible, amid all these questions, to give the idea of creation order a meaningful content within a theoretical context? Is there a future for creation order?

In my contribution to this volume, I take the liberty to interpret the expression *the future of creation order* in a different sense: What is the future for the order of creation itself? Understood this way, the idea of creation order is accepted, at least for the moment. Instead of asking the question about the vitality of the idea of creation order for contemporary and future discussions, I want to place the order of creation itself in the perspective of the future. Taking *creation* in the biblical sense, my question concerning "the future of creation order" is about the relation between the order of creation and the new creation.

Formulating my topic this way may raise the question of how it is possible in philosophy to speak about the new creation. Actually, several publications have recently appeared that I can make use of. They can be divided in two groups. The first group deals with science and eschatology. Some of these publications focus on the laws of nature in relation to new creation, others concentrate on human identity before and after the resurrection. They are not necessarily philosophical in a strict sense, but they clearly do imply philosophical issues. These issues will be discussed here. The second group of publications is in the field of political philosophy. Several contemporary philosophers show a surprising interest in the writings of the apostle Paul. One of the themes they discuss is "messianic hope." I will discuss this second group in part 2 of my "Creation Order in the Light of Redemption" (Geertsema forthcoming).

The current chapter is divided in two main sections: the first is entitled "Science, Eschatology, and the Natural World," and the second "Resurrection and Personal Identity." Each section starts with a characterization of the issues to be discussed, followed by some biblical and philosophical comments. The main emphasis is on the latter, though. My focus will be on two questions: (1) How should we understand the nature of creation order, and what should be the relationship between our everyday experience and the results of science, both in our interpretation of creation order as such and in relation to the new creation? (2) What kind of theoretical framework can we best make use of to interpret the continuity and discontinuity between the first and the new creation? Throughout the chapter the question concerning the relevance of the new creation for our understanding of creation order in the present time will be in the background of the discussion.

Evidently, I will be able to touch only the surface of the different issues. Yet, I hope it will become clear that important philosophical issues are involved when we reflect in a theoretical context on the biblical notion of the order of creation both in its original intention and in the perspective of the new creation as opened up by the Scriptures. My own philosophical approach is informed by what is usually called *reformational philosophy* (see Glas and De Ridder's introductory chapter in this volume).[1] I hope to show throughout the paper how fruitful this philosophy can be for the contemporary philosophical discussion of the issues at hand.

Science, Eschatology, and the Natural World

Recently some publications have appeared about science and biblical eschatology. I will discuss especially *The End of the World and the Ends of God*, a collection of papers edited by Polkinghorne and Welker (2000a), and *Christian Eschatology and the Physical Universe*, by David Wilkinson (2010). They are rich in content and very stimulating. I can recommend them to anyone interested in the subject. In the first section I focus on one specific subject: the ambiguity of our physical universe. In the biblical comments I relate this to the biblical perspective of creation, but I also broaden the perspective a little. In the philosophical comments an important issue is whether philosophical reflection on the order of creation in the light of the new creation should primarily focus on science or start with our integral experience. Another topic concerns continuity and discontinuity between creation and new creation. I suggest that some basic theories of Herman Dooyeweerd are taken as a theoretical framework to discuss this relationship.

Contemporary Discussion

The issue I want to mention here is the ambiguity of our physical universe. On the one hand, contemporary science sketches a development in which a rich diversity both of physical elements and living creatures has evolved in the course of time. The physical universe has developed from the Big Bang into a hardly imaginable manifold of galaxies, stars, and physical elements and structures. The latter are the necessary condition for the evolution of life on earth with its abundant diversity of forms. Within the perspective of creation the complexity of both can only increase the praise for the Creator.

On the other hand, though, the same contemporary scientific picture implies that the prospects for life and the physical universe in the long run are very dark. There is the likelihood of the impact of asteroids and comets with possibly disastrous consequences for life on earth. This could happen any time in the future. The occur-

---
[1] For further introduction, see Dooyeweerd (1960, 1979), and Clouser (2005).

rence of two other events, although expected for the far future, is not just a possibility but, according to present knowledge, a certainty. The first concerns the death of the sun, which will also imply the end of life on earth. The other expected event is the end of the universe itself. According to Wilkinson, the prospect for the universe is what could be called a cold death (Wilkinson 2010, 16). Of course, the time span for these events is almost beyond human imagination. Yet the question arises of how this bleak perspective relates to the perspective of creation. Is it possible that the end of the universe makes the whole project of creation ultimately futile and meaningless?[2]

The issue does not only concern Christian believers. Wilkinson mentions Freeman Dyson and Frank Tipler and the theories they have developed to overcome this bleak perspective with the means of science itself (Wilkinson 2010, 18–19). He himself takes a different approach, and so do the other authors in the books I mentioned. They point to the biblical perspective. All these authors agree on one point: in order for creation to fulfill the ultimate purposes of God, the present world with its laws and structures that point to death and decay as the final end cannot be its ultimate destination. That would be against the faithfulness of God in relation to what he has started. A new creation is necessary in which the present world is transformed in such a way that time and decay, life and death, are no longer intricately connected. This means that at least some laws of nature have to change. There needs to be continuity between the present world and the world to come; otherwise, the first creation would still be futile and ultimately meaningless. But there needs to be discontinuity as well, or death and decay will have the final word (Polkinghorne 2000, 30, 38–41; Wilkinson 2010, 83, 86, 109, 111ff.).

This, of course, raises all kinds of questions. What can we say about continuity and discontinuity between creation and new creation with the means of science? For now I leave these questions aside and move to my biblical comments.

Biblical Comments

The question for a biblical believer is not about the expectation of the new creation, also in a physical sense. The Bible is clear enough about this (cf., e.g., Wright 2007). But what about the ambiguity of the first creation? Does that fit the biblical picture? Should we not take all death and decay as the result of sin and evil? Is it legitimate to apply the scientific picture of the universe and of life as sketched above to creation as such?

---

[2] It is not just the end of the world that poses this problem. Rather, the ambiguity is already part of the development of life. From the very beginning it appears that life and death are intricately interwoven. The clearest example may be the togetherness of predator and prey. But the issue goes far deeper. Complex order as expressed in living organisms can only exist at the expense of the growth of disorder in the surrounding world. How can death figure in the good creation from the beginning? How is it possible that the temporal nature of the creation involves decay from the outset?

I cannot go deeply into this. But I do believe it is possible to connect the scientific picture of the universe with the biblical story. Obviously, the scientific picture has its limitations. It is based on present knowledge and theories. It remains a projection into the past from our present condition. Science can never take the position of an outside observer, a God's eye point of view. Yet, as it is, with the ambiguity mentioned, it does not necessarily contradict the biblical doctrine of creation. In Romans 8 (vv. 19–22) the apostle Paul writes about creation as subjected to frustration, as bound to decay, as groaning in the pains of childbirth. This could be interpreted as referring to the consequences of the curse mentioned in Genesis 3 after the fall. But I wonder whether this explains the positive implication of the expression "pains of childbirth." Moreover, the word translated here as "frustration" is, in the Greek translation of the Septuagint, the same word as used in Ecclesiastes 1, where it is translated either as "meaningless," as in the New International Version, or as "emptiness," as in the New English Bible. The illustrations given in Ecclesiastes of this meaninglessness or emptiness are mainly taken from the natural world as such, not necessarily as affected by the fall. And there are other indications in the biblical teaching that there is at least an edge of darkness and threat, maybe even of moral evil, to the creation as described in Genesis 1. The night and the sea mentioned in Genesis 1 refer back to the chaos described in verse 2 in terms of darkness and stormy waters. They are set boundaries, but they are still there. Against this background we can understand why we are told in Revelation 21 and 22 that in the new creation there will be no more sea, no more night. The edge of darkness, in the deep figurative sense, and the edge of threatening powers, will be no more. In the language of contemporary science: the ambiguity of the laws of nature will be gone forever. So, I have no basic biblical reservations about the ambiguity as pointed out by Polkinghorne and Wilkinson.[3]

I would like to add a second comment here to avoid the impression that all we can say about the new creation from the biblical perspective is that it will overcome both the ambiguity of the first creation and sin and its consequences. In the new creation, there will also be a new dimension in the sense that God will be all in all (1 Cor. 16:28; cf. Wilkinson 2010, 155–156). I connect this with Revelation where it is said that in the city of God, the new Jerusalem, there is no need for the light of the sun and the moon anymore, "for the glory of God is its light, and the Lamb is its lamp" (Rev. 21:23, cf. 22:5). This clearly refers back to Genesis 1 where it says the sun and the moon are given as light, the sun for the day and the moon for the night (vv. 14–18). In this creation we can see the world around us in the light of the sun, the moon, and the stars. And we know God by faith, both in relation to the world and to ourselves. This faith-knowledge may be confirmed by our experience and reflection, but it will always remain dependent on faith and trust. I take the words of Paul and of Revelation to mean that in the new creation we will see God directly in the world around us and experience him in ourselves; not in the sense of pantheism or panentheism, but in the sense that reality is immediately transparent for us to God.

---

[3] I will shortly come back to the issue of death and decay in relation to life in my discussion of the understanding of time.

The distinction between Creator and creature remains. But the creature as creature of God is directly experienced and seen. Interestingly, Revelation adds the Lamb as the lamp, implying that in the new creation the work of Christ as crucified and risen is also immediately visible. Also in relation to evil and suffering there is continuity and discontinuity: they are definitely overcome, but as such they remain visible. This can only be a reason for more praise and wonder, as the hymns in Revelation 4 and 5 testify.

Philosophical Comments

I limit myself to two topics. First, I focus on the view of science and relate this to our understanding of the order of creation. Then I discuss the nature of time. The focal point here is the relationship between creation and new creation in terms of continuity and discontinuity.

Science and Concrete Experience

In the books I mentioned earlier, the limitations of science are emphasized several times. Yet, there is still a tendency to take scientific knowledge as the model for all our knowledge. This becomes clear when in the discussion of theology and the natural sciences theological knowledge is supposed to be of the same nature as scientific knowledge (Polkinghorne and Welker 2000b, 4, 6; Polkinghorne 2000, 29, 38–39; Wilkinson 2010, 6, 24–25, 51). The impression is given that only the use of scientific method can give reliable knowledge. Implied is the suggestion that, over against our everyday knowledge, only scientific understanding may give us, however tentatively, the true picture of the world.

At this point I would like to call for caution. There is a long tradition in Western thought that aims at overcoming the limitations of our concrete experience by way of knowing through theoretical reason. Parmenides at the beginning of Greek philosophy and Descartes at the beginning of modern times could be mentioned as illustrations. Within philosophy the distinction between appearance and true reality has been very influential. Philosophy relates, at least tentatively, to true reality. Everyday knowledge and experience relate only to the appearances. At the basis of the distinction lies a deep distrust of the senses, at least in terms of their functioning in our concrete experience, and of all knowledge gained from tradition. However tentatively, over against the limitations of our everyday knowledge, a God's eye point of view is aimed at, even when it is acknowledged that this never can be fully achieved. Science runs the risk of continuing this tradition, aiming at a kind of certainty that is no longer dependent on the limitations of our everyday knowledge (cf. Putnam 1981, 49ff.; 1983, 210).

The point I want to make is not that our everyday knowledge is infallible. This evidently is not the case. Scientific research can not only deepen and broaden our everyday knowledge, but also correct it. My point is that it can never replace our everyday knowledge in a full sense. Our concrete experience of the world remains the starting point, and this also applies to scientific understanding, however far it may appear to be removed from our concrete experience. To avoid misunderstanding I should emphasize that by *concrete experience* I do not just mean our sense experience. What I think of is the full integral experience of reality with its many aspects that characterize our concrete existence. To this experience belong not just the senses and analytical distinction but also language and the diversity of human relationships with their basic intuitions, and even faith and worldview as guiding our ultimate understanding (cf. Geertsema 2008).

The reason why I mention this is not only that we should be aware of the limitations of scientific knowledge. It is also important for our understanding of the order of the universe. In contrast with an approach that claims to be guided by unbiased reason and theoretical or scientific analysis, we should take our starting point in our concrete experience as guided by faith. We should recognize the finite nature of human existence. More importantly, as Christians we find the ultimate horizon for understanding the world and ourselves in our relationship as creatures with the Creator, not in an abstract order in terms of the theoretical laws of science. This relationship encompasses all other relationships, all things (creatures!), and events. It determines the meaning and purpose of life, the ultimate nature of evil, and how this is overcome. It makes the universe deeply personal in the sense that its Origin can only be understood if personal categories are applied.

Because our concrete experience is foundational to all our knowing, and our relationship as creatures with the Creator is primarily a lived relationship, creation order should in the first place be understood as the concrete order which we live, experience and understand, ultimately, by faith. We should not take it primarily in terms of the laws we find by means of scientific research and theoretical models. The order of creation is, in the first place, lived and experienced in the concreteness of our being. We get to "know" it when we learn to walk, to speak, to relate to others. Philosophy and science relate to this order by distinguishing different aspects we can analyze theoretically, but these aspects by themselves are not the order of creation. They are, if analyzed correctly, abstract elements of the full order of creation. Philosophy and science can certainly open up and even correct our understanding both of the order we experience and the disorder, which is also part of the world. Yet, as theoretical knowledge can never replace our concrete experience of the world, the order of reality should never be identified with what we can say about it by means of theoretical analysis. The full order of creation cannot be grasped by us anyway. We do not have a God's eye point of view. Our first "contact" with it lies in our concrete being and living. Reflection, both practical and theoretical, can only clarify, open up, and sometimes also correct this primary understanding.

This distinction between concrete experiential and abstract theoretical knowledge should also be taken into account when we discuss the new creation. Whatever we say by means of science will remain abstract in itself. It may concern valid elements, but it will never express the fullness of the new creation. For both creation and new creation we will need our concrete experience, opened up by faith in the present life with all its limitations and misunderstandings, and opened up in the life to come to its full potential. Therefore, there is some risk involved when we take scientific knowledge as the model for all our knowledge. It may be that we limit our perspective to what can be approached by scientific method and thereby leave out information that is part of concrete experience or faith as based on biblical revelation. The next section may prove this point.

Time and New Creation

The second topic I want to address is the nature of time. I limit myself to some remarks concerning the view of Wilkinson. In Wilkinson (2010), he devotes a full chapter to space-time in creation and new creation. First I give a short summary of his view. Then I elaborate on some points and add my own comments. In fact, it will appear that Wilkinson's discussion of time can serve as an illustration of the first point I discussed, i.e., the tendency to give priority to scientific knowledge over our concrete experience.

The main question of the chapter is "whether time and space can provide the key to understanding the continuity and discontinuity of the relationship of creation and new creation" (116). Throughout his book Wilkinson emphasizes the togetherness of continuity and discontinuity concerning the relationship between creation and new creation. These two sides also appear to be crucial for his understanding of the nature of time. In contrast with the tradition which places the new life in heaven as eternal over against the temporal nature of our life here and now, Wilkinson emphasizes that "time is real both in creation and new creation." That is the continuity part. At this point, he takes a clear position in opposition to the Neoplatonic view of time which, because of its negative appraisal of time and change, has no place for them in new creation. The discontinuity he sees reflected in two elements: (1) "There is a decoupling of time and decay in new creation"; and (2) "Time is not limiting in new creation in the same way that it is limiting in this creation" (134). As I agree with the continuity part, I start my comments with these two elements and will then add a more general remark about the nature of time in relation to our concrete experience.

Our experience of time in the present world can be characterized by both growth and decay. There is a positive and a negative side to time. The negative side is especially connected with decay. According to Wilkinson, in new creation, only the positive side will remain. There will be time and change but only as "growth and flourishing" (132). Time will be "emancipated from decay" (131). There will be "persisting fulfilment rather than transient coming-to-be" (quoting Polkinghorne). The images of music ("change without repetition or loss") and dance (highlighting the "dynamic nature and relatedness of time and eternity") should give some idea of what time would then look like (132).

It strikes me that in this context Wilkinson does not mention plants and animals. Animals are discussed later, especially in relation to the carnivorous nature of predators. Plants are not mentioned at all. Regarding animals, Wilkinson suggests that in new creation there will be animals but with a transformed nature, although there is no indication what this transformation will imply (170). But the question as to whether animals will be born and die is not even mentioned. And even if we may be able to imagine that the same animals will stay with us forever, so that there will be no decay in the animal world either, then there is still the question of plant life. It is hard to imagine that there will be no vegetation on the new earth. Will the fruits still be enjoyed by humans (Gen. 1:29) and the plants eaten by animals (v. 30)? How could this happen without an element of decay?

I wonder whether there is not still some influence of the Neoplatonic negative view of time in this approach. Decay certainly can be negative. But does that mean that all decay should be understood that way? I think a distinction should be made between the flow of time as such—expressed in birth, growth, and death, but also in past, present, and future—and the element of decay in its negative sense. The flow of time is connected with finitude, but finitude in itself should not be understood as something negative. Transience need not be understood as futility. This is only necessary when it is contrasted with eternity and eternal presence is taken as the ideal.[4] The flow of time in terms of past, present, and future, in the sense of change and development, and in the sense of the phases of life, is itself a condition for finite existence and therefore a positive element. The phases of our human existence in birth, growth, and death will apply no more to us after the resurrection (cf. Luke 20:34–36), but this does not necessarily mean that they, as such, have a negative meaning, certainly not in relation to plants and animals. As far as I know, there is no indication in the Bible that in the new creation the life cycle of birth, growth, and death will no longer hold for animals and plants. Death in the plant life is not necessarily negative. Even for animals we can see the end of life as completion. If there will be animals and plants in the new creation, which I believe, they will not, I think, have a continuous life in an individual sense. Individually, their lives will be completed without the experience of evil and suffering. Death may be the end of a completed life. We could even imagine that there will be lions and other predators that will play with their former prey in such a way that both can enjoy the game (cf. Isa. 11:6–7). Carnivores will eat animals then (in spite of Isa. 65:25) only after they have died a peaceful natural death. This certainly would imply a transformation of their nature, but not to such an extent that we could not recognize them anymore in the continuity with their present life.

The second element of discontinuity that Wilkinson mentions is that time in new creation will not limit in the same way as in this creation. This requires some explanation. The basic idea is a multidimensional time. The idea comes from physics. In our concrete experience space has three dimensions. In physics, time is often seen

---

[4] In this respect it is interesting that Wilkinson compares his own view of multidimensionality, discussed later in the main text, with "the 'eternalist' position which traces its history back to Plotinus" (Wilkinson 2010, 128).

as a fourth dimension. But this space-time can also be thought of as having more dimensions. Wilkinson is suggesting that we can use this multidimensionality as an analogy to understand God's relation to time.[5] We experience time in one dimension. Time flows in one direction: from past through present to future. That is our limitation. For God, time is multidimensional. This means that he is not constrained by the arrow of time as we are. "He does interact at various points in this time" (128). Actually Wilkinson suggests "that our experience of time in the physical Universe is a small and limited part of an ontologically real time that we might call eternity." "Eternity," then, is the same as "the higher dimensions of time" that God inhabits (126). Evidently, God is not limited by time in the same way as we are in this creation. The suggestion is that in the new creation for us, too, this limitation is somehow overcome. Consequently, we would not be bound anymore by the flow of time in the same way as we are now.

Although the point of the multidimensionality of time is central to Wilkinson's discussion in this chapter, it remains tentative and open. Wilkinson is aware that the analogy of multidimensionality to understand God's relation to time implies some risks. It might lead to "a panentheistic understanding of the relationship between God and the world" (127, cf. 133), or even to "problems of physicalisation and depersonalisation" in relation to our understanding of God (129). As to the multidimensionality of time in the new creation, he admits that there are also problems of a scientific nature. Therefore, more research needs to be done (135). Yet, it seems to be this idea of the multidimensionality of time that, besides the connection between time and decay, should answer the main question of the chapter, to wit, whether "time and space can provide the key to understanding the continuity and discontinuity of the relationship between creation and new creation" (116).

Regarding time and decay I have already expressed my reservations. As to the multidimensionality of time, I find the result rather meager. Not only do many questions remain, as Wilkinson admits, but even more important is that it does not become clear what this multidimensionality could mean concretely and how it would enrich our lives.[6] What strikes me especially is the fact that in the beginning of the chapter, time and space are mentioned as the possible key to understanding the continuity and discontinuity between creation and new creation; but later it is

---

[5] Wilkinson seems to follow the argument that God needs to be in time because only in this way could he be a personal agent (Wilkinson 2010, 125–126). I do not believe in this kind of argument. It draws conclusions about the nature of God based on what we can think, forgetting that our thinking is both made possible and constrained by the fact that we are creatures. In fact, I suspect that this kind of argument is an illustration of the tendency towards a God's eye point of view, which I mentioned in the first section, in this case even including the being of God. Typical for this way of thinking is the view about the relationship between thinking and being: whatever can be thought is supposed also to be possible; what cannot be thought is supposed to be impossible. This line of thought goes back to Parmenides. Evidently, the thinking in terms of what can(not) be thought applies to theoretical or strictly logical thought.

[6] It is only later in his book that Wilkinson mentions the multidimensionality of time as a possible solution to certain questions. For the experience of the believer, resurrection could take place immediately after death without an intermediate period of waiting (2010, 142), with reference to Luke 32:43 (145).

explicitly stated that the resurrection of Jesus is "the key to understanding the continuity and discontinuity" (130). Actually, this is the argument throughout the book. Because of this one would expect Wilkinson to try relating the gospel stories about the resurrected Christ to his idea of multidimensionality, either as an argument or as a critical test. But Wilkinson does neither. It is only with respect to time as "emancipated from decay" that the connection between time in the new creation and the resurrection is made (131). It seems, therefore, that the leading question of the chapter about space and time as the key to understanding continuity and discontinuity between creation and new creation should receive a negative answer.

My third remark is of a more general nature. It concerns the relation of a theory of time to our integral experience. In his discussion of space-time in connection with creation and new creation, Wikinson's focus is on a scientific understanding of time. Yet, he tries to connect this scientific understanding with our concrete experience. "Science can give some insights, however, it is not enough to represent the whole of our experience of time, both objectively and subjectively" (124). We must also make use of narrative. This applies to the relationship between creation and new creation in general. It holds also for our understanding of time (124). Yet it is clear that Wilkinson's focus is on scientific understanding. Referring to the use of narrative by the biblical writers, he asks whether "the concepts of modern science allow us to translate those narratives in a way that is fruitful in thinking about new creation" (ibid.). In fact, one could say that this is what Wilkinson attempts to do: using scientific concepts of time to understand the biblical story about the difference between creation and new creation. The actual result appears to be rather meager. As to the emancipation of time from decay (131), no attempt is made to formulate this in scientific concepts, although in an earlier context Wilkinson has mentioned the second law of thermodynamics (119). The real basis for the view of the continuity of time in the new creation but without decay appears to be the resurrection of Christ and the biblical images of a new life "without suffering, tears and death" (130, 131). And as to the idea of the multidimensionality of time as an attempt to conceptualize time in the new creation, the actual result remained problematic, as we have seen.

It appears that Wilkinson's approach to science in relation to our concrete experience is ambivalent. On the one hand, he ascribes to science only a modest role, as far as our understanding of the new creation is concerned (115). On the other hand, the impression is given that scientific models are the means par excellence to understand the relationship between creation and new creation. Actually, it appears from our analysis that he does not really succeed in connecting the scientific understanding of time with our concrete experience. I suggest, therefore, that we take our starting point in our concrete experience of time and attempt to develop a theoretical approach from there. This may help us also to understand the continuity and discontinuity between this and the new creation in terms of time. In what follows I give an indication of what this could mean.

Our concrete experience of time is manifold. We observe all kinds of change in the world around us: in physical reality, in living nature, in society, in our own existence. The world around us is fully temporal. But our experience itself is no less "in time." Time penetrates our own existence as much as the world around us. But within this temporality we perceive diversity and order. There is day and night. There are seasons and years. The order of the universe is connected with that of nature, also in a temporal sense. Yet there is also distinction. Our biological clock is connected with day and night through light and darkness. But our speaking of *connection* implies that they are not the same. Biology is more than physics in a narrow sense. Furthermore, there are different life cycles for all kinds of living creatures; plants, animals, and humans. Diversity and order we also find in a cultural setting: we give shape to legal order in terms of time, for instance, when a crime can only be prosecuted for a limited period. But the actual way this is done within different legal systems may vary considerably. Another illustration is the fact that in different cultures there is a different sense of urge with respect to being on time. In most Western countries we live by the clock. Africans on the other hand may feel the need for social contact on the road as much more urgent. A theory of time should account both for this diversity in the experience of time and for the order which is implied.

Wilkinson mentions the idea of narrative as important for our full experience of time. And it certainly is, if only because it can bring us back to concrete experience. Yet, narrative should never be identified with concrete reality as such. If the idea of narrative is developed, narrative is understood as having a specific structure: a plot, a beginning, a development, and an end. But this structure can never serve as a characteristic of our human existence. We can tell stories about our life, but that does not make life itself into a story. The same is true for the world at large. We can tell stories and write history, both about the human and the natural world. But no story or historical study will ever be identical with what actually happened, including all kinds of developments. Therefore, I do not think the idea of narrative can account for both the diversity and the order of the temporal nature of the world, not even of ourselves.

A theory which does account for both the diversity and the order of time can be found in the work of the Dutch philosopher Herman Dooyeweerd (1894–1977). As explained in Glas and De Ridder's introduction to this volume, he distinguishes around 15 aspects ("modes of being") to account for a basic diversity in our world. These aspects are both epistemologically and ontologically irreducible; epistemologically because their basic concepts cannot be reduced to one another, ontologically because the same applies to the laws that characterize them. In a provisional analysis, Dooyeweerd distinguishes the following aspects: the numerical, spatial, kinematic, physical, biotic, psychic, logical, historical, lingual, social, economic, aesthetic, juridical, ethical, and pistical (Dooyeweerd 1955).[7] In his view, time expresses itself in a particular way in all these aspects (Dooyeweerd 1940). Complementary to the theory of modal aspects is his theory of individuality structures (also called typical structures), which provides a structural analysis of different kinds of things in terms of these modal aspects (Dooyeweerd 1957). In this structural analysis of entities, time also has its place. All these things exist with a typical temporal order, such as the life cycle with its different time spans for everything that lives.

---

[7] It is important to realize that it is a *theory* about order. It does not describe the concrete order of reality but abstracts different aspects from this concrete order in theoretical analysis.

My suggestion is that we take this theoretical framework also as a starting point to get some idea of the continuity and discontinuity between creation and new creation. Both the modal aspects and the typical structures may account for the continuity. The discontinuity then is expressed in the concrete characteristics and laws that hold within these aspects and typical structures. Within this theoretical framework, scientific analysis may have its place, including Wilkinson's suggestions about the multidimensionality of time. One should realize, though, that these suggestions apply only to physical time, not to time as such, however basic physical time may be for the other aspects.

At least as important for our theoretical reflection about creation and new creation is thinking about historical time in the sense of ongoing cultural development. Wilkinson's images of music and dance for time in the new creation may seem attractive because they suggest "persisting fulfilment rather than transient coming-to-be" (132). For me, however, these images, although they make use of cultural achievements, suggest a cyclical rather than a linear view of time, as within the linear view the coming-to-be of something new, if only relatively, is essential. These images, therefore, can hardly account for the cultural-historical aspects of time with their linear element. Actually, for an understanding of time we need both elements. The cyclical may relate more to the natural aspects of creation, although nature implies also linear development as expressed in the theory of evolution. The linear may relate more to the cultural aspects of creation, because they are intrinsically connected with historical development. Yet history cannot be understood without a cyclical element. For instance, without the rhythm of years, it would be impossible to commemorate and celebrate important events. Anyhow, it would be strange if the natural aspects of time with their cyclical element would dominate the new creation at the expense of the cultural-historical aspects with their emphasis on linearity. Therefore, I believe that linear time with its elements of coming-to-be and transience will characterize the new creation to the same extent as cyclical time.

Resurrection and Personal Identity

The basic question that concerns us in this part is connected with the hope for humankind beyond death. This issue is related to the question of what characterizes us as humans. The Christian hope, biblically understood, implies a life after death, including a resurrected body. What does this imply for our understanding of human nature? Traditionally, the doctrine of life after death is connected with some kind of dualism between an immortal soul and a material body that will die. Is this the view we should still defend, or do we have an alternative? Do we even need one? How are we to understand human identity in relation to death and resurrection?

Again, I will first point to a contemporary discussion and then proceed with some biblical and philosophical comments.

Contemporary Discussion

The book I will refer to is the collection of essays entitled *Resurrection. Theological and Scientific Assessments* (Peters et al. 2002). Similar to the books I mentioned before, these essays want to connect theology and contemporary scientific understanding. I will focus on the chapters by John Polkinghorne and Nancey Murphy and discuss their views on personal identity before and after the resurrection.

It is interesting to see how different the starting points are of these two authors while their final views are basically rather similar. Polkinghorne's position comes close to the traditional view in terms of a soul that is re-embodied after the resurrection. He rejects a dualism in the Platonic and Cartesian sense (Polkinghorne 2002, 50), but defends a kind of holism that interprets the soul as the form of the body as in the Aristotelian-Thomistic tradition. Polkinghorne describes this form in contemporary terms as an "almost infinitely complex, dynamic, information bearing pattern" (51). It is this "information bearing pattern" that through death remains preserved "in the divine memory" (52) and is re-embodied in the resurrection (50).

Over against Polkinghorne Murphy takes her starting point in physical monism—more precisely, in non-reductive physicalism (for Murphy's position in this respect, see Murphy [1998b]). She considers Polkinghorne's holistic approach a possibility for Christians (Murphy 1998a, 24–25), but the results of the empirical sciences, such as neurobiology and cognitive science, point in the direction of physicalism. The practice of science shows clearly that "there is no need to postulate such things as *substantial* souls or minds" (Murphy 2002, 203). With respect to the nature of the resurrection this implies that "if there is life after death, it depends on bodily resurrection alone." Resurrection should be understood as "resurrection of the whole person from death, not as the restoration of bodily existence to a surviving immortal soul" (ibid.). The emphasis on the unity of the person over against any kind of dualism is also expressed by the later statement that "humans are their bodies" (207). Yet, at the end of her discussion she comes close to the position of Polkinghorne. Personal identity before and after death appears to depend upon mental and moral characteristic—not upon the body. The argument goes as follows.

In a thorough discussion Murphy defines the criteria for personal identity, both for this life and the next, in terms of memory, self-consciousness, moral character, and human relations (208). All these characteristics need a body as their substrate. This leads to the question of how there can be continuity between the person before and after the resurrection, because the body clearly dissolves after death. Thus, there is no material continuity between this life and he next.

To solve this problem, Murphy distinguishes between the nature of the person and the body. The body is called a "material object" provided with all the characteristics and capacities needed to support the characteristics of the person but, as such, is not identical with it (215). In this way it becomes possible to see the continuity between this life and the next in the personal characteristics. The continuity of the material body is not necessary, even more so because in our present life the continuity of consciousness as required for the continuity of the person is only contingently connected with the continuity of the body (ibid.). The continuity between this life and the next now becomes possible "if God can create a new (transformed) body and provide it with my memories" (211). No continuity is needed between our present body and our body in the resurrection since the continuity is not in the material, but in the personal characteristics. What is required of the new body is only that it is "similar in all relevant aspects" (215). This new body will provide the "substrate for ... our mental life and moral character" after the resurrection. The body is called the same, but with the possible "temporal interval between the decay of the earthly body and what is then essentially the re-creation of a new body out of different *stuff*" (ibid.).

Still, according to Murphy, resurrection is not the "restoration of bodily existence to a surviving immortal soul" (203). But the difference between Murphy and Polkinghorne appears to lie primarily in the nature of the "soul." Murphy does not speak of the soul as an information pattern of the body as Polkinghorne does. She does not even mention the term. However, for both Murphy and Polkinghorne, the soul or the personal characteristics are preserved in the memory of God and can therefore be connected with a new body. So in fact there seems to be no reason why Murphy should not speak, like Polkinghorne, of re-embodiment of what is preserved. Actually, the unity between the personal or mental characteristics (the soul) and the body may be more intrinsic in Polkinghorne's approach, where the soul is the form of the body, than in Murphy's, for whom the body serves only as a substrate for the personal or mental characteristics. But both reject the view of the soul as a substance in itself.

On reading Polkinghorne and Murphy, it appears quite difficult to leave behind a language that somehow reminds of a dualism of body and soul, of material and mental characteristics (properties), the latter representing continuity, the former discontinuity. Let us see how far this goes.

Biblical Comments

Let me start again with some biblical comments. I will mention three points. My first point concerns the connection between our identity and the relationships we exist in. Both Polkinghorne (2002, 52) and Murphy (2002, 213) emphasize the importance of relations. They are part of our identity, all kinds of relationships, including the relation with God. The latter is not given special importance, though, in the sense that it is connected with the very core of our identity. Yet, should we not say, if we are created in the image of God, as the Bible teaches, that our deepest identity lies in this relationship with God? If we are created by God, this relationship will characterize our very being and encompass all other relationships. It may be helpful to characterize this relationship in line with Genesis 1 in terms of call and answer, or as promise-command to be and response, as I like to say (Geertsema 1993). All creatures, in fact, exist in this relationship. But we humans exist responsibly, responding to God who created us both by our very being as self-conscious creatures, even if only in a pre-reflective sense, and in the many ways we live. In this responding, I contend, lies our deepest identity.

In the second place, it struck me that, in the discussion about continuity and discontinuity between our life before and after the resurrection, both Polkinghorne and Murphy connect the discontinuity primarily with the body and the continuity with the person as distinct from the body. After the resurrection, different laws of nature will apply to the body. As to the person, God preserves its basic characteristics in his memory.[8] But is this in accordance with the biblical picture?

According to the New Testament death applies as much to us as persons, or souls, as to our bodies. Jesus teaches that "whoever finds his life will lose it, and whoever loses his life for my sake will find it" (Matt. 10:39).[9] And, "if anyone would come after me, he must deny himself and take up his cross and follow me" (Matt. 16:24). This may imply a physical death, but primarily it implies an inner conversion. Paul speaks of the death of the old and the birth of the new self: "I have been crucified with Christ and I no longer live, but Christ lives in me" (Gal. 2:20; cf. Rom. 6:6, 8:6ff.). What Paul refers to is the death of Christ at the cross, but this death certainly was not just a physical death—however deep the physical suffering may have been because of the kind of death. The suffering of Christ also consisted in being forsaken by God. This need not happen to us anymore because Christ died in our place. But somehow the death of Christ will affect us through our being in him (cf. Schuele 2002).[10]

To make use of the metaphor of our answering nature again, it is our very identity in living as responding to God's promise-command to be that is at stake. Since we have failed in our responding to God, Christ takes our place. But this means that we die with him to also be raised from the dead with him. We should give up our old nature, involving sin and evil, which goes beyond our moral character as it concerns the attitude of our hearts towards God. There is a discontinuity in the core of our being because of the attitude we have chosen towards God. Therefore I need to be crucified with Christ and "I" should no longer live, but "Christ in me," as Paul writes. But these very words of Paul imply also a continuity, because he writes that Christ lives in *me*. It should be clear, then, that we cannot relate discontinuity

---

[8] Neither Murphy nor Polkinghorne reflects upon our historical and cultural situatedness which shapes both our memory and our moral convictions, not to mention our sense of self. Think of people from different historical periods and cultural contexts that will meet in the new creation. Memory is certainly important (as are the other characteristics Murphy mentions) in relation to our identity in the resurrection, but the history that has shaped us goes beyond our memory. Our being as an answer to the promise-command to be through time goes beyond all the stories that could be told or remembered, even if they relate what is most important and most characteristic.

[9] The word *life* here is the translation of the Greek word for *soul*.

[10] Interestingly, Murphy asks to which extent "personal identity can be maintained through the elimination of negative characteristics" and refers in this context to "narratives of sinners transformed in this life" (2002, 214). Yet her emphasis is on the "preservation of human moral character" (218), including "virtues (or vices)" (212).

primarily to the body and continuity to the person or soul. In both respects there will be both discontinuity and continuity.[11]

My last point concerns the nature of the continuity between the present life and life after the resurrection. Both Polkinghorne (2002, 52) and Murphy (2002, 211) take the continuity between death and resurrection as a preservation of personal characteristics in the memory of God. Yet the impression given in the New Testament is different. I limit myself to the letters of Paul. Paul definitely does not support the idea of an immortal soul over against a perishable body. *We* are perishable, not just our bodies. Therefore, the "perishable must cloth itself with the imperishable, and the mortal with immortality" (1 Cor. 15:53). But Paul does speak about "being with Christ" immediately after death (Phil. 1:23; cf. 2 Cor. 5:1–10). Lampe (2002, 110) makes the important observation that Paul uses personal terms in this context and not words like "spirit" or "soul": "I desire to depart and be with Christ" (Phil. 1:23), and "We have a building from God, an eternal house in heaven.... We do not wish to be unclothed but to be clothed with our heavenly dwelling" (2 Cor. 5:1, 4).

The word Paul uses for being dead before the resurrection is also interesting. *Koimasthai* can mean both "being asleep" and "being dead." This is the word Paul uses in 1 Thessalonians 4 and 1 Corinthians 15 when he refers to those who have died before Christ comes back. The word does suggest continuity but not mental activity. "Sleep" concerns both the body and the mind. It could be understood as a peaceful rest which, within biblical language as a whole, may even be disturbed (cf. 1 Sam. 28:15). To be sure, we do not get a clear picture of how Paul imagined the "being with Christ" between death and resurrection, but what he does say suggests a continuity that is more than merely being preserved in the memory of God. At the same time we should be careful with taking his words as an argument for the existence of a substantial soul which, different from the material body, survives death, since his words do not necessarily imply such an interpretation.

Philosophical Comments

In my philosophical comments I will continue the discussion of my biblical comments. First, I will deal with what I see as the background of the position of Polkinghorne and Murphy, both in relation to the parallel between continuity and discontinuity of person and body and the denial of continuity of the person in the period between death and resurrection, apart from the memory of God. I find this background in the problems of substance dualism. Then I will attempt to develop a different approach based on my earlier contention that the core of our identity lies in the relationship with God. Starting with our concrete experience I move on to a more theoretical position. Finally, I will try to apply this theoretical framework to the issues of continuity and discontinuity between our present life and life after the resurrection and of the continuity in the interval between death and resurrection.

---

[11] This issue will be taken up in our discussion of Badiou in part 2 on political philosophy (cf. Geertsema forthcoming).

Substance Dualism Rejected

It is not hard to see why Polkinghorne and Murphy place the continuity of the person between death and resurrection exclusively in the memory of God. Both reject the idea of a substantial soul. Personal characteristics depend for their realization on their connection with the body. Therefore they cannot continue to exist when this material substrate has dissolved because of death. If there is no material continuity, there cannot be mental continuity either. There is no such thing as a substantial soul that could exist by itself. The continuity of personal characteristics is only possible, therefore, in the memory of God. The background of Polkinghorne's and Murphy's position lies in the problems of substance dualism. The framework within which they think is construed in terms of the two substances of body and soul, or matter and mind. Starting from there, they deny the independence of a substantial soul—Polkinghorne in terms of a kind of holism, Murphy in terms of physical monism. Therefore, there cannot be a continuity of personal ("soulish") or mental characteristics without a material substrate. But in this way, as we have seen, they cannot do full justice to biblical teaching.

The same background appears in relation to the point of the continuity of personal characteristics and the discontinuity of the body. Polkinghorne explicitly rejects a dualism "either of a Platonic or of a Cartesian kind" (2002, 50), but acknowledges the similarity of his position to "the old idea ... that the soul is the form of the body," referring to Aristotle and Thomas Aquinas (51). Murphy emphatically starts with the physicalist thesis that there are no "such things as *substantial* souls or minds" (2002, 203). The theoretical framework is set by substance dualism, which is then considered to be problematic. From here they develop their own positions.

The question is whether we should not take leave of substance dualism in a more radical way. Both Polkinghorne and Murphy want to relate their discussion of human identity to our concrete self-experience. Polkinghorne uses the term *soul* as an equivalent for "the core of the person," the "self," the "real me" (2002, 51). The characteristics of personal identity that Murphy mentions include self-recognition and relations (208). Her concept of a person "involves both a body and subjectivity" (210). Still, it is hard to recognize ourselves in a full sense in the descriptions they give of human identity. Ultimately, for both, the body becomes just a substrate. But that is not how I experience my bodily existence. Neither do I recognize myself as a person in Polkinghorne's characterization of a person as an "information bearing pattern." The description of Murphy comes closer to my understanding of self, yet it still does not express it fully. It mentions certain important characteristics but these by themselves do not express who I am as this unique person.

I wonder whether this problem does not arise because Polkinghorne and Murphy start from a theoretical position: they attempt to grasp human identity in terms of properties connected to one or two substantial natures. But this already presupposes a theoretical scheme of one or two substantial essences and related properties. It is possible, of course, in a loose way to relate the diversity of our concrete experience to *body* and *soul* as referring to outer and inner experience. But this does not imply thinking in terms of two substantial essences with related properties. That scheme does not reflect my concrete experience with its richness and diversity.

This applies even more to the dualism of material and mental properties. As a theoretical clarification it is not even very helpful. It ignores the diversity both within the material and the mental (Geertsema 2000) and cannot do justice to the unity of my "embodied self" (Geertsema 2011). Replacing this dualism either by a kind of holism or by physical monism does not help. The ultimate horizon from which we are to understand the world and ourselves as living within it cannot be the relation between the physical and the mental. In this approach, the theoretical background of thinking in terms of substances with essential properties remains predominant. We should start with our concrete experience, including our basic worldview perspective and the leading function of faith. As I claimed before, theoretical analysis can clarify, expand, even correct this concrete experience, but it can never replace it. We basically understand the world and our identity as persons from our concrete experience as guided by faith, even if both somehow will remain a mystery to us, because of our finiteness.

This brings me to my second point, to wit, the need for a different approach both with respect to our basic identity and with respect to the theoretical framework by which we try to account for the way we are structurally related to the diversity of the world around us.

A Different Approach

We have a sense of self in all our experiences and activities. It is *I* who experiences and who acts. This sense of self has developed since our childhood and is still developing. It is there in all our relations with people, with animals, and with things. Yet, if someone would ask us to give a definition of who we are, we would not be able to provide one. Our *self* appears to escape our conceptual grasp. Polkinghorne speaks of the "real me" as something "problematic" (51). This expression, though, betrays his focus on theoretical understanding: the real me is a problem for our theoretical analysis because we are unable to give a precise definition. When he uses the term *carrier* in relation to the question of what can account for the continuity of ourselves through all the material changes, this even reminds of the idea of a substance as the carrier of properties. I would rather speak of a mystery, which by its very nature escapes theoretical analysis. We live "our selves" in our relationship with God and our fellow creatures (cf. the great commandment in Matt. 22:36–40). In this way our selves are opened up (or closed down). Reflection can certainly be of help, and may even be indispensable, although it may also lead us in the wrong direction. But conceptual analysis will never be able to grasp fully who we are.

Earlier I mentioned the all-decisive nature of our relationship with God. I described it as the relation between promise-command to be and response. I characterized our deepest identity as our responding in this relationship. In philosophical terms this means that we should understand ourselves primarily from a second-person perspective; not in terms of the third-person perspective of theoretical analysis, and neither in terms of the first-person experience of subjectivity, but in the second-person perspective of answering in relationships. Only in this way can we do full justice to who we are. In other words, to express our nature as persons we need in the first place the use of personal pronouns—not properties, essences, or substances. We are *I*, *you*, *he*, or *she*; with others we are *us*, *you* (plural), and *they*. But before we respond as *I*, we are addressed as *you*. The first-person perspective follows the second-person perspective and remains dependent on it.

This is true in the first place for our relationship with God. He calls us to be and we respond by our very being. Secondarily it holds also for the relationships with other humans. Our sense of self develops in the context of being addressed as a little child. In fact, I think, our sense of self is rooted in the relationships in which we exist. Our theoretical reflection (third-person perspective) is only secondary. Both the third-person perspective of science and philosophy and the focus on the first-person perspective in much modern thought may have had a deep influence on the way we experience ourselves. In themselves they are not sufficient for us to really understand who we are. As such, these modern perspectives of third- and first-person may even close off our sense of self instead of opening it up in the relations with other creatures. From the biblical perspective this applies even more to our relationship with God as our Creator and Redeemer.

The real me cannot be grasped in a theoretical analysis nor in an inner reflection. It seems to me that this is the reason why the approach of Polkinghorne and Murphy ultimately fails. They try to grasp human identity in terms of properties and concepts (the third-person perspective), while at the same time making use of the first-person perspective of subjective experience. Actually, because they try to grasp the *me* of inner experience (first-person perspective) in terms of properties and concepts (third-person perspective), there remains a gap between the two. Polkinghorne comes close to the identification of *me* with a substantial soul, although he denies it an independent existence. But within substance dualism the soul-substance refers not only to the individual person but also to the universal essence that characterizes the first. The first-person perspective (*me*) is reduced to the third-person perspective of theoretical analysis, which means that in fact the first-person perspective is abandoned.

Murphy attempts to identify the *me* in terms of mental properties which are non-reducible to material ones but have no independence in an ontological sense. In this way she, too, necessarily leaves the first-person perspective behind. The second-person perspective by its very nature cannot be closed off (or grasped) in a first- or third-person perspective. Not only is it primarily lived—although it can be reflected upon, and opened up that way—but it also, by its very nature, points beyond itself. It implies a relationship of responding to the other/Other. Therefore it cannot be grasped in a theoretical concept.

This does not mean, though, that we cannot attempt to bring some clarity to the concrete lives we live by way of theoretical analysis. Earlier I mentioned Herman Dooyeweerd's theory of modal aspects and of individuality structures. I now want to apply these theories (third-person perspective) to our human being and then connect them with the first- and second-person perspectives.

As mentioned before, Dooyeweerd distinguishes 15 modal aspects to give an idea of the basic diversity of our world. All these aspects apply in principle to all creatures, either in an active or in a passive way. Dooyeweerd speaks of subject-functions in the first case and of object-functions in the second case. Object-functions of one entity relate to the corresponding subject-functions of other entities; for instance, the being visible of a stone or a tree relates to the ability to see of animals and humans. All modal aspects, therefore, are aspects of coherence. Because all things function in all the aspects they are, as such, all related to one another. Characteristic for human beings is that they function in all these aspects in an active way. This does not mean, though, that humans stand over against the other creatures. In all the modal aspects humans exist in relationships of coherence. Sometimes they exist in subject-subject relations, for instance, in the spatial and the physical aspects with material things, plants, and animals, in the biotic with plants and animals, and in the psychic aspect, concerning the senses and feeling, with animals, especially the higher ones. And sometimes humans exist in subject-object relations, for instance, in the lingual, the economic, or the juridical aspect, with material things, plants, and animals. The uniqueness of humans makes them distinct from the other creatures but does not separate them from them. One should rather speak of a special responsibility.

Although the human person has a distinct character in terms of the theory of individuality structures, there are also similarities with the other creatures. One can analyze different entities in terms of the modal aspects. Then the main question concerns the modal aspect that provides the qualifying function. One can also deepen the analysis in terms of individuality structures that are integrated within specific kinds of things. With respect to plants one can distinguish, for example, a physical structure of molecules (DNA) which are integrated in the overall biotic structure. In the case of the higher animals there is also the higher structure which is qualified by the psychic aspect. For the human person Dooyeweerd distinguishes four different structures: the physical-chemical, the biotic, the psychic, and what he calls the act-structure.[12] The last structure concerns all typical human activities which can have their qualification from the logical to the pistic aspect, such as acts of thinking, of prayer, but also buying and selling. The act-structure relates to typical human responsibility and therefore distinguishes the human person from animals, plants, and material things. The other structures are to some extent shared with other entities: the physical-chemical with material things, plants, and animals; the biotic with plants and animals; the psychic with animals. Yet because these structures are integrated within a higher unity, they do have typical traits in each case.

Before I move on to my last point I want to emphasize the difference between this Dooyeweerdian analysis and the approach in terms of substance dualism and its alternatives of holism and physical monism. In the first place, the basic diversity is

---

[12] Within this theoretical context of modal aspects and different structures, it makes sense to call one aspect or structure the substrate for another; for example, in the case of animals the physical-chemical and the biotic for the aspects of the senses.

much more nuanced than merely in terms of the relation between the physical and the mental. Over against an analysis which starts from a dualism or a monism, 15 irreducible aspects are pointed out. In the second place, the terminology of substance is completely avoided. The theory of individuality structures does concern the issues that the idea of substance relates to, especially in its Aristotelian form. It addresses both the general structures and the unique individuality of things (cf. Dooyeweerd 1957). But there is no ultimate essential nature, either in terms of dualism or in terms of monism. The diversity of kinds of things is analyzed in terms of the integration of the different modal aspects or of the intertwinement of part-structures within an all-encompassing structure. Yet these structures have no independent existence apart from the concrete entities they hold for. The givens of concrete experience—the diversity of things and their unique individuality—are taken as a starting point. To some extent they can be accounted for by means of theoretical analysis, but this analysis does not aim at theoretical reconstruction, let alone construction of an in itself disorderly world. Rather, it aims at clarification and deeper understanding of what is given in concrete experience.

The difference between the Dooyeweerdian approach and the approach in terms of substance dualism or monism becomes even stronger when we relate the structural analysis as sketched above to the first- and second-person perspectives mentioned before. One could call the theoretical analysis in terms of modal aspects and structural patterns a third-person perspective. But this perspective can easily be related to the first person: these aspects and structures are aspects and structures of *me*, of *us*. There is no gap, because *me* and *us* are not being defined in terms of these aspects and structures in a way similar to the approach of Polkinghorne and Murphy. The first-person perspective is not reduced to the third-person perspective, as in the case of substance dualism or physical monism. The aspects and structures concern our possibilities and limitations, our responsibilities and vulnerabilities. They can be used to characterize us but not to define us in the strict sense. Only with these qualifications in mind will we understand what is actually involved. Aspects and structures are not about abstract substances—they are about *us*.

But we need to take another step. The aspects and structures not only concern our possibilities and limitations, our responsibilities and vulnerabilities (first-person perspective), but they also belong to us as called to be by the creating word of God (second-person perspective). Our being as responding to the promise-command to be is given concrete form by these aspects and structures. Therefore all these aspects and structures can be called aspects and structures of answering. There is an inner connection between the first- and second-person perspectives: my possibilities and limitations, my responsibilities and longings are part of me as being called to be by God. There is also an inner connection between these two perspectives and the third-person perspective of theoretical analysis: the diversity of aspects and structures that shape my factual being is actually a characterization of the structural side of my being as responding to God. We keep speaking about ourselves in personal terms (first-person perspective), because aspects and structures which can be studied scientifically (third-person perspective) never determine us so fully as to leave no room for responsibility (except in extreme cases). They remain structures of answering. Therefore, ultimately these personal terms, including the structural possibilities and limitations, remain characterized by our relationship to God, who is personal himself.

Continuity and Discontinuity Between This Life and the Next

I now want to apply this sketch of our human nature to the issues of continuity and discontinuity between our present life and life after the resurrection and of a possible existence between our death and resurrection. For convenience sake I will call the central relationship of our responding to God's call to be our *personal identity*, and the diversity of aspects and structures that we can distinguish in our actual being our *structural identity*. It is important to realize that this distinction is not the same as the traditional one between soul and body. Structural identity includes all 15 aspects, so it covers both the material and the mental properties. Neither is the distinction identical with personal—in the sense of numerical—identity over against structural identity where *structural* refers to the covering concept which provides the criteria for speaking of a person (cf. Murphy 2002, 208). Because both personal and structural identity, as I use the terms, have a temporal or historical dimension with an element of irreversibility, they include what is unique for each person as well as what characterizes us as humans.[13] Personal identity connects with my deepest sense of self (first-person perspective) but is understood in terms of responding to God's promise-command to be (second-person perspective). It escapes theoretical analysis. Structural identity refers to the diversity of aspects and structures that can be studied by the empirical sciences, both the natural and the social (third-person perspective).

In relation to our understanding of continuity and discontinuity between our present life and life after the resurrection, my basic contention is that continuity and discontinuity apply both to personal and structural identity. As I pointed out in my biblical comments, the deepest mystery of death and resurrection is to have died with Christ and be raised from the dead with him. This concerns our central relationship with God, our personal identity. It is hard to fathom what it means to be cut off from God in the sense that Christ was forsaken by God at the cross. It is also hard to conceive of the implications of our being renewed in the sense that sin does not play a role anymore in our lives, and that our relationship with God is fully as it was meant to be. Both continuity and discontinuity are involved here. We will still be the same person as originally called to be by God, but we will have gone through a radical transformation.

There is also both continuity and discontinuity in relation to our structural identity. In a concrete sense one should say that my personal history in all its diversity will still be my history. Yet in this respect there certainly will be also transformation. There will be healing and restoration in a complete sense. Not only will there be no

---

[13] The distinction is similar to Dooyeweerd's distinction between the central-religious and the temporal sphere. Both have a law- and a subject-side; what holds for all and what is unique for each person. I differ from Dooyeweerd in that I do not take personal identity to be supra-temporal.

more suffering and evil in the resurrected life, but the suffering and evil of the past will also be healed and overcome. Not even the memory will be painful anymore. It will be related to the work of Christ, understood in the light of the lamb that is slaughtered (Rev. 21:23). To make use again of the characterization of our nature as responding to a promise-command to be, we could say that both the promise and the command in the promise-command will be fulfilled. We will have reached our destination in a deep sense. At the same time, the command of God will be fully interiorized in our heart. But there will also be a newness that transcends the first creation. New elements will be opened up which go beyond our deepest longings. There will be a closeness to God that surpasses our present understanding.

In a more theoretical approach we could think of the framework of aspects (and possibly structures) as being continued in the new creation. This applies to all 15 aspects and also to the four basic structures as Dooyeweerd has analyzed them. The framework will continue, I suggest. But the specific way the aspects will function may be different. For some this certainly will be the case. For instance, there will be no more death: this clearly affects the biotic aspect as it is now and probably also the physical. Faith, which characterizes the pistic aspect, will be replaced by sight (2 Cor. 5:7; cf. my second biblical comment in the first of the two sections entitled "Biblical Comments"). Trust will have a new dimension. But perhaps there will be transformation for all aspects, because they always function in coherence with one another. In that case, not only will there be no more pain (the psychic aspect) and no more injustice (the political and economic aspect), but these aspects will also function in a way that goes far beyond whatever we can imagine. As I mentioned before, all of reality will be transparent in relation to God, because he will be all in all. That will characterize all aspects—and structures.

Could we also use the distinction between personal and structural identity to understand a continued existence between death and resurrection? I think it can be helpful, at least in the sense of opening up some possibilities for thought that differ from the traditional understanding in terms of substance dualism and its alternatives of holism and physical monism. In the first place, the continuity that Paul assumes in his letters need not be related to a soul-substance as distinct from the body. I suggest that we take this continuity in the first place as related to what I have called *personal identity*. God's call to be is still valid after death, at least for those who live with Christ. My existence as a response to that call will therefore be continued in some way, however hard it may be for us to imagine this kind of existence. It certainly does not imply an immortal soul. We need personal terms, not theoretical ones like substance.

But it may also be helpful to apply the concept of what I have called *structural identity* to our existence between death and resurrection. In terms of the scheme of modal aspects, I see no reason why only immaterial aspects would be involved in this intermediate existence and not the material aspects. When Paul speaks about the "earthly tent" (2 Cor. 5:1), it appears from the context that he does not mean just his body in a restricted sense, but his overall present existence. The inner treasure of the light of Christ is kept in "jars of clay" (2 Cor. 4:7). This may suggest just the material body, but what Paul means is explained in the following verse: "We are hard pressed on every side, but not crushed; perplexed, but not in despair; persecuted, but not abandoned; struck down, but not destroyed' (vv. 8–9). It all relates to "carry[ing] around in our body the death of Jesus so that the life of Jesus may also be revealed in our body" (v. 10). The body, like the earthly tent, appears to refer to Paul's concrete existence in hard conditions. It is this concrete existence that will be cast off at death, not a material body over against an immaterial soul. And it is "he" that will "have a building from God, an eternal house in heaven" (5:1).

I for myself see no reason why there could not also be a material side to our existence with Christ between our death and the resurrection. Our body as we have it now will certainly be dissolved. But our whole earthly existence will have come to an end. Just as a possibility for thought, I suggest that, although we will not function in the biotic aspect during the intermediate state because we have died, there may be a spatial and a physical aspect. If that is the case, it will not be with a body as we have it now—but science has shown how complex physical reality is in terms of space-time and of the interconnection between matter, energy, and information. It may even be much more complex than we know now. We may also function in some other aspects, such as the psychic, the logical, the lingual, and the social. Most likely, in some way we will function in the aspect of faith, because we are with Christ. Anyhow, in all respects our existence will probably be rather inactive. Think of the image of sleep which is used by Paul to describe death. In any case, *we* in the sense of our personal identity (second- and first-person perspectives) will not be reducible to one or more aspects, whatever they may be, because that would imply a reduction of our personal to our structural identity.

Of course, what I suggest is all rather speculative. What I want to show is that it is possible to think about a continued existence between our death and resurrection without making use of the idea of substance dualism in terms of an immaterial soul over against a material body. Crucial to me seems to be the biblical belief that we will be with Christ immediately after death. For this we need only personal terms—what I have called personal identity—not a theoretical explanation. It is part of the second- and first-person perspectives of concrete experience and faith, not of the third-person perspective of theoretical analysis. And when we do use this third-person perspective, it appears to be possible to avoid substance dualism and its varieties. I even contend that my suggestions are closer to biblical language because they do not imply substance dualism.

Concluding Remarks

In our discussions we have taken our starting point in the biblical ideas of creation and new creation on the assumption that both creation and new creation imply some order. It has appeared that the idea of new creation does have implications for our understanding of the present world. The ambivalence of the scientific picture of the universe is placed in the perspective of God's faithfulness as Creator and, therefore, does not necessarily lead to a negative view of the future. At least some laws will be different in new creation. This applies also to our existence as human beings. Therefore, we should not take the present world as absolute. This applies to our own present life as well. Besides, looking at ourselves in the light of new creation and of the period between death and resurrection leads to specific questions concerning our identity, both in a personal and a structural sense.

Theoretical reflection about the new creation involves philosophical issues, both in relation to the natural world and to our human identity, as we have seen. In the first place, it appeared to make a difference whether the focus was placed on science by itself or that concrete experience was taken as starting point. I argued for the priority of our integral concrete experience as guided by faith or worldview. Science, therefore, should not be taken as the exclusive model for reliable knowledge. Concrete experience may also serve as a valid source. Starting with concrete experience, the results of science should be integrated as far as possible within this perspective.

In the second place, I suggested we take some basic theories of Herman Dooyeweerd as the framework for a theoretical interpretation of continuity and discontinuity between creation and new creation. In relation to the world as a whole the framework of modal aspects and structural patterns could help us understand continuity, while discontinuity would relate to specific laws and concrete characteristics of these aspects and structures. As to our human nature, I introduced the distinction between personal and structural identity to overcome the dilemma of substance dualism and its alternatives of physical monism and holism. In terms of substance dualism and its alternatives, continuity tends to be related to the soul or personal properties and discontinuity to the body or material characteristics. An interpretation in terms of personal and structural identity gives room for continuity and discontinuity in both respects and can stay closer to biblical language.

I should add a remark about my use of Dooyeweerdian philosophy in relation to understanding creation order. Within this philosophy the distinction between law-side and subject-side has an important place. The law-side refers to the order of reality, the subject-side to reality which is ordered. The law-side is easily connected with theoretical analysis in terms of the laws of modal aspects and structural patterns. The subject-side, then, applies to reality in its concreteness. This may lead to our understanding of the order of creation primarily in these (theoretical) terms of modal aspects and structural patterns. My emphasis on the primacy of concrete experience as guided by faith has also implications for the way we understand creation order. The order of creation should primarily be taken as the order of concrete reality. Making use of the distinction between law-side and subject-side, we should say that it is these two sides together that form the order of creation. It is not just the law-side of reality—understood in universal terms, as opposed to concrete entities—let alone a law for reality, which forms the order of creation. My contention is that the order of creation refers to an order that is established, or that will be and needs to be established, in the concrete actual world. Without its concrete implementation in unique situations, the order of creation is not complete.[14]

---

[14] In part 2 on political philosophy I will elaborate on this point in relation to the social-political world (cf. Geertsema forthcoming).

The order of creation concerns creatures in their concrete relationships with God and with each other. It can be understood as the answering nature of created reality, which responds to the promise-command of God. It encompasses the subjective answering and the law-like answering structures together. We can study elements or aspects of this order, especially its law-side, by theoretical means, both in science and philosophy. Yet, these elements should not be understood as the order itself, but indeed only as elements or aspects, abstracted from creation order in its concreteness. Creation order is the concrete order as we live and understand (or misunderstand) it in our actual lives. This limits the significance of science and philosophy, but also gives them room for the pursuit of research and analysis with their own means.

References

Clouser, Roy. 2005. *The Myth of Religious Neutrality: An Essay on the Hidden Role of Religious Belief in Theories*, rev. ed. Notre Dame: Notre Dame University Press.

Dooyeweerd, Herman. 1940. Het tijdsprobleem in de wijsbegeerte der wetsidee [The Problem of Time in the Philosophy of the Cosmonomic Idea]. *Philosophia Reformata* 5: 160–182 (part 1), 193–234 (part 2).

———. 1955. *A New Critique of Theoretical Thought*. Vol. 2, *The General Theory of the Modal Spheres*. Amsterdam/Philadelphia: Paris/Presbyterian and Reformed Publishing Company.

———. 1957. *A New Critique of Theoretical Thought*. Vol. 3, *The Structures of Individuality of Temporal Reality*. Amsterdam/Philadelphia: Paris/Presbyterian and Reformed Publishing Company.

———. 1960. *In the Twilight of Western Thought. Studies in the Pretended Autonomy of Philosophical Thought*. Philadelphia: Presbyterian and Reformed Publishing Company.

———. 1979. *Roots of Western Culture: Pagan, Secular and Christian Options*, ed. M. Vander Vennen and B. Zylstra, trans. J. Kraay. Toronto: Wedge.

Geertsema, Henk G. 1993. Homo Respondens. On the Historical Nature of Human Reason. *Philosophia Reformata* 58: 120–152.

———. 2000. Science and Person: Beyond the Cartesian Paradigm. In *Studies in Science and Theology*, ed. N.H. Gregersen, U. Görman, and W.B. Drees, vol. 7, 47–64. Aarhus: University of Aarhus.

———. 2008. Knowing Within the Context of Creation. *Faith and Philosophy* 25: 237–260.

———. 2011. Embodied Freedom. *Koers* 76: 33–57.

———. Forthcoming. Creation Order in the Light of Redemption (2): Political Philosophy. In *The Future of Creation Order*. Vol. 2, *Order Among Humans*, ed. Govert Buijs and Annette Mosher. New Approaches to the Scientific Study of Religion. New York: Springer.

Lampe, Peter. 2002. *Paul's Concept of a Spiritual Body*. In Peters et al. 2002, 103–114.

Murphy, Nancey. 1998a. Human Nature: Historical, Scientific, and Religious Issues. In *Whatever Happened to the Soul? Scientific and Theological Portraits of Human Nature*, ed. Warren S. Brown, Nancey Murphy, and H. Newton Maloney, 1–29. Minneapolis: Fortress Press.

———. 1998b. Nonreductive Physicalism: Philosophical Issues. In *Whatever Happened to the Soul? Scientific and Theological Portraits of Human Nature*, ed. Warren S. Brown, Nancey Murphy, and H. Newton Maloney, 127–148. Minneapolis: Fortress Press.

———. 2002. The Resurrection Body and Personal Identity: Possibilities and Limits of Eschatological Knowledge. In Peters et al. 2002, 202–218.

Peters, Ted, Robert John Russell, and Michael Welker, eds. 2002. *Resurrection. Theological and Scientific Assessments*. Grand Rapids: Eerdmans.

Polkinghorne, John. 2000. Eschatology: Some Questions and Some Insights from Science. In Polkinghorne and Welker 2000a, 29–41.

———. 2002. Eschatological Credibility: Emergent and Teleological Processes. In Peters et al. 2002, 43–55.

Polkinghorne, John, and Michael Welker, eds. 2000a. *The End of the World and the Ends of God. Science and Theology on Eschatology*. Harrisburg: Trinity Press International.

———. 2000b. *Introduction: Science and Theology on the End of the World and the Ends of God*, ed. Polkinghorne and Welker 2000a, 1–13.

Putnam, Hilary. 1981. *Reason, Truth and History*. Cambridge: Cambridge University Press.

———. 1983. *Realism and Reason. Philosophical Papers*. Vol. 3. Cambridge: Cambridge University Press.

Schuele, Andreas. 2002. Transformed into the Image of Christ: Identity, Personality, and Resurrection. In Peters et al. 2002, 219–235.

Wilkinson, David. 2010. *Christian Eschatology and the Physical Universe*. London/New York: T&T Clark International.

Wright, N.T. 2007. *Surprised by Hope*. London: SCPK.

Part II
Creation Order, Emergence, and the Sciences

Christianity and Mathematics

Danie Strauss

Abstract The trends discernible within the history of mathematics display a recurring one-sidedness. With an alternative non-reductionist ontology in mind this contribution commences by challenging the assumed objectivity and neutrality of mathematics. It is questioned by the history of mathematics, for in the latter Fraenkel et al. (*Foundations of Set Theory*, 2nd rev. ed., Amsterdam: North-Holland, 1973) distinguish three foundational crises: in ancient Greece with the discovery of incommensurability, after the invention of the calculus by Leibniz and Newton (problems entailed in the concept of a limit), and when it turned out that the idea of infinite totalities, employed to resolve the second foundational crisis, suffered from an inconsistent set concept. An alternative approach is to contemplate the persistent theme of discreteness and continuity further while distinguishing between the successive infinite and the at once infinite. Weierstrass, Dedekind, and Cantor define real numbers in terms of the idea of infinite totalities. Frege reverted to a geometrical source of knowledge while rejecting his own initial logicist position. Some theologians hold the view that infinity is a property of God and that theology therefore should mediate its introduction into mathematics. Avoiding the one-sidedness of arithmeticism (over-emphasizing number) and geometricism (over-emphasizing spatial continuity) will require that both the uniqueness of and mutual coherence between number and space is acknowledged. Two figures capture some of the essential features of such an alternative approach. Mediated by a Christian philosophy and a non-reductionist ontology, Christianity may therefore contribute to the inner development of mathematics.

Keywords Christian philosophy · Mathematics · Non-reductionist ontology · Successive infinite · At once infinite · Modal aspects · Herman Dooyeweerd · Limit concept · Irrational numbers

---

D. Strauss (✉)
North-West University, Potchefstroom, South Africa
e-mail: dfms@cknet.co.za

Introduction

Part of the fame of mathematics as a scholarly discipline is that it represents the acme of sound reasoning and a manifestation of what it means to be an "exact science." A brief look at the question what mathematics is as well as a brief account of the history of mathematics demonstrates the contrary. According to Fraenkel et al. (1973) mathematics went through three foundational crises. One of the key issues shared by all three is given in the nature of infinity. Oftentimes reflection on Christianity and mathematics assumes that infinity is an attribute of God introduced in mathematics via the discipline of theology. Interestingly, one of the requirements stipulated by Hersh as a test for every philosophy of mathematics is found in the following question: "Does the infinite exist?" (Hersh 1997, 24).

So reflecting on the meaning of infinity may play a mediating role between theology and mathematics as well as between Christianity and mathematics. But also from the perspective of mathematics as such the notion of infinity appears to be of central importance. David Hilbert states: "From time immemorial, the infinite has stirred men's *emotions* more than any other question. Hardly any other *idea* has stimulated the mind so fruitfully. Yet, no other *concept* needs clarification more than any other question." He holds that the "definitive clarification of the nature of the infinite … is needed for the dignity of the human intellect itself" (Hilbert 1925, 163). Hermann Weyl holds that a concise definition capturing the vital core of mathematics is that it is "the *science of the infinite*."[1]

Admiring the supposed "exactness" of mathematics as a scientific discipline faces two challenges: the one is derived from a number of contemporary assessments and the other from the history of mathematics.

Mathematics: The Acme of Sound Reasoning?

Positivism advanced the philosophical claim that science not only ought to be objective and neutral but must also be free from philosophical presuppositions. Sense data (positive facts) should be the ultimate judge. This view is but one among many other philosophical trends advocating what Dooyeweerd calls the *dogma of the autonomy of theoretical reason*. A recent admirer of the rationality of mathematics writes: "Mathematical calculations are paradigmatic instances of a universally accessible, rationally compelling argument. Anyone who fails to see 'two plus two equals four' denies the Pythagorean Theorem, or dismisses as nonsense the esoterics of infinitesimal calculus forfeits the crown of rationality" (Fern 2002, 96–97).

---

[1] "Will man zum Schluß ein kurzes Schlagwort, welches den lebendigen Mittelpunkt der Mathematik trifft, so darf man wohl sagen: sie ist die *Wissenschaft des Unendlichen*" (Weyl 1966, 89).

By contrast, Hersh mentions four myths regarding mathematics, namely, (i) its *unity*, (ii) its *universality*, (iii) its *certainty*, and (iv) its *objectivity* (Hersh 1997, 37–38). In this work his argumentation entails that these myths cannot account for the history of mathematics as scientific discipline. For example, the widely accepted assertion that mathematics *is* set theory leads into a cul-de-sac. Hersh asks:

> What does this assumption, that all mathematics is fundamentally set theory, do to Euclid, Archimedes, Newton, Leibniz, and Euler? No one dares to say they were thinking in terms of sets, hundreds of years before the set-theoretic reduction was invented. The only way out (implicit, never explicit) is that their own understanding of what they did must be ignored! We know better than they how to explicate their work! That claim obscures history, and obscures the present, which is rooted in history. (Hersh 1997, 27)

Without an ontic reference point mathematics cannot be defined effectively. Without such a point of reference the history of mathematics becomes meaningless. Hersh is therefore justified in holding that "an adequate philosophy of mathematics must be compatible with the history of mathematics. It should be capable of shedding light on that history" (ibid.).

Consider what a mathematician with a sense of the history of mathematics has to say about the recent history of mathematics. Morris Kline states:

> Developments in the foundations of mathematics since 1900 are bewildering, and the present state of mathematics is anomalous and deplorable. The light of truth no longer illuminates the road to follow. In place of the unique, universally admired and universally accepted body of mathematics whose proofs, though sometimes requiring emendation, were regarded as the acme of sound reasoning, we now have conflicting approaches to mathematics. Beyond the logicist, intuitionist, and formalist bases, the approach through set theory alone gives many options. Some divergent and even conflicting positions are possible even within the other schools. Thus the constructivist movement within the intuitionist philosophy has many splinter groups. Within formalism there are choices to be made about what principles of metamathematics may be employed. Non-standard analysis, though not a doctrine of any one school, permits an alternative approach to analysis which may also lead to conflicting views. At the very least what was considered to be illogical and to be banished is now accepted by some schools as logically sound. (Kline 1980, 275–276)

This proliferation of viewpoints does not merely occur within the *philosophy* of mathematics but also within the discipline itself. It is confirmed by Stegmüller:

> The special character of intuitionistic mathematics is expressed in a series of theorems that contradict the classical results. For instance, while in classical mathematics only a small part of the real functions are uniformly continuous, in intuitionistic mathematics the principle holds that any function that is definable at all is uniformly continuous. (Stegmüller 1970, 331; see also Brouwer 1964, 79)

Beth also highlights this point:

> It is clear that intuitionistic mathematics is not merely that part of classical mathematics which would remain if one removed certain methods not acceptable to the intuitionists. On the contrary, intuitionistic mathematics replaces those methods by other ones that lead to results which find no counterpart in classical mathematics. (Beth 1965, 89)

Such a widely diverging assessment of the status of mathematics as a scholarly discipline has deep historical roots. The question is whether we can discern persistent themes running through its development. In their standard work on the foundations of set theory Fraenkel et al. (1973) identify three foundational crises in the history of mathematics.

The First Two Foundational Crises in the History of Mathematics

They first of all mention that "the development of geometry as a rigorous deductive science" in ancient Greece was accompanied by two discoveries. The first one concerns the fact that geometrical entities lacked mutual commensurability, entailing, for example, that "the diagonal of a given square could not be measured by an aliquot part of its side" (that the square root of 2 is not a rational number—Fraenkel et al. [1973, 13]). This was paradoxical because the Pythagoreans were convinced that "everything is number" (see Reidemeister 1949, 15, 30). While it is possible to represent every numerical relationship in a geometrical way, not every relationship between lines could be represented arithmetically. The result of this switch from number to space as modes of explanation is that in the books of Euclid the theory of numbers appears as a part of geometry (see Laugwitz 1986, 9).

After the discovery of the calculus by Leibniz and Newton scant attention was given to its conceptual foundations, causing the second foundational crisis of mathematics emerging at the beginning of the nineteenth century. The weak point was the concept of a limit.

Struggling with the Mathematical Concept of a Limit

Boyer mentions attempts to avoid the *petitio principii* present in Cauchy's limit concept, such as those pursued by Heine and Cantor (in 1872—Boyer 1959, 289–290). The new way explored by Weierstrass contemplates the totality of all fractions smaller than $\sqrt{2}$. Fraenkel explains:

> So, for example, the "irrational number" $\sqrt{2}$ appears as the cut of which the upper class is constituted by the totality (*Gesammtheit*) of the positive rational numbers larger than 2. It should be pointed out that this reasoning does not contain a kind of existential proof or calculation method for irrational numbers, but a *definition* of new numbers on the basis of known numbers; every other conception is circular. (Fraenkel 1930, 283)

On the same page Fraenkel also points out that the square root of the number 2 cannot be obtained with the aid of the row of rational numbers 1; 1,4; 1,41; 1,414; 1,41,42, and so on, because $\sqrt{2}$ already has to be defined as a number in order to serve as a limit: "A statement such as that the just-mentioned row of rational numbers have the limit-value of $\sqrt{2}$ is only then meaningful when $\sqrt{2}$ has already been defined" (see also Fraenkel 1930, 293). Cantor and Weierstrass realized that they have to introduce the idea of *infinite totalities* in order to understand the nature of real numbers. Weierstrass simply defines $\sqrt{2}$ as the infinite totality of rational numbers smaller than $\sqrt{2}$.

Owing to the second foundational crisis, modern mathematics since 1872 realized that no real number could be generated by introducing converging sequences of rational numbers, such as viewing $\sqrt{2}$ as a number generated through a converging sequence of rational numbers. Whatever serves as a limit of a converging sequence of rational numbers already had to be a number in the first place—therefore real numbers cannot come into being through such a process of convergence. In his *Textbook of Analysis* (1821) it is clear that the French mathematician Cauchy still wrestles with this issue: "When the successive values assigned to a variable indefinitely approaches a fixed value to the extent that it eventually differs from it as little as one wishes, then this last [fixed value] can be characterized as the limit of all the others."[2]

Cauchy still thought that one can obtain an irrational (real) number with the aid of a convergent series of rational numbers, without recognizing the circularity entailed in this argument (see also Weyl 1919). Since 1872 Cantor and Heine made it clear that the existence of irrational (real) numbers is *presupposed* in the definition of a limit. In 1883 Cantor expressly rejected this circle in the definition of irrational real numbers (Cantor 1962, 187). The eventual description of a limit still found in textbooks today was only given in 1872 by Heine, who was a student of Karl Weierstrass with Cantor (cf. Heine 1872, 178, 182). Compare here the original explanation of Heine. He commences by defining an *elementary sequence*: "Every sequence in which the numbers a_n, with an increasing index n, shrink beneath every specifiable magnitude, is known as an elementary sequence" (Heine 1872, 174). In this first section of his article Heine points out that the word *number* consistently designates a "rational number." A more general number (or numeral) is the sign belonging to a sequence of numbers. He secures the existence of irrational numbers with the aid of the following theorem (*Lehrsatz*):

> When a is a positive, non-square integer, the equation $x^2 - a = 0$ does not have an integer as root and consequently also not a rational root. However, on the left hand side it contains, for specific distinct values of x, opposite signs such that the equation has an irrational root. Through this it has been proven that not all number signs can be reduced to rational numbers, *because there exist also irrational numbers*. (Ibid., 186)

This article contains the modern limit concept. However, later on Cantor pointed out that the ideas contained in this article of Heine were actually derived from him (see Cantor 1962, 186, 385).[3] Employing the at once infinite in defining limits (and real numbers) caused the following remark of Boyer: "In a sense, Weierstrass settles

---

[2] "Lorsque les valeurs successivement attribuées à une même variable s'approchent indéfiniment d'une valeur fixe, de manière à finir par en différer aussi peu que l'on voudra, cette dernière est. appelée la *limite* de toutes autres"—quoted in Robinson (1966, 269).

[3] In general, a number l is called the limit of the sequence (x_n) when for an arbitrary rational number $\epsilon > 0$ there exists a natural number n_0 such that $|x_n - l| < \epsilon$ holds for all $n \geq n_0$ (see Heine 1872, 178, 179, and in particular page 184 where a limit is described in slightly different terms). Consider the sequence $n/n + 1$ where $n = 1, 2, 3, \ldots$ (in other words the sequence 1/2, 2/3, 3/4, ...). This sequence converges to the limit 1.

the question of the *existence* of a limit of a convergent sequence by making the sequence (really he considers an unordered aggregate) itself the number or limit" (Boyer 1959, 286).

Two Kinds of Infinity

The concept of a converging sequence of rational numbers merely employs the most basic understanding of infinity: one, another one, and so on, indefinitely. Traditionally this form of the infinite is known as the *potential infinite* (preferably to be called the *successive infinite*). However, as we just noted, the limit concept which is employed in the calculus (in mathematical analysis) moves one step further by employing what traditionally was designated as the *actual infinite* (preferably to be labelled as the *at once infinite*). The employment of the at once infinite was driven by the urge to arrive at a complete arithmetization of mathematics. Cantor claims that he had no other choice but to employ the possibly most general concept of a purely arithmetical continuum of points.[4] This step accomplished the complete reversal of the geometrization of mathematics in Greek thought by employing the idea of infinite totalities. It also terminated the long-standing reign of the potential infinite (successive infinite). Becker writes:

> The decisive insight of Aristotle was that infinity just like continuity only exists potentially. They have no genuine actuality and therefore always remain uncompleted. Until Cantor opposed this thesis in the second half of the 19th century with his set theory in which actual infinite multiplicities were contemplated, the Aristotelian basic conception of infinity and continuity remained the unchallenged common legacy of all mathematicians (if not all philosophers).[5]

Nonetheless the acceptance or rejection of the at once infinite continued to separate mathematicians. While rejecting the at once infinite, intuitionism also questions the transfinite number theory of Cantor by describing it as a *phantasm* (see Heyting 1949, 4). In his rejection of the actual infinite Weyl even employs biblical imagery. He believes that "Brouwer opened our eyes and made us see how far classical mathematics, nourished by a belief in the 'absolute' that transcends all human possibilities of realization, goes beyond such statements as can claim real meaning and truth founded on evidence" (Weyl 1946, 9). On the next page he alleges that the precise meaning of the word *finite* (where the members of a finite set are "explicitly exhibited one by one") was removed from its limited origin and misinterpreted as something

---

[4] "Somit bleibt mir nichts Anderes übrig, als mit Hilfe der in §9 definierten reellen Zahlbegriffe einen möglichst allgemeinen rein arithmetischen Begriff eines Punktkontinuums zu versuchen" (Cantor 1962, 192).

[5] "Die entscheidende Erkenntnis des Aristoteles war, dass Unendlichkeit wie Kontinuität nur in der Potenz existieren, also keine eigentliche Aktualität besitzen und daher stets unvollendet bleiben. Bis auf Georg Cantor, der in der 2. Hälfte des 19. Jahrhunderts dieser Thesen mit seiner Mengenlehre entgegentrat, in der er aktual unendliche Mannigfaltigkeiten betrachtete, ist die aristotelische Grundkonzeption von Unendlichkeit und Kontinuität das niemals angefochtene Gemeingut aller Mathematiker (wenn auch nicht aller Philosophen) geblieben" (Becker 1964, 69).

"above and prior to all mathematics." It was "without justification" applied "to the mathematics of infinite sets." And then he adds the remark: "This is the Fall and original sin of set theory, for which it is justly punished by the antinomies" (ibid., 10).

From his orientation as an axiomatic formalist, David Hilbert appreciates it as the finest product of the human intellect and proclaimed that nobody will be able to drive us out of the paradise created by Cantor.[6]

The Solution as New Source of Trouble

However, what appeared to be a solution to the second foundational crisis of mathematics, namely the employment of the at once infinite, soon turned out to be a key element in the third foundational crisis of mathematics. In 1900, Russell and Zermelo (see Husserl 1979, xxii, 399ff.) independently discovered that Cantor's naïve set concept is inconsistent. Just consider the set C having as elements all those sets which do not contain themselves as elements. Then C is an element of itself if and only if it is not an element of itself.

In this context the fate of Gottlob Frege's logicism is rather tragic. In the appendix to the second volume of his work on the basic laws of arithmetic he had to concede that Russell's discovery (in 1900) of the antinomous character of Cantor's set theory for some time delayed the publication of this volume (1903) because one of the cornerstones of his approach had been shaken.

Close to the end of his life, in 1924–25, Frege not only reverted to a geometrical source of knowledge, but also explicitly rejected his own initial logicist position. In a sense he completed the circle—analogous to what happened in Greek mathematics after the discovery of irrational numbers. In the case of Greek mathematics this discovery prompted the geometrization of their mathematics, and in the case of Frege the discovery of the untenability of his *Grundlagen* also inspired him to hold that mathematics as a whole actually is *geometry*:

> So an *a priori* mode of cognition must be involved here. But this cognition does not have to flow from purely logical principles, as I originally assumed. There is the further possibility that it has a geometrical source.... The more I have thought the matter over, the more convinced I have become that arithmetic and geometry have developed on the same basis—a geometrical one in fact—so that mathematics in its entirety is really geometry. (Frege 1979, 277)

For those who still adhere to the arithmetization of the continuum it is important to account for the idea of infinite totalities. Unfortunately, as noticed by Lorenzen, "arithmetic does not contain any motive for introducing the actual infinite."[7] Heyting points out: "Difficulties arise only where the totality of integers is involved" (Heyting 1971, 14). Yet Weierstrass was convinced that his static view of all real

---

[6] "Aus dem Paradies, das Cantor uns geschaffen [hat], soll uns niemand vertreiben können" (see Hilbert 1925, 170).

[7] "In der Arithmetik ... liegt kein Motiv zur Einführung von Aktual-Unendlichen vor" (Lorenzen 1972, 159).

numbers as an infinite totality (from which anyone can be "picked") proceeds in a *purely* arithmetical way.

It is clear that a systematic account of the two kinds of infinity is required since implicitly they surfaced in all three foundational crises in the history of mathematics. From a theological angle, however, it is sometimes claimed that theology has to explain the use of the term *infinity* because it is an attribute of God.

Theology as Mediator Between Christianity and Mathematics?

When mention is made of God's infinity in a theological context, the (theo-ontological) assumption is that infinity originally (eminently) belongs to God. Whenever the notion of infinity is employed in mathematics it is therefore *derived* from the theological understanding of God's infinity. Since Cusanus it is also customary to say that infinity in the sense of endlessness belongs to mathematics, but that the actual infinite is reserved for God only.

This theological legacy does not start by analyzing the structural interconnections between a fiduciary mode of speech (the "language of faith") and the numerical (quantitative) aspect of reality. As a result, it does not realize that *infinity* is a mathematical notion that can only subsequently be employed theologically. The prevailing theo-ontological circle first lifts infinity from its cosmic "place" ("seated" within the arithmetical aspect and its interrelation with the spatial aspect) by assuming that it originally belongs to God. Once this shift is made, it is alleged that infinity can only be (re-!)introduced within the domain of number by taking it over from theology.

An instance of investigating alleged historical (genetic) relationships while ignoring the structural relationships between (the fields of investigation of) theology and mathematics is present in the thought of Chase.

Are Numerical Relations Intrinsically "Theological" in Nature?

No one can deny that both mathematicians and theologians use numerical terms like the numerals *one*, *two*, and *three*. The underlying philosophical issue is, are these notions originally (that is, in a structural-ontic sense) *numerical* notions that are analogically used within a different (faith) context when theologians say that there is but *one* God, or when they speak of God's "tri-unity"? An investigation of the basic concepts and ideas employed in theological parlance reveals a remarkable coherence between the certitudinal aspect of reality and the different other aspects of reality. They are related by means of moments of similarity that are analogically reflected within the structure of the fiduciary aspect of reality (see the explanation of Dooyeweerd's theory of modal aspects explained by Glas and De Ridder in their introductory chapter).

These connections between different aspects are designated as analogical basic concepts of the scholarly disciplines. The simple succession of one, another one, and so on (indefinitely), provides us with the most basic (primitive) awareness of infinity. When this numerical intuition is deepened by our spatial awareness of at once and of wholeness (totality), an infinite succession of numbers may be considered as if all its elements are given at once, that is, as a "completed totality." Our above-mentioned preference to designate these two forms of the infinite as the successive infinite and the at once infinite follows from the fact that both these notions of infinity have their original seat (location/place) within the numerical aspect of reality.

The primitive meaning of number comes to expression in distinctness (discreteness) and (an order of) succession. It provides the basis for the remark of Weyl that it is the principle of (complete) induction which safeguards mathematics from becoming an enormous tautology (Weyl 1966, 86). Continuity in a spatial sense entails both simultaneity (an order of at once) and the notion of wholeness, of being a totality. When a sequence of numbers points towards these two features the primitive numerical meaning of infinity is deepened towards the idea of infinite totalities.

It is therefore surprising that Chase is asking: "Could infinities such as a completed totality be brought into mathematics without a Christian theological foundation?" (Chase 1996, 209). He continues: "At the very least, some idea of God standing outside of our experience must have been necessary, since apart from God we have no experience of the infinite" (ibid.). Chase also mentions the following fact: "Some Scholastics in the Middle Ages and Cantor in the nineteenth century believed in an actual mathematical infinity, based on God's infinity" (ibid., 209–210).

Having failed to investigate any structural relationship between mathematics and theology, the historical analysis provided by Chase precludes the option of acknowledging infinity in both its forms as truly "mathematical." Once the two forms of infinity are recognized in terms of the interconnections between number and space, one can explore additional numerical analogies within the fiduciary aspect. In other words, instead of supposing that "infinities such as a completed totality" originally is a theological idea that is completely foreign and external to mathematics, one would rather account for the numerical meaning of the successive infinite (one, another one, and so on, indefinitely, endlessly, infinitely). Likewise, accounting for the deepening of this primitive meaning of infinity under the guidance of our intuition of simultaneity will point us towards the idea of the at once infinite.

If we proceed from a structural-genetic perspective as the basis of a historical analysis (something absent in Chase's article), one can rephrase the point he wants to make: theological reflection and speculation about the "infinity" of God indeed paved the way for and promoted the eventual mathematical development of a theory of transfinite numbers (Cantor), but in doing that, theology simply digressed into quasi-mathematical considerations which, in the first place, refer to purely mathematical notions related to the inter-modal coherence between number and space.

Chase actually defends a kind of "negative theology." He does not acknowledge the numerical and spatial source of the "potential" and the "actual" infinite, but argues that these terms are originally theological in nature. In the final analysis they are then sent back to the domain from which they were (implicitly) captured in the first place—in the form of theological notions allegedly fruitful for the further development of modern mathematics. This also explains why Chase does not enter into a discussion of the notion of infinity as it is traditionally employed in Christian theology. At least such an investigation might have taken note of the fact that the Bible nowhere explicitly attributes infinity to God. Theologians traditionally extrapolate God's infinity from his omnipresence and eternity. In contrast, recognizing the (deepened) numerical seat of the notion of infinity should rather start from the assumption that theologians could only use notions of infinity as mathematical analogies in their theological argumentation.

Clearly, in the absence of a truly encompassing Christian philosophy, those sincere Christians who want to establish a link between their Christian faith and what mathematics is all about easily fall prey to the long-standing effect of a circular (scholastic) theo-ontology. Such a theo-ontological view proceeds from a particular ontological conception not necessarily inspired by a biblical perspective on reality. A theo-ontological approach takes something that is created and positions it in the "essence" of God, after which it is then finally copied back to creation.

Theo-ontology

Another example of such a theo-ontological approach is found in the work of James Nickel, who claims that the tension between the one and the many "is resolved and answered in the nature of the ontological trinity, the eternal one and the many.... We can do mathematics *only* because the triune God exists. Only biblical Christianity can *account* for the ability to count" (Nickel 2001, 231).[8] One immediately thinks of the second *hypostase* in the thought of Plotinus, the *Nous*, which is also designated as the one-in-the-many (*hen-polla*). The *Nous* does not display an absolute Unity, for it exhibits a balance between unity and multiplicity, as unity-in-the-multiplicity (cf. Plotinus 1956, VI.2.2.2, VI.7.14.11–12). In both instances something from creation is elevated to become a part of God's nature and then afterwards projected back into creation.

---

[8] Nickel quotes Vern S. Poythress saying: "In exploring mathematics one is exploring the nature of God's rule over the universe; i.e. one is exploring the nature of God Himself" (see Poythress 1976, 184).

Acknowledging the Order of Creation

Alternatively, a Christian approach to God's creational law-order may commence by investigating the fact that throughout the history of mathematics various points of view are found, as briefly noted earlier. Such an undertaking immediately relativizes the claims of "universal reason" which precludes the possibility of genuine divergent (or, conflicting) views within the so-called exact sciences, such as mathematics and physics. In the philosophical legacy primarily operative within the English-speaking world the term *science* is even restricted to the disciplines known as mathematics and physics. What is needed is our acceptance of the primary (and indefinable) meaning of the numerical aspect (where, as *part of creation*, our awareness of the one and the many finds its seat). Everything is not number, but everything functions within the numerical aspect. This remark presupposes the ontic givenness of the various modal (functional) aspects of reality, including number and space. Existence is more than *entitary* existence. In a different context I have argued that the lifting out of a particular aspect while disregarding others constitutes the distinctive feature of scholarly thinking, to wit, *modal abstraction* (see Strauss 2011b).

From this it follows that a scholarly account of the nature, scope, and limits of any scientific discipline in principle exceeds the limits of such a discipline (special science). The moment a mathematician wants to define or describe what mathematics is all about, something is said that falls *outside* the universe of discourse of mathematics as a discipline. Suppose it is alleged that mathematics is the discipline investigating "formal structures" (Bernays), that it is the "science of order" (Russell), that it is the "science of order in progression" (Hamilton), or even that it is a discipline constituted by the two subdisciplines algebra and topology (Bourbaki). Then we have to realize that in all these (and many more similar) "definitions" something is said *about* mathematics without, in any way, getting involved in *doing mathematics*. It is therefore a simple fact of philosophy of science that talking about mathematics cannot be equated with being involved in doing mathematics. The decisive criterion in defining mathematics is not *who* defines mathematics but the *nature* of such a definition of mathematics. Defining mathematics belongs to the philosophy of mathematics.

The Discrete and the Continuous

We have seen that the history of mathematics toggled between the extremes of an arithmetized and a geometrized perspective and that one of the key issues is found in alternative views of the nature of the infinite. Fraenkel et al. connect this to the relationship between the "discrete and continuous." They even speak about a "gap" in this regard which has remained an "eternal spot of resistance and at the same time of overwhelming scientific importance in mathematics, philosophy, and even physics" (Fraenkel et al. 1973, 213).

Apparently the "discrete admits an easier access to logical analysis," explaining according to them why "the tendency of arithmetization, already underlying Zeno's paradoxes, may be perceived in axiomatic set theory." Yet since "intuition seems to comprehend the continuum at once" the Greeks believed that continuity could be comprehended at once, explaining their inclination "to consider continuity to be the simpler concept" (ibid.). Recently a number of French mathematicians returned to the priority of space (the continuum)—they are even known as mathematicians of the continuum (see Longo 2001).

Yet the mentioned urge to arithmetize mathematics inspired Cantor to employ the general concept of a purely arithmetical continuum of points.[9] This step accomplished the complete reversal of the geometrization of mathematics in Greek thought by employing the idea of infinite totalities.

Unfortunately, modern set theory turned out to be burdened by the troublesome presence of what Cantor called *inkonsistente Vielheiten* (inconsistent sets) (see Cantor 1962, 447). Zermelo introduced his axiomatization of set theory in order to avoid the derivation of "problematic" sets while Hilbert dedicated the greater part of his later mathematical life to develop a proof of the consistency of mathematics. But when Gödel demonstrated that in principle it is not possible to achieve this goal, Hilbert had to revert to intuitionistic methods in his proof theory ("meta-mathematics").

This orientation is still alive within mathematics. We have quoted Fraenkel et al. (1973, 213) regarding the basic position of continuity. More recently, Longo highlights the above-mentioned views of René Thom and other mathematicians: "For him, as for many mathematicians of the continuum, 'the continuum precedes ontologically the discrete,' for the latter is merely an 'accident coming out of the continuum background,' 'a broken line'" (Longo 2001, 6). He also remarks: "By contrast Leibniz and Thom consider the continuum as the original giving, central to all mathematical construction, while the discrete is only represented as a singularity, as a catastrophe" (ibid., 19). Of course Longo is quite aware of the fact that the set theory of Cantor and Dedekind assigns priority to notions of discreteness "and derive[s] the mathematical continuum from the integers" (ibid.; cf. 20).

The history of mathematics therefore opted at least for three different possibilities: (i) to attempt exclusively to use the quantitative aspect of reality as mode of explaining for the whole of mathematics—Pythagoreanism, modern set theory (Cantor, Weierstrass), and axiomatic set theory (axiomatic formalism—Zermelo, Fraenkel, von Neumann, and Ackermann); (ii) to explore the logical mode as point of entry (the logicism of Frege, Dedekind, and Russell); and (iii) to assert the geometrical nature of mathematics—an attempt which was once again taken up by Frege, now close to the end of his life, and by the mentioned mathematicians of the continuum.[10]

---

[9] See the quotation given in n. 4 above.

[10] Bernays also consistently defended the position that continuity belongs to the core meaning of space and that the modern approach of Cantor and Weierstrass to mathematical analysis did not accomplish a complete arithmetization of the continuum (see Bernays [1976, 188] and also Strauss [2011a]).

Christianity and Scholarship

The legacy of reformational philosophy, particularly in the thought of Herman Dooyeweerd, proceeds from the basic biblical conviction that within itself created reality does not find an ultimate or final mode of explanation. The moment a thinker attempts to pursue this path, the honour due to God as Creator, Sustainer, and ultimate *Eschaton* of created reality is dedicated to a mere creature. The distorting effect of this inclination is manifest in all the antinomous "isms" discernible within all the disciplines (not only within mathematics). The challenge is to explore the option of acknowledging the uniqueness and irreducibility of every aspect inevitably involved in practicing mathematics without attempting to reduce any of the modal aspects to another. Dooyeweerd claims that whenever this anti-reductionist approach is not followed, theoretical thought inescapably gets entangled in *theoretical antinomies*. His claim is, in addition, that the logical principle of non-contradiction finds its foundation in the more-than-logical (cosmic) principle of the excluded antinomy (*principium exclusae antinomiae*) (see Dooyeweerd 1997, 2:37ff.).

A Christian attitude within the domain of scholarship, while observing the *principium exclusae antinomiae*, will attempt to avoid every instance of a one-sided deification or reification of anything within creation. The biblical perspective that God is Creator and that everything within creation is dependent upon the sustaining power of God opens the way to acknowledging the life-encompassing consequences of the redemptive work of Christ, for in him we are in principle liberated from the sinful inclination to search for a substitute for God within creation. We are in principle liberated from this inclination in order to be able—albeit within this dispensation always in a provisional and fallible way—to respect the creational diversity and its dependence upon God with the required intellectual honesty.

The Meaning of an Aspect Comes to Expression in Its Coherence with Other Aspects

This depth dimension contains the supra-theoretical motivation for articulating a non-reductionist ontology. The dimension of cosmic time and that of modal aspects constitute essential elements in such an ontology. Discerning the uniqueness and coherence within the diversity of various aspects within created reality is guided by the philosophical hypothesis that no single aspect could ever be understood in its isolation from all the other aspects. Furthermore, the core meaning of an aspect is indefinable and this insight entails the indispensability of "primitive terms." Yourgrau explains that Gödel "insisted that to know the primitive concepts, one must not only understand their relationships to the other primitives but must grasp them on their own, by a kind of 'intuition'" (Yourgrau 2005, 169).

The crucial point is that the meaning of an aspect only comes to expression in its unbreakable coherence with other modes—exemplified in what is designated as the *modal analogies* within each modal aspect, reflecting the inter-modal coherence between a specific aspect and the other aspects. These analogies are retrocipatory (backward pointing) or anticipatory (forward pointing) in nature. They are therefore also known as *modal retrocipations* and *modal anticipations*. Within the quantitative aspect of reality no retrocipations are found and within the certitudinal aspect there are no anticipations.

Christianity impacts mathematics through the non-reductionist ontology of a biblically informed philosophy. Such a philosophical orientation aims at avoiding the one-sided conflicting "ismic" trends operative throughout the history of mathematics. What it entails is summarized in the following thesis: accept the *uniqueness* and *irreducibility* of the various aspects of created reality, including the aspects of quantity, space, movement, the physical, the logical-analytical, and the lingual (or sign) mode, while at the same time embarking upon a penetrating, non-reductionist analysis of the inter-modal connections between all these aspects.

This proposal crucially depends upon a more articulated account of the theory of modal aspects as such. Figure 1 captures the most important features of a modal aspect.

Although Christians and non-Christians are living in the same world and do the same things, they indeed do these *differently*. What does this mean for mathematics?

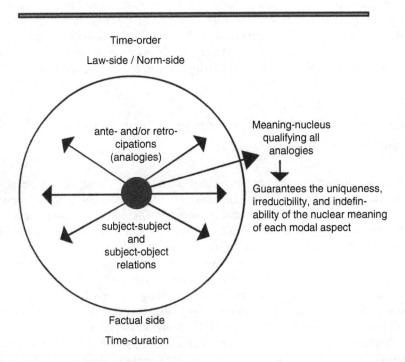

Fig. 1 The structure of a modal aspect

Infinity as Point of Entry

We will restrict ourselves to some general remarks regarding the inter-modal coherence between the aspects of number and space as it comes to expression in the earlier-mentioned two kinds of infinity, namely, the successive infinite and the at once infinite.

Since Descartes advanced the nominalistic conviction that "number and all universals are mere *modes of thought*" (*Principles of Philosophy* 1.57) only concrete entities were considered to be "real." Yet the question is whether there is not, prior to any human intervention, construction, or cognition, a *given* multiplicity of aspects or functions of reality. Hao Wang points out that Kurt Gödel[11] is very "fond of an observation that he attributes to Bernays"; to wit, "That the flower has five petals is as much part of objective reality as that its color is red" (quoted in Wang 1988, 202).

Surely the quantitative side or aspect of entities is not a product of thought, for at most human reflection can only explore this given (quantitative) trait of reality by analyzing what is entailed in the meaning of multiplicity. Bernays is a prominent mathematician who explicitly questions the dominant view that ascribes reality only to entities. He says that this mistaken conception acknowledges one kind of factuality only, namely that of the "concrete" (Bernays 1976, 122).[12]

Perhaps the most impressive advocate of the ontic status of modal aspects is found in Gödel's thought. He advances the idea of "semiperceptions" in connection with "mathematical objects." Next to a physical causal context within which something can be "given," Gödel refers to data of a second kind, which are open to semi-perceptions. Data of this second kind "cannot be associated with actions of certain things upon our sense organs" (quoted in Wang 1988, 304). In terms of a Dooyeweerdian approach these semiperceptions relate to the functional aspects of reality. Gödel says: "It by no means follows, however, [that they] are something purely subjective as Kant says. Rather they, too, may represent '*an aspect of objective reality*' but, as opposed to the sensations, their presence in us may be due to another kind of relationship between ourselves and reality" (ibid.; my italics).[13]

Theoretical and non-theoretical thought can explore the *given* meaning of this quantitative aspect in various ways. It is first of all done by forming *numerals* or *number symbols*, such as "1," "2," "3," and so on. The simplest act of counting

---

[11] At the young age of 25 Gödel astounded the mathematical world in 1931 by showing that no system of axioms is capable—merely by employing its own axioms—of demonstrating its own consistency (see Gödel 1931). Yourgrau remarks: "Not only was truth not fully representable in a formal theory, consistency, too, could not be formally represented" (Yourgrau 2005, 68). The devastating effect of Gödel's proof is strikingly captured in the assessment of Hermann Weyl: "It must have been hard on Hilbert, the axiomatist, to acknowledge that the insight of consistency is rather to be attained by *intuitive reasoning* which is based on evidence and not on axioms" (Weyl 1970, 269).

[12] Note that Bernays employs the term *factual* in the sense in which we want to employ it—referring to what is given in reality prior to human cognition.

[13] Kattsoff defends a similar view where he discusses "intellectual objects" which are also characterized by him as "quasi-empirical" in nature (Kattsoff 1973, 33, 40).

already explores the original meaning of the quantitative aspect of reality. It occurs in a twofold way, because every successive number symbol ("1," "2," "3," etc.) is correlated with whatever is counted.[14]

Figure 1 highlights the correlation between the law-side and factual side of an aspect and it also captures the way in which the dimension of cosmic time comes to expression in the first two modal aspects distinguished by Dooyeweerd—by acknowledging the numerical time-order of succession and the spatial time-order of simultaneity.

The Attempt to Arithmetize Continuity

We shall now briefly focus on the attempt to arithmetize continuity by calling upon the non-denumerability of the real numbers. When the elements of a set can be correlated one-to-one with the natural numbers [0, 1, 2, 3, ...] then it is called *denumerable (countable)*. Cantor has shown that the integers and fractions are denumerable. So the question remained, what about the real numbers?

In his famous diagonal proof Cantor only considered the closed interval [0,1] (i.e., $0 \leq x_n \leq 1$) which represents all the real numbers. In his diagonal proof Cantor commences by assuming that there is an enumeration x_1, x_2, x_3, ... of real numbers satisfying $0 \leq x \leq 1$ given by

$$x_1 = 0.a_1 a_2 a_3 \ldots$$

$$x_2 = 0.b_1 b_2 b_3 \ldots$$

$$x_3 = 0.c_1 c_2 c_3 \ldots$$

Now another real number is considered, namely, y conforming to the stipulation that y_1 is different from a_1; y_2 from b_2; y_3 from c_3, and so on. This number y certainly also belongs to the closed interval [0, 1] because $0 \leq y \leq 1$.

Nonetheless it differs from every x_n in a least one decimal position, from which Cantor concludes that every attempt to enumerate the real numbers within this interval will leave out at least one real number. The conclusion is that the real numbers are uncountable or non-denumerable. Grünbaum attempted to use this outcome in developing a consistent conception of the extended linear continuum as an aggregate of unextended elements. His entire argument crucially depends upon the non-denumerability of the real numbers and he is fully aware of the necessity of this assumption: "The consistency of the metrical analysis which I have given depends crucially on the non-denumerability of the infinite point-sets constituting the intervals on the line" (Grünbaum 1952, 302).

---

[14] Frege correctly remarks "that counting itself rests on a one-one correlation, namely between the number-words from 1 to n and the objects of the set" (quoted in Dummett 1995, 144).

Cantor starts his diagonal proof with the assumption that all real numbers in the closed interval [0, 1] are arranged in a countable succession. They are given as an infinite totality. He then shows that another real number (y) could be specified in such a way that it differs from every real number in their assumed denumerable succession (each time at least in the nth decimal place).

The crucial point is that the entire diagonal proof presupposes the assumption of an infinite *totality* containing all its elements *at once* (that is, conforming to the spatial time-order of simultaneity).

Sometimes the idea of an infinite totality is not accepted, such as one finds in intuitionistic mathematics (this school of thought only accepts the successive [potential] infinite). When the at once infinite is rejected, Cantor's proof does not yield non-denumerability, because there is no constructive transition from the successive infinite to the at once infinite (see Heyting 1971, 40; Fraenkel et al. 1973, 256, 272; and Fraenkel 1928, 239n1). Without the idea of an infinite totality Cantor's diagonal proof merely demonstrates that for a given sequence of successively infinite sequences of numbers it will always be possible to find another sequence of numbers differing from the given sequence of successively infinite sequences of numbers at least in one decimal place. In this interpretation non-denumerability nowhere surfaces.

Proceeding from a non-reductionist ontology, no contradiction is at stake, because the two kinds of infinity merely reflect inter-modal connections between number and space. We noted earlier that the primitive meaning of infinity is given in the successive infinite: one, another one, and so on, endlessly, infinitely. Once this primitive meaning of infinity is deepened under the guidance of the spatial time-order of *simultaneity* (at once—on the law-side) and the (factual) spatial *whole-parts relation*, it is possible to employ the idea of the at once infinite. Therefore, Cantor's diagonal proof is not contradictory even though an undisclosed (not-yet-deepened) use of the infinite does not yield non-denumerability. The latter can only be inferred when the at once infinite is assumed (see Strauss [2014] for a more extensive treatment of this issue).

The diagonal proof demonstrates a couple of things that are of central importance for the relationship between Christianity and mathematics. First of all, it shows that an apparently exact mathematical proof may lead to opposing results depending upon the underlying philosophical view on the relationship between number and space. Secondly, it highlights the implications of a non-reductionist ontology for mathematics as an academic discipline by suggesting an alternative for the historical movement to and fro between the aspects of number and space. Instead of reducing space to number or number to space—twice accomplished in the history of mathematics—one should accept their uniqueness (sphere sovereignty) and their mutual coherence (sphere universality).

It should be noted that the rejection of the at once infinite by intuitionistic mathematics resulted in a different theory of the real numbers (normally referred to by mathematicians as *the continuum*) instead of merely imitating a spatial feature in a numerical way—see Fig. 2.

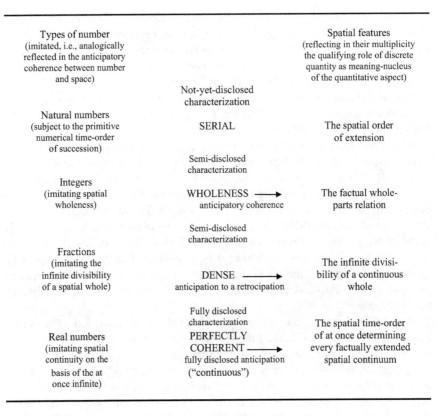

Fig. 2 The uniqueness of and interconnections between number and space

The Intuitionism of Dooyeweerd and Dummett

Interestingly, Dooyeweerd followed intuitionism in its rejection of the at once infinite (see Dooyeweerd 1997, 1:98–99n1, 2:92, 340n1). He did not realize that his own theory of modal aspects provides a point of departure for a unique and meaningful account of the nature of the at once infinite, namely as a regulative hypothesis in which the meaning of the numerical time-order of succession is deepened by anticipating the spatial time-order of at once. This deepened hypothesis makes it possible to view any successively infinite sequence of numbers as if it is given at once, as an infinite whole or an infinite totality.

It is worth mentioning that when Dummett explains the intuitionistic notion of infinity, a striking dialectic surfaces. He both uses and discards notions that are essentially spatial in nature, for without any hesitation he speaks about "infinite *totalities* of mathematical objects" (my italics). On the next page the expression "infinite *domain*" is used as a substitute for "infinite totality" (cf. similar usages—Dummett 1978, 22, 24, 57, 58, 59, 63, and so on). Sometimes the phrase "infinite *structures*" is used (ibid., 56, 62). At the same time Dummett holds on to the successive infinite since according to him any infinite structure is never something given as a set of "completed objects" (cf. Dummett 1978, 62).

Because both the time-order of at once and the whole-parts relation are embedded in the irreducibility of the spatial aspect, it begs the question if an attempt is made to reduce continuity to discreteness on the basis of the non-denumerability of the real numbers since this feature of the real numbers, as we noted, presupposes the at once infinite while the latter presupposes the spatial time-order of at once. The upshot of such an attempted arithmetization is antinomous and comes to expression in the following contradiction: continuity (space) could be reduced to number if and only if it cannot be reduced to number.

Approximating the Theory of Modal Aspects

Although Paul Bernays did not develop a theory of modal aspects, he has a clear understanding of the cul-de-sac entailed in arithmeticistic claims. He writes:

> It should be conceded that the classical foundation of the theory of real numbers by Cantor and Dedekind does not constitute a complete arithmetization.... The arithmetizing monism in mathematics is an arbitrary thesis. The claim that the field of investigation of mathematics purely emerges from the representation of number is not at all shown. Much rather, it is presumably the case that concepts such as a continuous curve and an area, and in particular the concepts used in topology, are not reducible to notions of number [*Zahlvorstellungen*]. (Bernays 1976, 187–188)[15]

Concluding Remark

Exploring an analysis of more interrelations between number and space exceeds the scope of this contribution. However, the preceding discussion sufficiently supports the claim that Christianity—via a Christian philosophy and a non-reductionist ontology—does have a meaningful contribution to make to the discipline of mathematics.

---

[15] "Die hier gewonnenen Ergebnisse wird man auch dann würdigen, wenn man nicht der Meinung ist, daß die üblichen Methoden der klassischen Analysis durch andere ersetzt werden sollen. Zuzugeben ist, daß die klassische Begründung der Theorie der reellen Zahlen durch Cantor und Dedekind keine restlose Arithmetisierung bildet. Jedoch, es ist sehr zweifelhaft, ob eine restlose Arithmetisierung der Idee des Kontinuums voll gerecht werden kann. Die Idee des Kontinuums ist, jedenfalls ursprünglich, eine geometrische Idee. Der arithmetisierende Monismus in der Mathematik ist eine willkürliche These. Daß die mathematische Gegenständlichkeit lediglich aus der Zahlenvorstellung erwächst, ist keineswegs erwiesen. Vielmehr lassen sich vermutlich Begriffe wie diejenigen der stetigen Kurve und der Fläche, die ja insbesondere in der Topologie zur Entfaltung kommen, nicht auf die Zahlvorstellungen zurückführen."

References

Becker, Oscar. 1964. *Grundlagen der Mathematik in geschichtlicher Entwicklung*. Freiburg: Alber.
Bernays, Paul. 1976. *Abhandlungen zur Philosophie der Mathematik*. Darmstadt: Wissenschaftliche Buchgesellschaft.
Beth, Evert. 1965. *Mathematical Thought*. Dordrecht: D. Reidel Publishing Company.
Boyer, Carl B. 1959. *The History of the Calculus and Its Conceptual Development*. New York: Dover.
Brouwer, Luitzen Egbertus Jan. 1964. Consciousness, Philosophy, and Mathematics. In *Philosophy of Mathematics, Selected Readings*, ed. Paul Benacerraf and Hillary Putnam, 78–84. Oxford: Basil Blackwell.
Cantor, Georg. 1962. *Gesammelte Abhandlungen Mathematischen und Philosophischen Inhalts*. Hildesheim: Georg Olms Verlagsbuchhandlung.
Chase, Gene B. 1996. How Has Christian Theology Furthered Mathematics? In *Facets of Faith and Science, The Role of Beliefs in Mathematics and the Natural Sciences: An Augustinian Perspective*, ed. Jitse M. van der Meer, vol. 2, 193–216. New York: University of America Press.
Dooyeweerd, Herman. 1997. *A New Critique of Theoretical Thought*, The Collected Works of Herman Dooyeweerd, Series A. Vols. 1–4. Lewiston: Edwin Mellen Press.
Dummett, Michael A.E. 1978. *Elements of Intuitionism*. Oxford: Clarendon Press.
———. 1995. *Philosophy of Mathematics*. 2nd ed. Cambridge, MA: Harvard University Press.
Fern, Richard L. 2002. *Nature, God and Humanity*. Cambridge: Cambridge University Press.
Fraenkel, Abraham. 1928. *Einleitung in die Mengenlehre*. 3rd ed. Heidelberg: Springer.
———. 1930. Das Problem des Unendlichen in der neueren Mathematik. *Blätter für deutsche Philosophie* 4 (1930/31): 279–297.
Fraenkel, Abraham A., Yehoshua Bar-Hillel, and Azriel Levy. 1973. *Foundations of Set Theory*. With the Collaboration of Dirk van Dalen. 2nd rev. ed. Amsterdam: North-Holland.
Frege, Gottlob. 1903. *Grundgesetze der Arithmetik*. Vol. 2. Jena: Verlag Hermann Pohle.
———. 1979. *Posthumous Writings*. Oxford: Basil Blackwell.
Gödel, Kurt. 1931. Über formal unentscheidbare Sätze der Principia Mathematica und verwandter Systeme 1. *Monatshefte für Mathematik und Physik* 38: 173–198.
Grünbaum, Adolf. 1952. A Consistent Conception of the Extended Linear Continuum as an Aggregate of Unextended Elements. *Philosophy of Science* 19 (2): 288–306.
Heine, Eduard. 1872. Die Elemente der Functionenlehre. *Journal für reine und angewandte Mathematik* 74: 172–188.
Hersh, Reuben. 1997. *What Is Mathematics Really?* Oxford: Oxford University Press.
Heyting, Arend. 1949. *Spanningen in de wiskunde*. Groningen/Batavia: P. Noordhoff.
———. 1971. *Intuitionism: An Introduction*, 3rd rev. ed. Amsterdam: North-Holland.
Hilbert, David. 1925. Über das Unendliche. *Mathematische Annalen* 95: 161–190.
Husserl, Edmund. 1979. *Aufsätze und Rezensionen (1890–1910)*. Husserliana: Edmund Husserl – Gesammelte Werke 22. Den Haag: Martinus Nijhoff.
Kattsoff, Louis O. 1973. On the Nature of Mathematical Entities. *International Logic Review* 7: 29–45.
Kline, Morris. 1980. *Mathematics: The Loss of Certainty*. New York: Oxford University Press.
Laugwitz, Detlef. 1986. *Zahlen und Kontinuum: Eine Einführung in die Infinitesimalmathematik*. Mannheim: B.I.-Wissenschaftsverlag.
Longo, Giuseppe. 2001. *The Mathematical Continuum: From Intuition to Logic*. ftp://ftp.di.ens.fr/pub/users/longo/PhilosophyAndCognition/the-continuum.pdf. Accessed 19 Oct 2010.
Lorenzen, Paul. 1972. Das Aktual-Unendliche in der Mathematik. In *Grundlagen der modernen Mathematik*, ed. Herbert Meschkowski, 157–165. Darmstadt: Wissenschaftliche Buchgesellschaft.
Nickel, James D. 2001. *Mathematics: Is God Silent?* Vallecito: Ross House Books.

Plotinus. 1956. *The Enneads,* 2nd ed. Trans. Stephen MacKenna, Revised by D.S. Page. London: Faber & Faber.
Poythress, Vern S. 1976. A Biblical View of Mathematics. In *Foundations of Christian Scholarship: Essays in the Van Til Perspective*, ed. Gary North, 158–188. Vallecito: Ross House Books.
Reidemeister, Kurt. 1949. *Das exakte Denken der Griechen. Beiträge zur Deutung von Euklid, Plato und Aristoteles.* Hamburg: Reihe Libelli.
Robinson, Abraham. 1966. *Non-Standard Analysis.* Amsterdam: North-Holland.
Stegmüller, Wolfgang. 1970. *Main Currents in Contemporary German, British and American Philosophy.* Dordrecht: D. Reidel Publishing Company.
Strauss, Danie. 2011a. Bernays, Dooyeweerd and Gödel – The Remarkable Convergence in Their Reflections on the Foundations of Mathematics. *South African Journal of Philosophy* 30 (1): 70–94.
———. 2011b. Defining Mathematics. *Acta Academica* 4: 1–28.
———. 2014. What Is a Line? *Axiomathes* 24: 181–205. https://doi.org/10.1007/s10516-013-9224-5.
Wang, Hao. 1988. *Reflections on Gödel.* Cambridge, MA: MIT Press.
Weyl, Hermann. 1919. Der circulus vitiosus in der heutigen Begründung der Analysis. *Jahresberichte der Deutschen Mathematiker-Vereinigung* 28: 85–92.
———. 1946. Mathematics and Logic: A Brief Survey Serving as Preface to a Review of *The Philosophy of Bertrand Russell. The American Mathematical Monthly* 53: 2–13.
———. 1966. *Philosophie der Mathematik und Naturwissenschaft*, 3rd rev. and exp. ed. Munich/Vienna: R. Oldenbourg.
———. 1970. David Hilbert and His Mathematical Work. In *Hilbert. With an Appreciation of Hilbert's Mathematical Work by Hermann Weyl*, ed. Constance Reid, 243–285. New York: George Allen & Unwin.
Yourgrau, Palle. 2005. *A World Without Time. The Forgotten Legacy of Gödel and Einstein.* London: Penguin Books.

Properties, Propensities, and Challenges: Emergence in and from the Natural World

Marinus Dirk Stafleu

Abstract Applying the basic distinctions of the philosophy of the cosmonomic idea, it may be possible to suggest some contributions to the development of the concept of emergence, of new properties and new propensities, of new characters, and even of humanity. I will first discuss successively emergence within the physical world, emergence of the physical world, and emergence from the physical world. Next, I will review some implications of the theory of evolution for the challenging emergence of mankind from the animal world. This chapter ends with a discussion of the meaning of *emergence*.

Keywords Characters · Evolution · History · Normativity · Randomness · Reductionism · Relation frames

Introduction

In his *Purposein the Living World? Creation and Emergent Evolution*, Jacob Klapwijk discusses the theory of emergent evolution (Klapwijk 2008, 2011). He describes the creation as a system of layers with increasing complexity, ignoring the mathematical aspects as well as Dooyeweerd's distinction of modal aspects and structures of individuality (Klapwijk 2008, 106–118). His book is intended to show that the biotic or living world of bacteria emerges from the physical world; the vegetative world of plants from the living one; the sensitive or animal world from the vegetative world; and finally humanity from the animal world. He concludes that no theory of emergence is available that could be considered explanatory. At most it is a philosophical framework theory (Klapwijk 2008, 162; 2011, 23). In particular, Klapwijk is unable to present a mechanism for the emergence of, for instance, the first living cells. But then, nobody else knows of such a mechanism, including the staunchest defenders of naturalistic evolutionism. I intend to present a small but perhaps significant contribution to understanding the phenomenon of emergence.

M. D. Stafleu (✉)
Independent Scholar, Zeist, The Netherlands
e-mail: m.d.stafleu@freeler.nl

The common view is that at a higher level of complexity new properties emerge that do not occur at a lower level. The whole is more than the sum of its parts (Popper 1972, 242–244, 289–295; Popper and Eccles 1977, 14–31; Mayr 1982, 63–64). In suit of Dobzhansky and using a term also applied by Dooyeweerd (Geertsema 2011, 69), Stebbins speaks of *transcendence*:

> In living systems, organization is more important than substance. Newly organized arrangements of pre-existing molecules, cells, or tissues can give rise to emergent or transcendent properties that often become the most important attributes of the system. (Stebbins 1982, 167)

Besides the emergence of the first living beings (prokaryote bacteria) and of humanity, Stebbins mentions the following examples: the first occurrence of eukaryotes (cells with a nucleus and other cell bodies, probably of prokaryote origin); of multicellular animals and fungi; of invertebrates and vertebrates; of warm-blooded birds and mammals; of the higher plants and of flowering plants. According to Stebbins, reductionism and holism are contrary approximations in the study of living beings, with equal and complementary values.

I interpret the modal aspects or relation frames (including the mathematical ones) to be primarily aspects of time (Stafleu 2002a, 2011). Each aspect expresses a directive order of time, determining subject-subject relations and subject-object relations. Because everything is related to everything else, and human experience depends on relations, each relation frame is also an aspect of being and of human experience. In this interpretation, the relation frames are aspects of becoming and of change as well. According to Dooyeweerd, this comes about by the opening up of the anticipations in each modal aspect. For historical developments, this process is founded in the so-called historical modal aspect, and guided by the aspect of faith. In turn, the opening up of the aspect of faith (which has no anticipations) is guided by religion. Because several religious ground motives are operative, this historical process can occur both according to and contrary to God's laws (Dooyeweerd 1953–1958, 2:181–365).

Dooyeweerd never came to terms with the theory of natural evolution (Dooyeweerd 1959; Verburg 1989, 350–360; Stafleu 2002b; Wearne 2011, 88–100; van der Meer, chapter "Is there a Created Order for Cosmic Evolution in the Philosophy of Herman Dooyeweerd?," this volume). One reason may be the tension between evolution and his philosophy of history. In Dooyeweerd's philosophy there is no place for a modal aspect having the same function for evolution as the historical aspect has for history, and Dooyeweerd never suggests that natural evolution is guided by religion, faith, or any other aspect.

I do not believe that it makes sense to call the technical or formative aspect *historical*, and I do not believe that the opening process occurs mainly in the modal aspects (Stafleu 2008, 2011). Evolution is only manifest in the coming into being of natural things and events at the subject- and object-side of reality. In contrast, historical development also occurs at the law-side, in the human development of normative principles into variable norms. In addition, at the subject- and object-side one finds the historical development of acts, artifacts, and associations, in which people take an active part (Stafleu 2011).

For understanding this, the distinction of relation frames and characters is crucial—these are Dooyeweerd's modal aspects and structures of individuality (Stafleu 2002a, chap. 1). Both have a law-side and a subject- and object-side (Stafleu 2014). At the law-side of the relation frames we find universal laws for temporal intersubjective relations and for subject-object relations. For instance, we find that all physical things are subject to the temporal order of irreversibility, valid for any kind of interaction, and that all living beings are genetically related, subject to the genetic law that any living being descends from another one. Characters are sets of universal laws and type laws, determining a subjective class of things or events, and an objective ensemble of possible states. The character of electrons differs from the character of hydrogen molecules, not because of universal physical laws, but because of different specific laws. At the subject-side of natural characters, one finds individual things and events, plants and animals; at the object-side, one finds specific properties. At the subject-side of normative characters, individual human beings and their associations function; at the object-side, acts and artifacts are operative.

Coming into being refers to the realization of characters of things such as molecules. *Change* refers to typical events such as radioactive decay. In these processes, the temporal order and temporal relations of several modal aspects are involved, besides the typical properties of the structures concerned. Even if we assume that natural laws are invariant, the realization of the characters of things and events, with their properties and propensities, is a temporal process. Sometimes this is called *emergence*.

In the following section, I propose to distinguish the emergence of physical things *within* the physical world from the emergence of the physical world itself (popularly called the Big Bang) and the emergence of living beings *from* the physical world. From an analysis of the first kind we may learn a few points that can help us to understand the other two. Also, we shall see how humanity, emerging from the animal world, is challenged to activate its own history. For a discussion of emergence within the living world, see Jitse van der Meer's contribution to this volume.

Properties and Propensities

Emergence Within the Physical World

Let me start with a familiar example. Suppose you have a container with a gas mixture of two parts of hydrogen and one part of oxygen. The mixture has properties that do not differ much from a combination of the properties of its components. The properties of a mixture form a mixture of the properties. But if you apply a spark, the mixture will explode and you get water vapor. If you cool this, you will get liquid water and eventually ice, the crystal form of water.

In this example, we see both the emergence of new structures and of new properties. Water has completely different properties from hydrogen and oxygen, and the properties of vaporous, liquid, and solid water differ too. In the framework of the philosophy of the cosmonomic idea, each characteristic whole has the following threefold character (Stafleu 2002a, chap. 1).

1. Each typical whole is *primarily* characterized by one of the relation frames. Reversely, all modal aspects (including the mathematical ones) qualify various characters. The structures of hydrogen, oxygen, and water have in common that they are primarily physically qualified. This means that they are mainly characterized by *interaction*, which is the universal physical subject-subject relation. Something is physical if it interacts with something else.
2. The mentioned structures differ because of their having *secondary retrocipatory properties* (i.e., referring the qualifying relation frame to earlier frames). For instance, mass is a quantitative objective magnitude of physical subjects, having different values for hydrogen, oxygen, and water molecules. Symmetry is an emergent spatial property. Whereas free atoms have the symmetry of a sphere, hydrogen and oxygen molecules have the symmetry of a dumbbell. A water molecule has a triangular shape giving rise to an electric dipole, which partly explains the peculiar properties of water and ice. Solids are characterized by their long-distance crystal symmetry, completely absent in the composing atoms or molecules. The parts of each physical thing perform periodic motions, contributing to the thing's stability.

One may recognize these two features of being qualified by one modal aspect and founded in a preceding one from Dooyeweerd's theory of individuality structures. I have shown that the secondary characteristic, based in a preceding relation frame, leads to secondary types of things and events. Physical things and processes occur in three secondary types, based in the quantitative, spatial, and kinetic frames. Similarly, there are four secondary types of biotic characters (Stafleu 2002a, chaps. 5 and 6). Therefore, I have no need for a vegetative layer besides the biotic one, as Klapwijk postulates, after omitting the mathematical relation frames (Klapwijk 2008, 106–115). In my analysis, bacteria do not constitute a primary type, but belong to a secondary type of living beings, which are primarily biotic (Stafleu 2002a, 186–191).

Klapwijk does not recognize the relevance of retrocipations for typical structures, and both Klapwijk and Dooyeweerd do not recognize a third, a tertiary characteristic of any typical whole. It is sometimes related to anticipations, looking forward in the order of the relation frames.

3. Hydrogen has the *tertiary disposition* or *propensity* of being combustible. Its structure anticipates that of water. In turn, water has many more or less unique propensities, such as being a solvent of many chemical compounds. In particular, water is a prerequisite of all kinds of life as we know it. Clearly, with the emergence of new structures, both new *properties* and new *propensities* emerge. With the formation of a new physical thing, possibilities are realized and new

possibilities are disclosed. For our understanding of evolutionary processes, the tertiary dispositions are paramount. Because of their tertiary propensities, in the course of physical time a hierarchy of structures came into existence. For example, a crystal consists of molecules (or ions), a molecule consists of atoms, and an atom has a nucleus and a number of electrons. A nucleus consists of protons and neutrons, which are combinations of quarks.

Reductionists (such as many high-energy physicists, naturalist philosophers, and popular science writers) believe that understanding interaction at the deepest level in a so-called theory of everything will inevitably lead to the understanding of interaction at all higher levels of complexity, including biotic, animal, and human life. Antireductionists (such as several solid-state physicists, biologists, and holist philosophers) point out that in general the emergence of new structures, new properties, and new propensities cannot (or cannot easily) be explained from knowledge of lower levels. A reductionist theory may be able to explain properties but is often unable to explain or predict propensities. Some people consider this merely a practical difference; others see it as a matter of principle.

Whereas the stable structure of systems like nuclei, atoms, and molecules is fairly well known, the *processes* leading to their formation are far less understood. For instance, the crystal structure of solid matter is widely investigated, but up till now there is no explanation available of why below a fixed temperature a fluid becomes a crystal, showing a typical long-distance ordering of the composing atoms, ions, or molecules which is absent in gases and fluids.

Whereas the structure of a stable physical system is largely *determined* by general and typical laws, transitions between unstable states—as in radioactivity, or the emission of light—are to a large extent *random* processes, subject to stochastic laws, i.e., probability laws. This is even more the case with the emergence of new structures like the formation of molecules from other molecules. In general, stable structures are less accidental than processes. In our example of the formation of water from hydrogen and oxygen, only water molecules can be formed (besides some related molecules, such as hydrogen peroxide), but it is largely accidental which pair of hydrogen molecules will bind with which oxygen molecule into two water molecules. Statistical laws for random or stochastic processes were first introduced in statistical mechanics for the explanation of irreversible processes and of Brownian motion, and were next discovered in radioactive decay at the turn of the twentieth century. Quantum physics has confirmed the occurrence of randomness in every physical or chemical process (e.g., the formation of a water molecule from hydrogen and oxygen), later followed by chaos theory. In biology, genetic drift is a random process in the theory of evolution, and the process of fertilization in plants and animals is very much random. Nevertheless, many reductionist philosophers and scientists maintain their unshaken belief in nineteenth-century determinism. Sometimes it leads them to the empirically unsubstantiated hypothesis of universes parallel to the observable one. If something seems to be the random realization of a possibility (as in radioactivity), they suppose that the other possibilities occur in some other universe, such that determinism is saved.

For the understanding of evolution, the distinction of the law-side and the subject- and object-side of reality is very important (Geertsema 2011, 59–64). Realist physicists readily assume that physical laws are invariant. Otherwise they could not apply the laws they have found from observations and experiments in the present to past and distant events in astrophysics. Even positivist physicists do that. They also assume that mathematical laws are invariantly valid. Biologists, however, appear to be less confident about the invariance of natural laws, apparently because they seem to believe that something can only be called a *natural law* if it has a mathematical form. Nevertheless, in the theory of evolution they apply whatever patterns they discover in the present to events in the past. Hence they implicitly acknowledge the persistence of natural laws, also in the field of biology.

One should always keep in mind the distinction between general modal laws and type laws, which are valid only for a restricted set of things or events. If we assume that natural evolution takes place at the subject- and object-side of typical laws, we can hypothetically accept that species as natural kinds are invariant characters, whereas their temporally and spatially limited realizations (which biologists call *populations*) are subject to evolution (Stafleu 2000, 2002a, chap. 6). Evolutionary processes are always directed by the modal temporal orders in the natural relation frames. Temporal order, as a set of general, universal laws, determines the relations between subjects, between objects, and between subjects and objects, and this is highly relevant for evolutionary processes.

The invariance of natural laws is in accord with the view of Christian philosophy that God created and sustains the world according to his laws. These are not rational, not *intelligent*, and structures are not *designed* according to some intellectual idea, the way technical artifacts are designed by people taking into account natural laws and human-made norms. Yet both modal laws and type laws are *intelligible*, meaning that science can investigate these and come to understand them and that people can apply them when designing artificial things and processes. Recognizing that the technical relation frame like all normative frames is irreducible to the natural ones, Christian philosophy should reject Plato's idea (developed in his dialogue *Timaeus*) of a semidivine demiurge, an anthropomorphic craftsman making imperfect copies according to a perfect, intellectual design. Therefore, we should not contemplate the idea of intelligent design of natural structures (Klapwijk 2008, 12–28, 128–136; 2011, 12–14). It appears that the idea of intelligent design is irrefutable (a trait it would share with, e.g., solipsism, or with the above-mentioned multi-universe hypothesis). This means that it is logically consistent, but empirically empty.

The Emergence of the Physical World

The combination of the general theory of relativity with insights gained of the various typical kinds of interactions led in the past century to the so-called standard model. This is a now generally accepted theory of the structure of sub-atomic particles and their interactions and of the development of the physical universe *after* it emerged in what is popularly called the Big Bang. Referring to a distinction made above, it explains many properties of sub-atomic particles, but hardly the emergence of any propensity. The fact that one needs the general theory of relativity shows that the relation frames preceding the physical one were crucial in this event. It is remarkable that these relation frames immediately occur in their "opened up" form, already anticipating the physical aspect (Stafleu 2002a, chap. 5). Only in the context of the physical aspect and the preceding aspects anticipating it could physical things and events emerge. Whereas the *theoretical* opening up of the modal aspects (including both their anticipations and their retrocations) is a historical process, in an ontological sense the modal aspects always include both their retrocations and anticipations. (The only and very important exception concerns the positivization of normative principles into historically developed norms—see below.) Therefore, evolution cannot be simply interpreted as the opening up of earlier modal aspects.

Sometimes the process occurring after the Big Bang is called *astrophysical evolution*. During this time, the initially very high temperature (meaning a high concentration of energy) gradually decreased when physical space expanded, allowing for the formation of protons and neutrons from quarks; nuclei from nucleons; atoms from nuclei and electrons; molecules from atoms; stars, comets and planets from interstellar dust; and then more atoms from exploding stars. This evolution means an enormously increasing diversity. Astrophysicists believe that they are able to give a fairly accurate (though not complete) description of what happened *after* the Big Bang, stressing that the emergence of the universe itself lies beyond the horizon of present-day theory. The reason is not difficult to understand: the physical world is characterized by interaction between existing things according to physical laws, and before it emerged, there was simply nothing to interact with.

Let us now see whether we can learn something from our summary of physical states of affairs for our third problem, the emergence *from* the physical world.

Emergence from the Physical World

Christian philosophy assumes that the aspect of life is irreducible to the physical aspect. This is a fruitful hypothesis, which should nevertheless not be taken as an unshakable dogma. But there are many reasons to stick to it, all of which I shall not mention. With my earlier comments in mind, I offer a few considerations.

In my view, each relation frame is characterized by one or more general laws determining subject-subject relations and subject-object relations. Quantitative relations are relations between numbers; spatial relations determine relative positions, and kinetic relations concern relative motions. Physical relations are always concerned with some kind of interaction: general relations, such as energy, force, and current, and typical relations, such as gravitational and electromagnetic interaction, and weak and strong nuclear interaction in all their typical manifestations, such as glue. I have argued that relations in the biotic world are always determined by genetic laws (Stafleu 2002a, chap. 6). The emergence of living beings is accompanied by the emergence of genetic relations between living beings as a universal trait of the living world.

The universality of interaction is expressed by the statement that *everything interacts with everything else*. Similarly, biologists argue that a living being always descends from another living being. Both statements are not trivial. The understanding that everything interacts with everything else became common sense in physics only in the seventeenth century with Newton's laws of motion and in particular the universality of gravity. The insight that a living being can only emerge from another living being (or a pair of them) dates from the nineteenth century, when the idea of spontaneous generation was abandoned. It played a part in Darwin's theory of evolution, but that theory only became fruitful in the twentieth century after the incorporation of genetics, particularly molecular genetics. Other general principles operative in Darwin's theory of evolution of living beings are competition, variation, selection, and genetic transmission. These point to universal biotic laws, which, like the later discovered laws of genetics, are not of a physical nature. These laws only operate within the realm of living beings, and can therefore not explain how living beings arise from the material world (Klapwijk 2008, 40). In fact, neither Darwin nor any other competent biologist pretends to know how life originated.

Everybody agrees that the emergence of the first living beings is a highly improbable process. Even naturalists, believing that it is a completely natural process, admit that the successive emergence of living beings, animals, primates, and human beings has such a low probability that we might very well be unique in the universe. The search for an extraterrestrial intelligent civilization is not very promising. Yet improbability is not the same as impossibility. The fact that we exist is the best and perhaps the only evidence that such a process is possible, and that the animal world has the propensity to give rise to humanity. The fact that chance plays a part does not mean that the emergence of a new thing is purely a chance process. The evidence is abundant that all processes in the natural world occur according to laws that delimit the possibilities. No less than properties, propensities are bound by laws. Without laws we would not even be able to estimate the chance of a certain process.

The assumption that the genetic relation characterizes all living beings implies that we cannot understand the emergence of the first living beings by biotic laws, because these were not operative before this emergence. This is similar to the inadequacy of physical laws to explain the emergence of the physical world. On the other hand, if we assume that physical things are only related by physical interaction, we cannot understand how living beings emerge from the physical world. In that case, like the emergence of the physical world, the emergence of the biotic world, the animal world, and the human world would forever remain behind the horizon of human experience and understanding. It would be tempting to assume that we have here what insurance brokers call an act of God, an unexplainable special act of creation. In contrast, one may also stress that God's hand is operative in any process, and that the emergence of the living world is neither excluded nor special.

However, this may be too fast. In this reasoning we overlook what Christian philosophy calls *anticipation*. Molecular biologists have been able to identify propensities in physical systems such as DNA molecules to play a part in genetic relations. What we know about that is a far cry from understanding the emergence of living beings, but we should not exclude its possibility. In this case, *anticipation* means a relationship between physical and biotic laws, structures and processes, our understanding of which is gradually growing. An example of a physical structure anticipating life could be a so-called self-replicating molecule like DNA or RNA, although it is really some other thing which undertakes a process to replicate the molecule.

It is quite clear that the biotic world presupposes the physical one. It can hardly be denied that the biotic world emerged from the physical one, the animal world from the living one, and humanity from the animal world. Reductionists believe that basically there is only one world, with various levels of complexity and diversity, but ultimately completely determined by physical laws only. Christian philosophy supposes that mathematical, physical, biotic, psychic, and normative relation frames are mutually irreducible. *Irreducibility* does not mean that the relation frames are independent of each other. In scientific investigation, the so-called retrocipatory connections between the relation frames usually receive more attention than the anticipatory ones. As observed above, the *properties* of structures are better understood than their *propensities* and the processes that realize them. For the understanding of the emergence of new structures and their properties and propensities we need to pay attention to both retrocipations and anticipations, not merely in the relation frames, but in particular in the typical characters of natural things and events.

It appears that all living beings have the same unique ancestor. That is not something that follows from any theory or philosophical point of view, but it might point to the improbability of this emergence. It has also been observed that the emergence of living beings and other major transitions happened in a relatively short time (measured on a geological scale). The sudden emergence of all phyla of animals, some 530 million years ago, is even called the *Cambrian explosion*. In the above-discussed example of a mixture of hydrogen and oxygen gas, one may ask what causes an explosion. Is it the spark, or is it the composition of the gas mixture? We have a tendency to blame the spark. However, the spark appears to be only the accidental, though sufficient, cause, whereas the presence of a mixture of hydrogen and oxygen is the necessary cause. The spark is not necessarily electric. Any other accidental disturbance could act as a trigger. Similarly, one may assume that at a certain moment the circumstances necessary for the emergence of living beings or the emergence of the animal world were present. Maybe we shall never know what the trigger was, because it is accidental and could have been different with the same result.

The laws of biological evolution—about adaption, natural selection, and common descent—have a general (i.e., modal) character. They show some resemblance to the physical laws of mechanics and thermodynamics, with the law of conservation of energy as a paradigm. In the nineteenth century, during Darwin's life time, positivist and materialist energeticists such as Ostwald, Mach, and (initially) Planck believed that all of physics should be explained from these general laws, which were assumed to be deterministic. They scorned Boltzmann for applying statistics to physical problems. They even rejected the reality of atoms and molecules. The development of physics during the twentieth century made clear that the general laws act as constraints, not showing what is possible but rather what is impossible. For instance, processes violating the law of energy conservation are prohibited. In the twentieth century it became clear that these general laws are not sufficient. Physicists discovered typical conservation laws (e.g., the law of conservation of electric charge) besides symmetry laws. Both kinds of laws do not determine but prohibit certain processes, giving room for other processes that might happen without determining which processes that would be.

Similarly, Darwin's theory may be able to explain which circumstances allow species to come into being or force them to be extinguished. But it does not explain why some species correspond with stable organisms in these circumstances and others do not. Since the twentieth-century synthesis of Darwin's theory with genetics and molecular biology, biologists have similarly become aware that the general laws of evolution should be complemented with specific laws for the enormous variety of living beings (Miller 1999; Cunningham 2010, chap. 4).

Challenges

Biology and the Emergence of Mankind

One of the basic assumptions of the standard Darwinian theory of evolution is that every living being descends from another one. For this reason alone, this theory cannot explain the emergence of the first living beings from non-living matter. There are more unexplained transitions, for instance, the emergence of the first eukaryotic cells, of sexual reproduction, and of the first animals. Finally, there is the emergence of mankind, for which I shall argue that the theory of evolution may be able to give an initial necessary explanation, but not a sufficient one.

In one way or another, history should be connected to evolution. Like most people, I assume that mankind has evolved from the animal world. A Christian view would be that humanity has been *called* from the animal realm, in a challenge to shape its own world. This appears to supplement the biblical testimony saying that Adam was made out of dust. Examination of fossils has sufficiently shown that human beings descend from hominids, sharing common forebears with other primates. Investigation of DNA and of behavior has shown much resemblance with primates, including patterns of genetic deletions and insertions, etc. In short, I think there is convincing evidence that humans descend from animals in a biological sense. Nevertheless, this provides at most a necessary part of a scientific explanation, not a sufficient one. In particular, it cannot explain how humanity became challenged to take an active part in its own history.

The Diversity of Humankind

From a biological point of view, human beings belong to a single species, *homo sapiens*, and a single subspecies, *homo sapiens sapiens*. If we distinguish between natural evolution and human history, it may be assumed that the latter started with the appearance of *homo sapiens,* which would include *homo neanderthalensis*. The main difference between humans and animals appears to be normative behavior, and the freedom and responsibility to actualize universal normative principles into culturally and historically different norms (Stafleu 2011, chap. 1). In contrast, natural things and events, plants and animals are subject to natural laws, which I assume to be invariant and universally valid.

All animals are specialized in one way or another. Each species occupies its own niche in which it can survive. In contrast, mankind shows an enormous individual diversity and cannot be characterized in a single way. One can easily think of ten different normative principles or conditions for being human (Stafleu 2011, chap. 1). People are called *homo faber*, or tool-making animals, developing all kinds of skills. Johan Huizinga introduced the term *homo ludens*, the playing man, who has pleasure in beauty. Some philosophers prefer *homo symbolicus*, pointing to the unique ability of mankind to communicate with symbols, in particular, language. *Homo sapiens* refers to the ability of mankind to reason, to make logical distinctions and connections. Another condition of being human is *to trust* each other, to have faith (which should be distinguished from religion) and to be faithful. People seek each other's *company*. They are called *homo economicus*, or *homo politicus*. They strive after *justice* and ought to *love* their neighbors. Of course, people often act contrary to these conditions, which otherwise would not be called normative.

In order to account for this diversity of human activity, besides six natural aspects I conjecture ten mutually irreducible normative modal aspects or relation frames as conditions for being human. These are the technical, aesthetic, semiotic, and logical aspects, which I take together with that of trust as being *cultural*; and companionship, economics, politics, justice, and loving care, which are aspects of *civilization*. In each of these relation frames, one finds normative principles, including a temporal order giving normative direction to the history of mankind. For instance, the technical frame is directed by the normative principle of progress. The semiotic or communicative relation frame is directed by the normative principle of clarity. *Culture* means the development of technical skills, of aesthetic expression, of the exchange of information by communication, of reasoning, and of trust. *Civilization* means the development of companionship, of economic services, of authority and discipline, of justice, and of love. Each of these is an expression of human experience, and each directs a kind of transfer of experience, an engine of history such as is unknown in the animal world (Stafleu 2011, chap. 2). By giving direction to historical events, the temporal order in each relation frame determines the *normative meaning* of history.

All these conditions for being human do not constitute properties of people, something people would have as a matter of course, but propensities, i.e., possibilities to realize, or rather challenges to develop the natural world into a cultural and civilized one.

The Experience of Normative Principles

These normative principles appear to be as universal as the natural laws. However, at the beginning of history, human beings have discovered that they are to a certain extent free to obey or to disobey these principles; a certain freedom that animals and human beings do not have with natural laws. Moreover, they have discovered that the normative principles are not sufficient. In particular, the organization of human societies requires the introduction of human-made *norms* as implementation or positivization of normative *principles*. Therefore, human freedom and responsibility have two sides. At the law-side they mean the development of norms from the normative principles; these norms are different at historical times and places and vary in widely different cultures and civilizations. At the subject-side, individual persons and their associations are required to act according to these laws, which ought to warrant the execution of their freedom and responsibility.

All people appear to have a sense of justice. Normative principles such as justice may be assumed to be universal and should therefore be recognizable in the whole of history (as far as we know it), in all cultures and civilizations. Human skills, aesthetic experience, and language may widely differ but are always present and recognizable in any human society.

From Evolution to History

This has led many scholars to assume that human history can be described as biological evolution, and they apply in particular Darwin's ideas of adaptation and natural selection to human history. They overlook the fact that Darwin's theory necessarily presupposes inheritance. Biological selection is a slow process. The evolution of hominids to mankind took at least 7 million years, which is not long on a geological scale. But human history is at most two hundred thousand years old. It happens much faster than biological evolution, and is even accelerating. Moreover, human experience cannot be inherited. In each relation frame, the transfer of experience is recognizable as an engine of history (Stafleu 2011, chap. 2).

As to the normative relation frames, concerning the cultural transfer of experience one might think of instruction and learning as the engine of technical progress, players and spectators as actors in aesthetic renewal, transfer of information as a source for the collective memory of mankind, reasoning as the engine of logical extrapolation, and faith as the motive force of reform. For the propagation of civilization, the following engines may be considered: education in keeping company, commerce in economic rendering of services, leadership, justice, and finally friendship and marriage. In this multifaceted transfer of experience, people attribute *subjective meaning* to history.

The transfer of experience is as diverse as human experience itself. It is completely absent in the animal world. The transfer of experience as the engine of history in each normative relation frame replaces inheritance as the engine of evolution. This is the nucleus of truth in the hypothesis that *memes* are the units of cultural transmission, comparable to inheritable *genes* in biotic evolution (Cunningham 2010, 206–212). Or, a bit more old-fashioned, it reminds of the distinction between nature and nurture.

Artifacts

The technical relation frame is the first cultural one. *Culture* means first of all *cultivating*, i.e., bringing nature into culture, the opening up of the natural relation frames, which therefore are no less aspects of human experience than the normative frames. People do that opening up by working. Labor is a cultural activity characterized by ability, skill, or command, in which people make use of instruments. Ability to perform labor is a universal value, a condition for progress as the historical temporal order for the technical relation frame. An event, process, artifact, or association, and even a personality may be called *historical* as far as they contribute positively or negatively to historical progress. The *intuition* of progress as a challenging value is not due to the Renaissance or the Enlightenment, but it is a condition for being human. The later *belief* in progress identified it with the factual history of seventeenth- to nineteenth-century science and technology. The Eurocentric belief in progress considered the technical and scientific progress even as characteristic for the entire history of mankind. This became a deep disappointment at the outbreak of the great European war in 1914, when science and technology turned out to be instruments of mass destruction. In 1931 Herbert Butterfield criticized the "Whig interpretation of history," which describes history as a continuous progress after the model of the British Empire.

Progress does not have the compulsory law-conformity of a natural law but is a normative challenging principle. As far as one can speak of progress in natural evolution as pointing, for instance, to an increasing complexity, this is not a value or norm but the effect of a natural process which has, as such, no normative character. Herein plants and animals do not play an active part comparable to that of human beings in technical progress.

As a normative principle, progress acts as the temporal order for the technical relation frame, as directive meaning of skilful labor. The history of technology concerns the elaboration of objects and the invention of artifacts besides training and education as the engines of technical progress.

History concerns the world as people have made it. For this purpose, I distinguish two kinds of experience (Stafleu 2011, chap. 3). The *natural* or *intuitive experience* which people to a certain extent share with animals is directly founded in sensory observation. In addition, there is an *indirect, detached form of experience*, in which people make use of some kind of instrument. This may be a material expedient, such as a microscope, with which people may reinforce their visual power. It may also be a logical device, e.g., a theory used to think about a problem. I apply the word *arti-*

fact as the generic name for any human-made object of human conduct primarily characterized by one of the normative relation frames (Stafleu 2014). This is a much wider definition than that applied in technology, where artifacts are technical products, or in archaeology, where artifacts are human-made material remains. Artifacts, or constructions, are often not primarily technical and by no means always material. In each relation frame artifacts are distinguished from other objects which are not characterized by that relation frame. A painting, for instance, is a material aesthetic artifact. It is an object characterized by the aesthetic relation frame, an instrument in one's aesthetic experience. As such it is not an economic artifact, though it can clearly be an economic *object*. Only its proceeds at an auction form an economic artifact that is established by people and is economically, not materially, typified. The price of a painting is *primarily* not characterized by aesthetic but by economic relations, and only *secondarily* by its aesthetic quality, rarity, and so on. Therefore, besides its being primarily aesthetic and secondarily technical, the painting also has an economic function, and the price of a painting has a history of its own.

Not only thing-like objects are artifacts. Events and processes playing an objective part in history can also be considered as artifacts if produced or interpreted. Therefore, artifacts show an enormous diversity, allowing mankind to leave the animal world and to explore nature. All artifacts (not only written sources) witness the *objective meaning* of history.

The Network Society

Humans live in societies. Evolutionists point to beehives and herds of cows as arguments to show that human behavior can be reduced to biotic or sensitive needs. In order to study the *social* or *communal meaning* of history, philosophy of history cannot neglect social philosophy. In particular the distinction between organized and unorganized social connections is relevant for understanding their influence on historical developments (Stafleu 2004, 2011, chaps. 4 and 5).

An unorganized group of people without leadership I call a *community*. Instances are a lingual community, a nation, or a people, a social class or caste, a culture or a civilization, but also a party during a reception or the public during a concert. These have a certain social coherence, forming a *network*, but not an organization with a governing board.

Alongside individual people, organized associations or corporations are active *actors* in the normative relation frames, contrary to unorganized communities. A lingual community and the public opinion are no more active than Christianity or the market. Communities cannot work, talk, act, or show respect for each other. They do not bear responsibility and are not answerable. Sometimes a community is objectively determined by an artifact, such as a lingual community by a language; sometimes by a common ideology, for instance, communism; sometimes by a connection with an association, the way a nation or a people is connected to a state; and sometimes it is related to an event, e.g., a birthday party.

In contrast, organized associations, having some kind of government with authority and members with discipline, are able to act in all relation frames, not unlike persons. Moreover, they do so increasingly, partaking in objective networks such as railroads and subjective networks such as the markets. These networks constitute the public domain with the state having the function of guarding the freedom and responsibility of the users of the networks. Nothing of that kind can be found in the animal world. The so-called *Umwelt* of a population of animals is fully restricted by the animal's experience. In the human world, this is replaced by the entirely open and multifaceted public domain.

The Meaning of Emergence

With Dooyeweerd we may state that *meaning* is the mode of being of all that is created. It becomes manifest whenever something new comes into being. The tertiary characteristics of natural things and events point to the possibility of the emergence of new structures with emerging new properties and propensities. It provides the original characters with meaning (Klapwijk [2008] prefers "purpose"), their proper position in the creation. The phenomenon of *disposition* shows that material things such as molecules have meaning for living organisms. It shows that organisms have meaning for animal life. The assumption that God's people are called from the animal world gives meaning to the existence of animals. Both evolution and history display the meaningful development of the creation, the coming into being of ever more structures. Artifacts, in particular written texts, are the most important witnesses of history (Stafleu 2011, chap. 3). They provide history with an *objective* basis. This objective basis complements the *normative* meaning of history as provided by the directive time-order in the relation frames (ibid., chap. 1). It also sustains the *subjective* attribution of meaning by individuals, associations, and unorganized communities, shaping their history by the transfer of experience (ibid., chap. 2).

As a starting point this implies a realist religious view, confessing that God created the world according to laws which are invariant because he sustains them. We know God through Jesus Christ, who submitted himself to God's laws. Partial knowledge of his laws can be achieved by studying the law-conformity of the creation. This also implies a dynamic view of the creation as developing continuously, in the natural evolution and in particular in human history.

The arguments in this chapter show that the theory of evolution may be able to provide necessary conditions for the emergence of human affairs but by no means sufficient conditions. As a religious statement, we may take the biblical message to reveal that humanity was led out of the animal world, called to be free and challenged to take responsibility for the development of God's creation. This may truly be called the *cultural mandate* of humanity. The above-mentioned relation frames include the relations of anyone with their true or imagined God. For Christians, these relations exist through Jesus Christ, who came into the world fulfilling God's laws for the creation, leading his people out of the animal world.

References

Cunningham, Conor. 2010. *Darwin's Pious Idea: Why the Ultra-Darwinists and Creationists Both Get It Wrong*. Grand Rapids: Eerdmans.
Dooyeweerd, Herman. 1953–1958. *A New Critique of Theoretical Thought*. 4 vols. Amsterdam: Paris.
———. 1959. Schepping en evolutie. *Philosophia Reformata* 24: 113–159.
Geertsema, Henk G. 2011. Emergent Evolution? Klapwijk and Dooyeweerd. *Philosophia Reformata* 76: 50–76.
Klapwijk, Jabob. 2008. *Purpose in the Living World? Creation and Emergent Evolution*. Cambridge: Cambridge University Press.
———. 2011. Creation Belief and the Paradigm of Emergent Evolution. *Philosophia Reformata* 76: 11–31.
Mayr, Ernst. 1982. *The Growth of Biological Thought*. Cambridge, MA: Harvard University Press.
Miller, Kenneth R. 1999. *Finding Darwin's God*. New York: HarperCollins.
Popper, Karl R. 1972. *Objective Knowledge*. Oxford: Clarendon Press.
Popper, Karl R., and John C. Eccles. 1977. *The Self and Its Brain: An Argument for Interactionism*. Berlin: Springer.
Stafleu, Marinus Dirk. 2000. The Idionomy of Natural Kinds and the Biological Concept of a Species. *Philosophia Reformata* 65: 11–31.
———. 2002a. Een wereld vol relaties: Karakter en zin van natuurlijke dingen en processen. Amsterdam: Buijten & Schipperheijn. English translation available at: www.scribd.com/doc/29057727/M-D-Stafleu-A-World-Full-of-Relations.
———. 2002b. Evolution, History and the Individual Character of a Person. *Philosophia Reformata* 67: 3–18.
———. 2004. On the Character of Social Communities, the State and the Public Domain. *Philosophia Reformata* 69: 125–139.
———. 2008. Time and History in the Philosophy of the Cosmonomic Idea. *Philosophia Reformata* 73: 154–169.
———. 2011. *Chronos & Clio: De tijd in de geschiedenis*. Amsterdam: Buijten & Schipperheijn.
———. 2014. Nuances in the Philosophy of the Cosmonomic Idea. *Koers* 79 (3).
Stebbins, George L. 1982. *Darwin to DNA, Molecules to Humanity*. San Francisco: Freeman.
Verburg, Marcel E. 1989. *Herman Dooyeweerd. Leven en werk van een Nederlands christenwijsgeer*. Baarn: Ten Have.
Wearne, Bruce C. 2011. Some Contextual Reflections on *Purpose in the Living World? Philosophia Reformata* 76: 84–102.

Nuancing Emergentist Claims: Lessons from Physics

Arnold E. Sikkema

Abstract From a consideration of emergence in physics, I outline how reformational philosophical concepts such as *idionomy*, *encapsis*, and *anticipation* can help nuance the claims of emergentism, whether within or beyond the discipline of physics. The methodological reductionist project has given physics significant success from Democritus through Newton to Hawking. Other sciences seek to employ, extend, and emulate physics with its theoretical precision and verisimilitudinous mathematical laws. Triumphalistic practitioners in disciplines from biology through psychology to sociology—hoping to position their theories as inexorable consequences of physics, touted for its firm foundation, solid knowledge, and clear vision—are applauded by public spokespersons of thoroughgoing ontological and naturalistic reductionism. Such optimism persists even when the so-called stratified nature of reality is acknowledged, especially if the concept of emergence is brought into view. But in addition to being poorly defined, emergence is used in exactly opposite senses: claims of unproblematic scientific explanation for a multileveled reality *and* claims of the intractable impossibility of such explanation. Sometimes enlisted in support of the former is the notion that emergence within physics is fully understood. A sober assessment of predictability and critical realism in physics, however, demonstrates that the nature of emergence within physics renders physics incapable of bearing its supposed grand foundational responsibility. Examples in various physics subfields are analyzed, demonstrating common themes and principles. Collective physical phenomena are strikingly characterized by robustness of the ordered macroscopic whole relative to variations in microscopic parts, universality near phase transitions, and symmetry breaking, but most importantly by surprise and incalculability.

Keywords Physics · Emergence · Phase transitions · Dooyeweerd · Idionomy · Encapsis · Anticipation · Approximation · Reductionism · Whole-part

---

A. E. Sikkema (✉)
Trinity Western University, Langley, BC, Canada
e-mail: arnold.sikkema@twu.ca

Introduction

This paper discusses what we know about emergence in physics with the hope that this could trigger, especially when common themes encounter Dooyeweerd's philosophy, some lessons about emergence in more general circumstances. In my work in theoretical biophysics, I explore collective motion in biological systems using methods of theoretical physics developed in the fields of thermodynamics, statistical mechanics, quantum physics, and condensed matter physics. My general working hypothesis is that phase transitions, for which there is a substantial network of experimental and theoretical understanding, bear many resemblances to putative examples of emergence. I do not attempt to be philosophically precise, but will speak as a physicist with an appreciation of Dooyeweerdian philosophy.

Methodological Reductionism in Physics

Reductionism as a method has served the discipline of physics well. Consider the examples of Democritus, Newton, and Weinberg. Democritus advanced the notion of the atom, the indivisible constituent of all matter, to explain many observations, such as the smell of fresh bread being due to the atoms of bread breaking free and traveling into our nostrils. Isaac Newton using a glass prism was able to demonstrate that white light consists of a visible spectrum of colors, and using observation and theory showed that the moon accelerates to the earth due to the same force of gravity as experienced by an apple. Steven Weinberg and others succeeded in unifying the electromagnetic and weak nuclear forces as facets of one electroweak interaction. These are relatively simple examples of reductionism as a method of describing and explaining physical phenomena. (One can identify a variety of forms of reductionism, e.g., theory reduction and composition reduction, as well as distinguish the philosophers' and the physicists' senses of reduction,[1] but that will not be the focus here.)

Because of its significant success, including especially its achievements via reductionism, physics is known for being theoretically precise, for having increasingly verisimilitudinous mathematical laws, for being a firm foundation upon which further study in physics, chemistry, and biology can be based, for providing solid and reliable knowledge of the world's basic constituents, and for having a clear vision by way of its methods of experimental and theoretical analysis. As such, physics is sometimes envied by practitioners of other disciplines, who seek to apply our methods in their own, be it biology (modeling the interactions of the parts of a

---

[1] On the physicists' vs. the philosophers' senses of reduction, see the helpful discussion in Batterman (2002), especially in the early pages and the conclusions.

cell), psychology (e.g., neural networks), sociology (e.g., computational simulations of social dynamics and crowd movement), or even religious studies.[2]

Ontological Reductionism

This last example—that of materialist religion—is an example of making the turn from the methodological value of reductionism to ontological reductionism, especially (but not only) in cases when it is believed, as in the liturgical *credo* of Sagan (1985, 1), that "The Cosmos is all that is or ever was or ever will be." Consider the following statements of Weinberg, Atkins, and Dawkins, respectively.

Nobel physicist Steven Weinberg (2001, 115; emphasis in original) writes, referring to Mary Midgley's fictitious character:

> the nervous systems of George and his friends have evolved to what they are *entirely* because of the principles of macroscopic physics and chemistry, which in turn are what they are *entirely* because of the principles of the Standard Model of elementary particles.

As Stuart Kauffman (2010, 43) correctly points out, "Weinberg rests his reductionism on the claim that the explanatory arrows always point downward, from society to cells to chemistry to physics," which Kauffman brilliantly demonstrates to be incorrect by an analysis of the utter inability of the physicist *qua* physicist to identify the function of the heart from among its various causal effects. This is because *function* is biologically, not physically, assessed.

Oxford chemist P.W. Atkins (1992, 34) writes, "Humanity should accept that science has eliminated the justification for believing in cosmic purpose, and that any survival of purpose is inspired solely by sentiment." Atkins (1995, 125, 129) goes on: "Science has never encountered a barrier, and the only grounds for supposing that reductionism will fail are pessimism in the minds of scientists and fear in the minds of the religious.... Reductionist science is omnicompetent." Richard Dawkins (1995, 133) writes, "The universe we observe has precisely the properties we should expect if there is, at bottom, no design, no purpose, no evil and no good, nothing but blind, pitiless indifference."

Over against ontological reductionism, Dooyeweerdian philosophy asserts, for good reasons, that reality is multiaspectual. And more and more scholars unfamiliar with Dooyeweerd (such as Roy Bhaskar and Alister McGrath) have also been discussing a so-called stratification, or layering, or reality.

Emergence has re-emerged as an argument in the refutation of ontological reductionism. Now, I will not attempt to define *emergence* precisely here. After all, definitions do vary significantly, and most have problems, large or small. In fact, I believe a precise definition is not possible, though perhaps one will yet emerge. But, roughly, the idea of emergence covers situations where a collective has properties

---

[2] For an example, see Vasquez (2010).

which its parts do not. A glass of water exhibits liquidity, while no molecule can. There is no conscious neuron, but you are conscious. A living cell has no living parts.

However, the picture is not as clear-cut as this. Instead, perspectives on emergence vary greatly, including its use *in support of* ontological reductionism. At one end of the spectrum, often based on assumptions of naturalism and materialism, is the claim that our multileveled reality is scientifically explicable in practice or principle, perhaps based on the supposition that emergence in and from physics (like the liquidity of water[3]) is understood. Just before the above quote by Weinberg (2001), he wrote, "Phenomena like mind and life do emerge [from biology and chemistry respectively]. The rules they obey are not independent truths, but *follow from* scientific principles at a deeper level" (2001, 115; emphasis added). Consider also this subsequent passage:

> I can readily believe that at least in principle we will one day be able to explain all of George's behavior reductively, including what he says about how he feels, and that consciousness will be one of the emergent higher-level concepts appearing in this explanation.... Of course, everything is at bottom quantum-mechanical.... [Research] moves us closer to the reductionist goal of finding the laws of nature that lie at the starting point of all chains of explanation. (Weinberg 2001, 117, 118, 121)

Weinberg is motivated in large part by his claim that high-energy particle research is uniquely fundamental and thus deserving of a great share of public funds.

The opposing claim, though not as broadly articulated, is that our multileveled reality cannot be scientifically explained even in principle, either because of or in spite of emergence. Now, in Dooyeweerd's philosophy one could distinguish emergence within a modal aspect (sometimes called *weak emergence*) and between different modal aspects (sometimes called *strong emergence*), even if these aspects are mutually irreducible.

I will leave aside intermodal emergence, and focus on one type of intramodal emergence, namely, physical emergence. That is, I shall consider emergence *within* physics, and discuss in particular the notion of predictability of properties of collectives. While twentieth-century quantum physics and chaos theory have shown that prediction in physics is problematic, the examples given below demonstrate other features of unpredictability in physics.

Emergence in Physics

Crystal Structure

Consider first the crystal structure of carbon. Let us ask what arrangement carbon atoms will take when in close proximity to one another, in the solid state. Well, diamond is a hard, clear, colorless crystal composed entirely of carbon. Thus one

---

[3] I have demonstrated that this supposition is false in Sikkema (2011).

could say that from the combination of carbon atoms a diamond crystal emerges. Furthermore, one might look to condensed matter theory and/or quantum chemistry to ascertain whether such a result might be predictable, namely, whether one could mathematically determine that the crystal structure of carbon is that of diamond, and whether diamond would be expected to have the optical and material properties it does. And as it turns out, the tetrahedral bonding is expected from the fact that the carbon atom has four valence electrons and from symmetry considerations.

That is, given all we know about carbon atoms and how they interact with one another, we are able to determine the crystal structure of carbon and a good number of its properties, including the fact that electrically it is an insulator. So it appears that the emergent structure and properties of carbon are predictable.

However, upon closer inspection, two significant matters arise and require us to recognize that such a conclusion would be premature. The first is observational, and the second theoretical.

First, diamond is not the only crystal structure of carbon. Graphite, used in pencil "lead," is also entirely composed of carbon atoms arranged in planes which exhibit a hexagonal lattice structure. Quite opposite to diamond, it is a soft, opaque, black material which conducts electricity. Going back to the drawing board (so to speak), we can also mathematically recognize and justify this crystal structure and explain it and its properties in terms of the carbon atom. But there is yet a third crystal structure, known as lonsdaleite; it is intermediate between that of graphite and diamond in that it has the tetragonal bonding of diamond while retaining the hexagonal structure of graphite. There are in fact a large number of different crystal structures of carbon; under different conditions of formation, quite a variety of periodic lattice forms can be taken (Krüger 2010). Among these are the recently discovered individual "two-dimensional" sheets of graphene (Novoselov 2011; Geim 2011). And furthermore, carbon atoms combine in a multitude of yet other ways in the solid form, including amorphous or glassy carbon, as well as any number of fullerenes (Curl 2003; Kroto 2003; Smalley 2003) and nanotubes.

Second, such calculations of crystal structure are not actually a priori, first-principles, straightforward calculations based only upon the previously known properties of carbon. These are instead retrodictions, possible for the diamond, graphite, and some other crystal and non-crystal structures of carbon, and quite common in the study of physical emergence. It is actually impossible to truly predict or calculate crystal structures; what is done instead is the *rationalization* of crystal structures. It is only after we observe and characterize a particular collective structure or phenomenon that we can bring to bear the tools and techniques of theoretical and computational physics to explain how the properties of the collective are found to be connected to the properties of its parts. The rationalizations rely upon what Laughlin and Pines (2000) call "art keyed to experiment," and not just a priori knowledge. The methods of condensed matter theory and quantum chemistry incorporate approximations and limits which have been found to be empirically satisfactory in many cases.[4] Having experience with the details of the known crystal structure, or at least similar structures, the scientists' approach assumes that certain things can be neglected, that periodicity will result, that the crystal is large enough, that the temperature is low enough, that certain symmetries are preserved and others broken, etc. Knowing that such methods have worked in the past for other crystals, one forges ahead in the new case, and hopes it will work. If it works, good! If not, one can make subsequent attempts with different approximations.

In addition, it is worth noting that the concept of crystal as a periodic lattice, a translationally invariant set of points in space, is an abstraction from our experience with crystals as found in creation. *Periodicity* means that each site of a lattice is identical in every way. The periodic lattice is qualified by the spatial aspect, and is of necessity infinite in extent. Any true material will exhibit interactions between the component parts (modeled as particles: nuclei and electrons) resulting in kinematics (such as crystal vibrations and optical effects) so that the crystal we find in creation is qualified by the physical aspect and is also active in the kinematic aspect. And of course a real crystal is only finite in extent; it has only limited regions ("domains") of approximate periodicity, often separated by dislocations and containing impurities, and has surfaces which are often faceted but can be rough and/or oxidized.

Correlated Electron Systems

There is growing recognition in the physics community that many condensed matter systems (not only crystal structure) ought to be considered examples of emergence.[5] Some, but not all of these, involve systems in which electrons are strongly correlated in quite unexpected ways. In "high-temperature" superconductivity, materials with very poor electrical conductivity when cooled below a critical temperature spontaneously develop exactly zero resistance and expel magnetic fields. While certain aspects of the general nature of superconductivity, first discovered in 1911 (Kamerlingh Onnes 1967) and explained in 1957 (Bardeen 1992; Cooper 1992; Schrieffer 1992), appear to somewhat apply to the new varieties discovered in 1986 (Bednorz and Müller 1988) and later, no satisfactory explanation has yet been accepted by the community. In general, when such novel behavior does get explained, it requires a great deal of insight, ingenuity, creativity, and imagination, rather than the straightforward application of basic principles and established techniques.

Furthermore, details of the basic compositional "foundation" of the material turn out, surprisingly, to not matter much. Clearly, the details ought to matter—such as

---

[4] It might be suggested that the reason for the inability to predict crystal structure is that, while the laws of physics are generally applicable ("modal laws"), the details of their application to particular systems ("typical laws") is the reason for the unpredictability discussed in this paper. But since many physical systems are predictable, it is doubtful that this "modal–typical" distinction is helpful in analyzing emergence.

[5] There are dissenting voices to this claim, such as Howard (2007).

the number of electrons each atom has, the geometrical arrangements of the atoms, the masses and charges of the nuclei—since the phenomenon occurs in certain materials and not in others. But the detailed explanatory theory in quantum critical phenomena tends to employ overarching concepts such as electron pairing, Bose-Einstein condensation, joint electronic wave functions, the Fermi sea, and excitation gaps. This feature of strongly correlated electron systems (found also in other examples of physical emergence) is known as robustness and universality, which is theoretically understood in terms of renormalization group flows. This approach develops an "effective" model which examines how the parameters vary as a function of a growing scale of analysis from the atomic level to the "infinite" whole, and there are basins of attraction toward stable fixed points.[6]

Besides the example of superconductivity mentioned above, other strongly correlated electron systems include the closely related superfluidity, varieties of magnetism (especially frustrated antiferromagnetic systems, where the local interactions can never all be equivalently satisfied in terms of generating what would locally be a lowest energy form), as well as the fractional quantum Hall effect (FQHE). In the FQHE, one is confronted with the observation that while each electron's charge is precisely $-e$, in certain two-dimensional systems subjected to large magnetic fields one finds novel collectives emerging which have exact fractional charges, such as $(2/7)e$. Robert Laughlin (1999, 863), who, together with Horst Störmer and Daniel Tsui, won the 1998 Nobel Prize in Physics for their explanation of the FQHE, wrote at length about the failure of reductionism in condensed matter theory. It is worthwhile to consider his introductory comments in full:

> One of my favorite times in the academic year occurs in early spring when I give my class of extremely bright graduate students, who have mastered quantum mechanics but are otherwise unsuspecting and innocent, a take-home exam in which they are asked to deduce superfluidity from first principles. There is no doubt a special place in hell being reserved for me at this very moment for this mean trick, for the task is impossible. Superfluidity, like the fractional quantum Hall effect, is an emergent phenomenon—a low-energy collective effect of huge numbers of particles that cannot be deduced from the microscopic equations of motion in a rigorous way and that disappears completely when the system is taken apart. There are prototypes for superfluids, of course, and students who memorize them have taken the first step down the long road to understanding the phenomenon, but these are all approximate and in the end not deductive at all, but fits to experiment. The students feel betrayed and hurt by this experience because they have been trained to think in reductionist terms and thus to believe that everything not amenable to such thinking is unimportant. But nature is much more heartless than I am, and those students who stay in physics long enough to seriously confront the experimental record eventually come to understand that the reductionist idea is wrong a great deal of the time, and perhaps always.

---

[6] An accessible treatment of these subjects and their relation to reduction and emergence is given in Batterman (2011).

Other Examples of Physical Emergence

In fluid dynamics, a classic example of emergence is the Rayleigh-Bénard convection cell. A liquid between two horizontal plates, with the bottom plate at a higher temperature than the top plate, undergoes a transition when the temperature difference is large enough. Before the transition, each point on the plates was equivalent (that is, there was horizontal translational symmetry). After the transition, an array of convection cells has formed. Now the motion of each particle of the fluid has two aspects: one is the random Brownian motion, and the other is the highly ordered pattern of motion established on the larger scale.

Other examples of emergence in physics, not further explored here, include nuclear physics (such as the fact that the carbon nucleus ^{12}C is not simply a combination of its six neutrons and six protons), fluid thermodynamics, sand ripples, and the relation between classical and quantum worlds.

Common Themes and Principles

It appears that examples of physical emergence have in common at least the following themes and ideas: incalculability, surprise, robustness of the ordered macroscopic whole relative to variations in microscopic parts, universality near phase transitions, and symmetry breaking. Some of these have been discussed in particular reference to specific cases above; here more will be said about the first two.

In terms of calculating the properties of a system whose parts are quite well characterized, we are thinking of both developing understanding of the new behavior and predicting it. Laughlin et al. (2000, 32–33) write,

> Although behavior of atoms and small molecules can be predicted with reasonable accuracy starting from the underlying laws of quantum mechanics, the behavior of large ones cannot.... Very large aggregations of particles have some astonishing properties ... that are commonly acknowledged to be "understood."... Some kinds of inanimate mesoscopic self-organization can be easily visualized, and perhaps not coincidentally are identified as understood.

It is important to note that the authors are here pointing out that even many in the physics community incorrectly believe that emergent physical behaviors are calculable and understood; the reason is that they are ignoring the fact that the calculations are not actually exact but rely upon approximations whose validity cannot be known in advance. Standard calculation methods "tend to be the least reliable precisely when reliability is most needed, i.e., when experimental information is scarce, the physical behavior has no precedent, and the key questions have not yet been identified" (Laughlin and Pines 2000, 28).

Perhaps more than anything, surprise is a feature of emergent systems. As Hamlet says, "as a stranger give it welcome. / There are more things in heaven and earth, Horatio, / Than are dreamt of in your philosophy." I will relate this to the contingency of the creation upon the Creator's will below. As expressed by theoretical physicists Laughlin et al. (2000, 33), "[mesoscopic ordering] rules that are dreamt up without the benefit of physical insight are nearly always wrong, for correct rules are really natural phenomena and therefore must be discovered, not invented."

Physics research continues to routinely unveil phenomena that were hitherto completely unexpected.

Difficulties with "Parts and Wholes"

A common way of characterizing emergence (including in this chapter!) is to note that a whole system has properties that its parts do not. From the point of view of physics, the notion of parts and wholes is seen as problematic. There are at least four important qualifications which can be given, each of which implies the impossibility of a simple demarcation of what the parts of a whole are and thus renders part-whole characterizations of emergence (as well as of reductionism) problematic. These have to do with isolatability, dynamics, delimiting, and material composition.

First, many nonlinear systems, at least for some range of parameter space, are affected in their qualitative large-scale behavior by even the weakest possible interactions across far reaches of intergalactic space. A clear example cited by John Polkinghorne (1999, 14) involves something as common as the motion of air molecules. If we seek to predict very broadly the direction of motion of a single air molecule after only 10^{-10} s, we must "take into account the effect of an electron (which is the smallest particle of matter) on the other side of the observable universe (which is just about the farthest away you can get) interacting with the air in this room through the force of gravity (which is the weakest of the fundamental forces of nature)." Clearly this universal influence can thus not be quantified, as that would require us to, in Polkinghorne's terms, "literally possess universal knowledge." This is an implication of the "sensitive dependence on initial conditions" of chaos theory. Thus, what we might think of as a closed system, such as the air in a closed container, is always unisolatable. That is, any finite "system" is always larger than we think. Thus the question of which "parts" are "in" any given "whole" is intractable. Thermodynamically closed systems are an abstraction; all real systems are thermodynamically open.

Second, dynamic systems generally do not retain specific parts. Consider three examples: an organism, a tornado, and the electromagnetic field. An organism digests, respires, and grows; the human body replaces 98% of its atoms each year (Aebersold 1954). It can also entirely lose parts while retaining identity, such as in accidental dismemberment or surgical removal.

Above I discussed Rayleigh-Bénard convection cells. Consider the relation of the "parts" to the "whole" in a more familiar but closely related example, that of a tornado. The vortex of a tornado consists materially of a swirling ensemble of air particles which are racing around at speeds like 180 km/h, or 50 m/s. However, given that there is a continual inrush of air along the ground toward the vortex and up through the vortex into the air above and that the typical speeds of the air particles themselves are about 10 times this, on the order of 500 m/s (even when air is "still"), this vortex is not composed of the same air particles for very long. (A detailed calculation includes the fact that these air particles bump into one another, and so considers the mean free path and the random walk of Brownian motion.) So given these two points, we know that large amounts of matter and energy are being transferred through the stable recognizable physical system. That is, the question of the "parts" of the tornado is somewhat problematic: there is not any way to specify at what moment a given particle is "part" of the vortex.

The steady magnetic field of an electromagnet (or the infrared radiation in a hot oven) can be thought of as an extremely dynamic collection of photons, each of which are being created and destroyed at all possible rates, involving spontaneous particle-antiparticle production and annihilation; the particles themselves have no enduring existence but are "fleeting." Yet the pattern (the magnetic field or, respectively, the infrared radiation) remains, while Heisenberg's uncertainty principle establishes limits of fluctuations of the number of photons.

Third, except for more general systems (e.g., social networks), being part of a whole naturally includes the concept of mutual spatial proximity. This is usually understood in terms of a boundary which can be drawn around the system, and this boundary will necessarily encompass all the parts of the system by virtue of the part-whole relation. However, as understood since the early twentieth century, nothing is truly spatially delimitable. For quantum theory indicates that no wave function has only finite extent. Thus any part of which a whole is made has a non-zero probability of being found arbitrarily far away from what is otherwise ordinarily thought of as the spatial extent of the whole. So the notion that any parts are spatially *within* the whole is not without its difficulties. This, of course, will be more evident for cases in which quantum physics is more important; however, with the increasing scales at which quantum effects have been demonstrated, this factor may be crucial in surprisingly large systems.

Fourth, material composition is itself a problematic concept, as seen from both physics and philosophy. In addition to increasingly recognizing that a physical theory of everything is not realizable (including down to, as well as from, "fundamental" material constituents), twentieth-century physics oversaw the "de-materialization of matter" (Sikkema 2005; see also McMullin 1963 & 1978 and Hanson 1962). Special relativity showed that matter and energy are not distinct. Quantum mechanics showed that particles have infinite extent. Chaos theory showed that universal influence exists. Quantum physics furthermore demonstrated entanglement. Quantum field theory showed that relations are inextricable from components.

Philosophically, composition is not only physical or material; that is, material composition does not exhaust composition. For example, relations themselves are constitutive; each relationship I have with another person is *part* of who I am, arguably more fundamentally so than my atoms and molecules. In fact, Neidhardt (1989), drawing on Torrance, suggests that Maxwell's development of the notion of the electromagnetic field was heuristically related to his appreciation of biblical anthropology and how the integrality of human interrelationships stems from the Trinitarian character of God.

Connection with Reformational Philosophical Concepts

Certain conceptual tools of reformational philosophy are valuable in the analysis of emergence in physics, and possibly emergence more generally. Thus I will conclude with some brief remarks on idionomy, encapsis, and anticipation.

Idionomy

The concept of idionomy, at least as articulated by Klapwijk (2008),[7] expresses the notion that new kinds of entities respond to new kinds of laws. While not thinking in these reformational philosophical terms, theoretical physicists approach such a conclusion as well (Laughlin et al. 2000, 32; emphasis added):

> Mesoscopic organization in soft, hard, and biological matter is examined in the context of our present understanding of the principles responsible for emergent organized behavior (crystallinity, ferromagnetism, superconductivity, etc.) at long wavelengths in very large aggregations of particles. Particular attention is paid to the possibility that as-yet-undiscovered *organizing principles might be at work at the mesoscopic scale*, intermediate between atomic and macroscopic dimensions, and the implications of their discovery for biology and the physical sciences. The search for the *existence and universality of such rules*, the proof or disproof of organizing principles appropriate to the mesoscopic domain, is called the middle way.

When we try to understand how new kinds of entities respond to new kinds of laws (or have new kinds of properties), Klapwijk (2011, 27) points out the limits of scientific theorizing, writing that "a believer has good reason to confess that the idionomy that we encounter in distinct levels of being … is, in the final analysis, grounded in theonomy, i.e. in laws of the creator God…. We see a world that is open to its Creator, [which] shows a fundamental receptivity to laws of a higher order…. The world of becoming … is responding to divine orderings."

The idea that knowing all that can be known about a given level or order is insufficient to determine the being or behavior of a higher level or order relates strongly to the notion of the contingency of the created order upon the free will of the Creator. It has been well argued by many that this recognition of contingency, such as in Bishop Tempier's Condemnation of 1277, is a crucial component of the Christian worldview which helped spur western European culture to develop modern science.[8]

---

[7] See also my essay review of Klapwijk's treatment of physics and the physical aspect (Sikkema 2011).

[8] See, for example, Pearccy and Thaxton (1994, 82).

Mark Noll (2011, 117) notes—in a chapter entitled "'Come and See': A Christological Invitation for Science"—"As described in the Gospels, individuals who wanted to learn the truth about Jesus had to 'come and see.' Likewise, to find out what might be true in nature, it is necessary to 'come and see.'" That is, it should not be surprising that one cannot derive, from the nature of things qualified by one aspect of reality, the character of something which is qualified by a higher aspect. Interestingly, Laughlin et al. (2000), quoted above, claim that even if we remain *within* the physical aspect, underivability is pervasive.

Encapsis

Jitse van der Meer (chapter "Is There a Created Order for Cosmic Evolution in the Philosophy of Herman Dooyeweerd?," this volume, section entitled "Differentiation") discusses as an example of diachronic emergence the development over time of encaptic relations between biotically qualified wholes and their physically qualified parts. Encapsis is thus not a prior condition that leads to evolutionary emergence, but a characteristic of at least some emergent systems. In the case of synchronic emergence, encapsis may be of even more value as an analytical aid in studying whole-part relations. For example, the nature of a collective influences the behavior of its parts. It is worth noting that Eleonore Stump's identification of the significant influence of *form* can be deepened through Dooyeweerdian analysis (Stump, chapter "Natural Law, Metaphysics, and the Creator," this volume). For this would reflect the importance of the (lower) spatial modal aspect. And additionally, other features of a collective are important, such as boundary conditions. With these brief though tantalizing points in mind, it appears that Dooyeweerd's conception of encapsis has potential to prove valuable in an ongoing research project analyzing emergence in physics and elsewhere.

Anticipation

Anticipation, in Dooyeweerd's philosophy, refers to features of one modal aspect which in some sense look ahead in order to allow for the opening up of a later (or higher) aspect.

First, within physics itself, one of the things which occur in many examples of emergence is *symmetry breaking*.[9] Consider the familiar example of water freezing. In the liquid phase, any given water molecule could be oriented in any direction, and (more importantly) the other molecules in its vicinity (or further away) could be found in any direction as well, especially when one considers the way in which the

---

[9] This term is also mentioned in Stump (chapter "Natural Law, Metaphysics, and the Creator," this volume, section entitled "Reductionism").

molecules are all moving about in their Brownian motion. In physics, this indicates a high degree of symmetry, for all directions are equivalent. In the freezing transition, we enter the solid state, whereupon the water molecules line up in a crystal structure and are now locked together in particular directions, and so the final state has a lower degree of symmetry. In addition, in the liquid, water molecules could be found at any location, while in the solid, water molecules are found only at (and vibrating near) the crystal lattice sites. This process of moving from high symmetry (all directions and locations equivalent) to low symmetry (specific directions and locations selected) is called *symmetry breaking*. The emergent (the crystal) does not pick up a new symmetry, but in a sense the liquid's initial symmetry is "prepared for" the emergent transition. It appears that the connection between this "preparation" and the Dooyeweerdian concept of anticipation for emergence has the potential for fruitful analysis.

For example, there is something in physics which gives a clue as to how a physically qualified thing can contribute to and be a "part of" a biotically qualified thing. This relates to both the synchronic and diachronic emergence of living things from non-living things. How, precisely, are electrons open to the biotic? Klapwijk (2008, 213–220) discusses enzymes, neurons, and emotion in terms of the Dooyeweerdian conception of anticipation. Similarly, electrons have been created with the properties they have so that they can be fruitfully parts of a greater whole with supraphysical properties. Preeminent among these physical properties is their indeterministic nature as revealed by quantum physics. That is, the initial state of an electron does not determine its final state. It seems to me that this "freedom," or "openness," of physically qualified things is fruitful in the "opening up" of the biotic world from the physical world. After all, the electronic, atomic, and molecular scales are small enough to experience this quantum openness and simultaneously large enough to be intimately involved in the biochemical processes important in the biotic realm.

Concluding Remarks

This paper has only scratched the surface as to how further explication and application of Dooyeweerdian and reformational philosophical concepts, particularly idionomy, encapsis, and anticipation, promise to be valuable in the exploration of the notion of emergence, both within the physical aspect and beyond. The future of creation order, as an interdisciplinary field of study, is bright.[10]

---

[10] I thank Harry Cook, Dick Stafleu, and Jitse van der Meer for valuable discussions, and the anonymous reviewers for their helpful critique.

References

Aebersold, Paul C. 1954. Radioisotopes: New Keys to Knowledge. In *Annual Report of the Board of Regents of the Smithsonian Institution, Publication 4149, Showing the Operations, Expenditures, and Condition of the Institution for the Year Ended June 30, 1953*, 219–241. Washington: United States Government Printing Office.

Atkins, Peter. 1992. Will Science Ever Fail? *New Scientist* 135 (1833): 32–35.

———. 1995. The Limitless Power of Science. In *Nature's Imagination: The Frontiers of Scientific Vision*, ed. John Cornwell, 122–132. Oxford: Oxford University Press.

Bardeen, John. 1992. Electron-phonon Interactions and Superconductivity. In *Nobel Lectures, Physics 1971–1980*, ed. Stig Lundqvist, 54–69. Singapore: World Scientific Publishing Co.

Batterman, Robert W. 2002. *The Devil in the Details: Asymptotic Reasoning in Explanation, Reduction, and Emergence*. Oxford: Oxford University Press.

———. 2011. Emergence, Singularities, and Symmetry Breaking. *Foundations of Physics* 41 (6): 1031–1050.

Bednorz, J. Georg, and K. Alex Müller. 1988. Perovskite-type oxide—The New Approach to High-T_c Superconductivity. *Reviews of Modern Physics* 60 (8): 585–600.

Cooper, Leon N. 1992. Microscopic Quantum Interference Effects in the Theory of Superconductivity. In *Nobel Lectures, Physics 1971–1980*, ed. Stig Lundqvist, 73–93. Singapore: World Scientific Publishing Co.

Curl, Robert F., Jr. 2003. Dawn of the Fullerenes: Experiment and Conjecture. In *Nobel Lectures, Chemistry 1996–2000*, ed. Ingmar Grenthe, 11–32. Singapore: World Scientific Publishing Co.

Dawkins, Richard. 1995. *River Out of Eden: A Darwinian View of Life*. New York: Basic Books.

Geim, Andre K. 2011. Nobel Lecture: Random Walk to Graphene. *Reviews of Modern Physics* 83 (3): 851–862.

Hanson, N.R. 1962. The Dematerialization of Matter. *Philosophy of Science* 29 (1): 27–38.

Howard, Don. 2007. Reduction and Emergence in the Physical Sciences: Some Lessons from the Particle Physics and Condensed Matter Debate. In *Evolution and Emergence: Systems, Organisms, Persons*, ed. Nancey Murphy and William R. Stoeger, 141–157. Oxford: Oxford University Press.

Kamerlingh Onnes, Heike. 1967. Investigations into the Properties of Substances at Low Temperatures, Which Have Led, Amongst Other Things, to the Preparation of Liquid Helium. In *Nobel Lectures, Physics 1901–1921*, 306–336. Amsterdam: Elsevier Publishing Company.

Kauffman, Stuart A. 2010. *Reinventing the Sacred: A New View of Science, Reason, and Religion*. New York: Basic Books.

Klapwijk, Jacob. 2008. *Purpose in the Living World? Creation and Emergent Evolution*. Cambridge: Cambridge University Press.

———. 2011. Creation Belief and the Paradigm of Emergent Evolution. *Philosophia Reformata* 76 (1): 11–31.

Kroto, Harold W. 2003. Symmetry, Space, Stars and C_{60}. In *Nobel Lectures, Chemistry 1996–2000*, ed. Ingmar Grenthe, 44–79. Singapore: World Scientific Publishing Co.

Krüger, Anke. 2010. *Carbon Materials and Nanotechnology*. Weinheim: Wiley-VCH Verlag.

Laughlin, R.B. 1999. Nobel Lecture: Fractional Quantization. *Reviews of Modern Physics* 71 (4): 863–874.

Laughlin, R.B., and David Pines. 2000. The Theory of Everything. *Proceedings of the National Academy of Sciences of the United States of America* 97 (1): 28–31.

Laughlin, R.B., David Pines, Joerg Schmalian, Branko P. Stojković, and Peter Wolynes. 2000. The Middle Way. *Proceedings of the National Academy of Sciences of the United States of America* 97 (1): 32–37.

McMullin, Ernan, ed. 1963 & 1978. *The Concept of Matter in Modern Philosophy*. Notre Dame: University of Notre Dame Press.

Neidhardt, W. Jim. 1989. Biblical Humanism: The Tacit Grounding of James Clerk Maxwell's Creativity. *Perspectives on Science and Christian Faith* 41 (3): 137–142.

Noll, Mark A. 2011. *Jesus Christ and the Life of the Mind*. Grand Rapids: Eerdmans.

Novoselov, K.S. 2011. Nobel Lecture: Graphene: Materials in the Flatland. *Reviews of Modern Physics* 83 (3): 837–849.

Pearcey, Nancy R., and Charles B. Thaxton. 1994. *The Soul of Science*. Wheaton: Crossway Books.

Polkinghorne, John. 1999. Can a Scientist Pray? *Logos: A Journal of Catholic Thought and Culture* 2 (2): 9–27.

Sagan, Carl. 1985. *Cosmos*. New York: Ballantine Books.

Schrieffer, J.R. 1992. Macroscopic Quantum Phenomena from Pairing in Superconductors. In *Nobel Lectures, Physics 1971–1980*, ed. Stig Lundqvist, 97–108. Singapore: World Scientific Publishing Co.

Sikkema, Arnold E. 2005. A Physicist's Reformed Critique of Nonreductive Physicalism and Emergence. *Pro Rege* 33 (4): 20–32.

———. 2011. Nuancing the Place and Purpose of the Physical Aspect in Biology and Emergence. *International Journal of Multi Aspectual Practice* 1: 29–39.

Smalley, Richard E. 2003. Discovering the Fullerenes. In *Nobel Lectures, Chemistry 1996–2000*, ed. Ingmar Grenthe, 89–103. Singapore: World Scientific Publishing Co.

Vasquez, Manuel. 2010. *More Than Belief: A Materialist Theory of Religion*. Oxford: Oxford University Press.

Weinberg, Steven. 2001. *Facing Up: Science and Its Cultural Adversaries*. Cambridge, MA: Harvard University Press.

Order and Emergence in Biological Evolution

Denis Alexander

Abstract In this chapter I provide a brief overview of the idea of progress that has so often been attached to the narrative of the biological theory of evolution and I argue that the outcome of that particular discussion makes relatively little difference to Christian theology. Second, I refer to more recent commentators who have suggested that Darwinian evolution is incompatible with either the ideas of progress or of purpose, or of both. Third, I suggest that in contrast to these commentators, recent biological insights point to evolution as a highly constrained process, consistent with the idea of a God who has purposes and intentions for all of his created order. In the dynamic interplay between chance and necessity that characterizes the evolutionary process, it is necessity that has the upper hand. Therefore, whatever one might think about the discussion about progress, as far as purpose is concerned, the biological data do not support those who suggest that evolution is incompatible with the idea of purpose.

Keywords Order · Emergence · Evolution · Progress · Purpose · Christian theology

Introduction

First, I will give a brief overview of the idea of progress that has so often been attached to the narrative of the biological theory of evolution and I will argue that the outcome of that particular discussion makes relatively little difference to Christian theology. Second, I will refer to more recent commentators who have suggested that Darwinian evolution is incompatible with either the ideas of progress or of purpose, or of both. Third, I will suggest that in contrast to these commentators, recent biological insights point to evolution as a highly constrained process,

---

The original version of this paper was first given at the Darwin Festival, Cambridge, on July 6, 2009. A second version was given at Vrije Universiteit Amsterdam on August 17, 2011.

D. Alexander (✉)
The Faraday Institute for Science and Religion, St Edmund's College, Cambridge, UK
e-mail: dra24@hermes.cam.ac.uk

consistent with the idea of a God who has purposes and intentions for all of his created order. In the dynamic interplay between chance and necessity that characterizes the evolutionary process, it is necessity that has the upper hand. Therefore, whatever one might think about the discussion about progress, as far as purpose is concerned, the biological data do not support those who suggest that evolution is incompatible with the idea of purpose.

The Question of Progress in Evolutionary History

First, progress. The idea already had a long and complex history for more than two millennia before biological evolution came on the scene, with many nuances in meaning, then as now. At the societal level it generally expressed the assumption that humanity was making social, economic, scientific, and political advances and that these trends were likely to continue on into the future. Within the context of natural history, the idea of progress was focused more on the upward climb of organisms, from the simplest to the complex, from "monad to man." The problem is that these two ideas of progress have often become inextricably entangled, with either the more social and political sense framing the narrative within which the biological story has been told, or the biological story generating or justifying accounts of human social and political progress, or a synergistic interaction between the two in which the social and biological narratives became interdependent.

A Christian vision for the idea of progress was famously stated by Augustine in what is arguably the first full-blown book on the philosophy of world history, *The City of God:* "The education of the human race, represented by the people of God, has advanced, like that of an individual, through certain epochs, or, as it were, ages, so that it might gradually rise from earthly to heavenly things, and from the visible to the invisible" (Augustine and Dods 2009, vol. 2, chap. 14). That same Christian vision framed the scientific revolution of the seventeenth century, which was strongly progressionist in tone, even though it was often presented as a concerted attempt to recover the knowledge that had been lost at the fall (Harrison 2007). As Francis Bacon proclaimed: "When he [Aristotle] had made nature pregnant with final causes, laying it down that 'Nature does nothing in vain, and always effects her will when free from impediments,' and many other things of the same kind [he] had no further need of a God" (Robertson 2011, 472). In that case, Bacon argued, "There [is] but one course left"; namely, "to try the whole thing anew upon a better plan, and to commence a total reconstruction of sciences, arts, and all human knowledge, raised upon the proper foundations" (Shapin 1998, 66). Bacon's *New Atlantis* published in 1624 (Bacon and Smith 1900) was a Utopian vision of life in which the benefits of science are used for the common good, established "for finding out the true nature of all things, whereby God might have the more glory in the workmanship of them" whilst the practical application of knowledge was "for the comfort of men" (Delbourgo and Dew 2008, 2).

The optimism displayed by the seventeenth-century Christian natural philosophers is, in retrospect, startling. John Wilkins, a founder member of the Royal Society in Britain, wrote a book in 1638 on the discovery of a new world, remarking that "without any doubt some means of conveyance to the moon cannot seem more incredible to us, than overseas navigation to the ancients, and that therefore there is no good reason to be discouraged in our hope of the like success" (Hooykaas 1977, 69). Kepler was quite sure that as soon as man mastered the art of flying, human colonies would be established on the moon.

Meanwhile the discoveries of the microscope were opening up visions of the minutiae of the biological world in remarkable detail. But it was a world that still remained framed within Aristotle's Great Chain of Being (*Scala Naturae*) in which the hierarchical order of creation, created by God at the top, was arranged in a vast systematic classification of angels, then man, then animals, then plants, then minerals at the bottom. When Linnaeus published his great classification system in 1737, it was arranged according to the three familiar lowest classes of the *Scala Naturae*: animals, plants, and minerals.

The static *Scala Naturae* hardly seemed a recipe for progress, but it was eventually this scheme that became transformed into Darwin's tree of life. As Wallace, Darwin's co-discoverer of natural selection, was to write in his 1865 paper entitled "On the Law which has Regulated the Introduction of New Species": "Every species has come into coexistence coincident in both space and time with a pre-existing closely allied species" (Berry 2013, 163). It was the brilliance of both Wallace and Darwin to bring history into biology, so that historical links now began to appear in the Great Chain of Being, joining it up as evolutionary history rather than as mere classification.

But first there had to be a readiness to accept change. The seeds were sown by the willingness of the seventeenth-century natural philosophers to question the ancients and "to commence a total reconstruction of sciences" (Shapin 1998, 66). In natural history, however, change began to take on some alarmingly materialistic connotations. This was seen most dramatically in the debates about reproduction between the so-called preformationists and epigeneticists that characterized the first half of the eighteenth century (Roe 2010). *Preformationism* referred to the belief that new organisms came from "germs" which represented preformed organized matter that had originally been brought into being by divine creation, and the idea gained popularity as a reaction against Cartesian mechanical views of reproduction. As Roe reports, "Encased within one another (the theory of *emboîtement*), all 'germs' existed in the first member of each species. By involving God's creative power in every future instance of reproduction, preformation came to be seen by many as a bulwark against the immorality to which atheism would inevitably lead" (Roe 2010, 36–37). By contrast, the epigeneticists saw the generation of each new organism during reproduction as entailing the formation of new order out of disorganized matter, but this was thought by many to open the door to materialism and atheism. Matter began to be seen by some as self-creating and self-energising, an idea that gained further traction with Abraham Tremble's discovery in late 1740 that the small freshwater polyp could self-regenerate after being cut up into separate pieces.

Furthermore, in the early 1740s, John Turberville Needham observed that when he added water to blighted wheat, fibres appeared to come to life like tiny worms that moved in a twisting motion for several hours. These observations created a sensation and in France the Encyclopedists, such as Denis Diderot (1713–1784) and Paul-Henri Thiry d'Holbach (1723–1789), embraced such ideas of self-organising matter and incorporated them within their radical anticlerical narrative in which a new natural basis for morality and society would emerge. The theme of change and progress, which was so central to the writings of the French *philosophes*, therefore became associated with a materialistic natural history in which matter acquired new powers. Animals developed from the moist earth heated by the sun. There was no need for God. For Diderot, Needham's microscopic observations provided the model for a world based on ceaseless activity and change, rather than preordained stability, and it was change associated with progress, progress in which the old static social order would be destroyed and a new world order ushered in that would lead to toleration and justice.

Unfortunately, the great chemist Antoine Lavoisier (1743–1794), one of the founders of the modern sciences of both chemistry and biochemistry, did not experience much toleration in the revolution that followed, losing his head in the process. Someone who disdained Lavoisier's chemistry, though in a way quite unconnected with his abrupt demise, was Jean-Baptiste Lamarck (1744–1829), who was more fortunate, and it is with Lamarck that the story of evolution in the form of a systematic theory really begins. Lamarck was much influenced by the *philosophes* and by their idea that physical laws produce complexity and progress. His mentor was the Comte de Buffon, one of the top French natural philosophers of his day, and Buffon's support earlier in Lamarck's career eventually led to his appointment in 1793 as one of the 12 professors in the newly organized Musée National d'Histoire Naturelle (National Museum of Natural History). Lamarck drew the short straw and was appointed professor of lower animals (insects and worms), a topic of which Lamarck knew little, his own previous expertise being more in the area of botany. But he set about the study of his new field with great enthusiasm, inventing a new word—"invertebrate"—to describe the objects of his study, and it was these investigations that led him to forsake his previous belief in the fixity of species and to become an evolutionist, as he stated in his introductory lecture for his new post given in 1800. Lamarck followed this up by three major publications in which evolution is presented in strongly progressionist terms, but ironically Lamarck himself, an ardent materialist, was also a convinced uniformitarian. To bring these two apparently incompatible ideas together, Lamarck envisaged the continuous spontaneous generation of the simplest organisms at the bottom, which then move up the escalator of life, with all steps occupied at all moments. In the newly minted Newtonian universe, there was a force that perpetually tends to make order (*le pouvoir de la vie* or *la force qui tend sans cesse à composer l'organisation*). As Lamarck himself expressed his theory:

> Ascend from the simplest to the most complex; leave from the simplest animalcule and go up along the scale to the animal richest in organization and facilities; conserve everywhere the order of relation in the masses; then you will have hold of the true thread that ties

together all of nature's productions, you will have a just idea of her pace, and you will be convinced that the simplest of her living productions have successively given rise to all the others. (Ruse 2010, 253)

Lamarck, unlike his colleague Georges Cuvier (1769–1832), did not believe in the extinction of species, and so in the progressive evolution of life, space was made available on the next step of the escalator by everything moving up in turn. When species disappeared it was because they had evolved to something different. When the environment changed, different organisms gradually adjusted their behaviours to thrive better in their new environments, and their adaptations were then inherited. This entailed a gradual move upwards towards increasing complexity and, in the end, perfection. As Lamarck wrote in *Philosophie zoologique*, "Nature, in producing in succession every species of animal, and beginning with the least perfect or simplest to end her work with the most perfect, has gradually complicated their structure" (Spencer 2014, 47). The Great Chain of Being had acquired an engine.

Charles Darwin's own grandfather, Erasmus Darwin (1731–1802), poet, rationalist, and botanist, to some degree foreshadowed Lamarck, in his work *Zoonomia* (1794–1796) envisaging that all living animals had arisen millions of years before man from one "living filament" which the great First Cause had endowed with the potential for delivering "improvements by generation to its posterity, world without end!" In Erasmus' poetry we start with "Organic Life beneath the shoreless waves" and finish with "Imperious man, who rules the bestial crowd." Life is always on the up. Yet as far as is known, Erasmus Darwin and Lamarck's evolutionary theories arose independently. Perhaps it is just that progress was in the air.

Although Lamarck died a pauper and the significance of his work lay unrecognised during his lifetime, Lamarckian themes gained great notoriety in later decades of the nineteenth century, not least when they were picked up by the Scottish publisher Robert Chambers in his then anonymous *Vestiges of the Natural History of Creation* (1844). Chambers presented his readers with a developmental hierarchy, which he termed the "universal gestation of nature." It was basically a story of the evolution of everything. In the sky a swirling fire-mist evolved into nebulae, solar systems, and planets; on the ground invertebrates, fish, reptiles, mammals, and man followed in order up life's great escalator; and in society there was development in civilization as Negro, Malay, American, Indian, Mongolian, and Caucasian gave way one to the other. The book was a sensation and it was not until the 1890s that the sales of Darwin's *Origin of Species* began to catch up with Chambers' popular work, despite, or perhaps because of, being lambasted by all the leading natural philosophers of his time.

If we score Lamarck, Erasmus Darwin, and Chambers 9 or 10 on a scale of 1–10 in the progressionist stakes, then I suspect that Charles Darwin himself would score around 5. As always he was temperate in his comments, at least by the standards of his time, balancing one comment off with another. On the one hand, in the *Origin of Species* we find Darwin writing that "as natural selection works solely by and for the good of each being, all corporeal and mental endowments will tend to progress towards perfection." Progress for Darwin was a consequence of biotic competition. In crowded ecosystems full of competing life-forms, the constant removal of inferior by superior life-forms would impart a progressive direction to evolutionary change in the long run. But then Darwin writes in his letter to the American progressionist palaeontologist Alpheus Hyatt on December 4, 1872: "After long reflection I cannot avoid the conviction that no innate tendency to progressive development exists." And we find Darwin scribbling in the margins of a progressionist text: "Never say higher or lower."

Darwin's most enthusiastic supporters simply dispensed with his caution and propounded a robustly progressionist view of evolutionary history. Herbert Spencer, arguably the most famous philosopher of his age, had already started working out a great developmental Lamarckian scheme for the evolution of nearly everything in *Progress: Its Law and Cause* published in 1857 and simply absorbed bits of Darwinism into his scheme as they came along, but remained more Lamarckian than Darwinian for the rest of his life. In his *Social Statics*, Spencer proclaimed that "Progress, therefore, is not an accident, but a necessity. Instead of civilization being artificial, it is a part of nature; all of a piece with the development of the embryo or the unfolding of a flower" (Spencer 1851, 65). Spencer maintained that the end point of the evolutionary process would be the creation of "the perfect man in the perfect society" with human beings becoming completely adapted to social life. Darwin did not use the word *evolution* at all in the first edition of the *Origin* because it carried the sense of "unfolding" with a strong connotation of inevitable progress, but Spencer pushed the word heavily and Darwin first starts using it in *The Descent of Man* in 1871.

Darwin's bulldog Thomas Henry Huxley was, unlike Spencer, a proper scientist, but was also a moderate progressionist, seeing evolution as the inexorable working out of natural laws, "a wider teleology which is not touched by the doctrine of Evolution" that "does not even come into contact with Theism, considered as a philosophical doctrine." But in later life we see Huxley reacting more strongly against progressionist views in his Romanes Lecture of 1893 entitled "Evolution and Ethics" where he critiques the idea that there is any order or purpose in evolution, and so moral values should be developed in defiance of nature's laws.

Despite the caution of Huxley and indeed of Darwin himself, the progressionist tradition continues on unabated right through the twentieth century, with evolutionary thinkers in the first half, such as T.H. Huxley's grandson Julian Huxley, and R.A. Fisher, together with the Catholic Lamarckian palaeontologist and theologian Teilhard de Chardin, in their very different ways keeping alive a progressionist stance. But as Michael Ruse points out, following the development of the neo-Darwinian synthesis in the 1920s and '30s, it now became much less respectable to talk about progression in scientific publications (Ruse 2010). Instead such material was relegated to the popular writings of the evolutionary biologists, as in Julian Huxley's hugely prolific output during the 1920s to 1950s, Huxley being attracted to vitalism and the writings of Henri Bergson.

Writing in the midst of the Second World War we find Julian Huxley extolling progress in the conclusion of his book *Evolution: The Modern Synthesis* (1942), but now tempered with the kind of realism that is rare in writers such as Spencer:

"Progress is a major fact of past evolution; but it is limited to a few selected stocks. It may continue in the future, but it is not inevitable; man, by now become the trustee of evolution, must work and plan if he is to achieve further progress for himself and so for life" (Huxley 1942, 578). So evolution passes on the baton of progress to man, who must keep up the good work. Top marks for optimism to someone publishing in 1942.

Many of the great evolutionary biologists of the latter half of the twentieth century, such as Ernst Mayr and E.O. Wilson, were likewise convinced progressionists, Wilson writing that "Progress, then, is a property of the evolution of life as a whole by almost any intuitive standard, including the acquisition of goals and intentions in the behaviour of animals." Ernst Mayr said the following to Michael Ruse in a taped interview in 1993 towards the end of his very long life, for he lived to be 100 (and published his last paper when he was 100, an example to us all):

> In your treatment, about half the time you talk about Progressionism with a sneer, always illustrated with some detestable examples of racism or male chauvinism. I think much of your writing would be improved if you would admit that much of Progressionism was a rather noble philosophy. In fact a very good case could be made for the claim that the current mess is the result of the loss of this philosophy. (Ruse 2010, 269)

Not until the writings of Stephen Jay Gould do we find a really vigorous all-out onslaught on progressionism, Gould proclaiming in 1988 that it is a "noxious, culturally embedded, untestable, non-operational, intractable idea that must be replaced if we wish to understand the patterns of history" (Gould 1988, 319). So, he does seem to have been against it. According to Gould, we are a "momentary cosmic accident," albeit a "glorious accident." Summing up his view, Gould writes: "Wind back the tape of life to the early days of the Burgess Shale; let it play again from an identical starting point, and the chance becomes vanishingly small that anything like human intelligence would grace the replay" (Gould 1991, 14).

And yet even Gould seems to have moderated his position in later life in *The Structure of Evolutionary Theory* published in 2002, the year of his death, and 1433 pages really do allow an author to add plenty of "ifs," "ands," and "buts." And so on page 468 we find Gould commenting: "But the history of life includes some manifestly directional properties—and we have never been satisfied with evolutionary theories that do not take this feature of life into account." And it turns out that the progressionist accounts that Gould liked to attack most heartily were, if not exactly windmills, then at least items that can leave other forms of progressionist narrative untouched.

No one could judge Richard Dawkins to be less than enthusiastic about evolution, and he comes across as rather a strong progressionist in his 1997 review of Gould's book *Full House*:

> progress [means] an increase, not in complexity, intelligence or some other anthropocentric value, but in the accumulating number of features contributing towards whatever adaptation the lineage in question exemplifies. By this definition, adaptive evolution is not just incidentally progressive, it is deeply, dyed-in-the wool, indispensably progressive. (Dawkins 1997, 1017)

And then in his *Ancestor's Tale* we find Dawkins writing that "the cumulative build-up of complex adaptations like eyes, strongly suggests a version of progress—especially when coupled in imagination with some of the wonderful products of convergent evolution" (Dawkins and Wong 2016, 681).

This sampling of progressionist and anti-progressionist narratives from the past few centuries makes one point clear: there is no overarching "grand narrative" that allows the philosophically inclined historian to categorise the various understandings of *progress* into some neat classification system. Biology, and life, are too messy for that. Nevertheless it is equally clear that at one end of the spectrum of meanings lie biologists such as Dawkins who see *progress* as defined within narrowly defined biological terms, entailing increased complexity associated with new adaptations, whereas at the other end lie the expansive visions of a Lamarck or a Julian Huxley, for whom the escalator of life is certainly moving towards something better, even though the delineation of the pot at the end of the rainbow remains rather ill-defined for these, as for other enthusiastic progressionists.

Evolutionary Progress and Christian Theology

This brief summary has at least reminded us of the immensely long and complex debate that has surrounded the whole issue of progression in the evolutionary literature and beyond. And the question I now wish to ask is whether the outcome of this discussion makes much difference to Christian theology.

The traditional Christian understanding of evolution, which starts with thinkers like Charles Kingsley and Frederick Temple as soon as the *Origin* is published, and continues in an unbroken lineage since that time, is that it represents the creative process that God uses to bring about his intentions and purposes for biological diversity in general and for humankind in particular. And the phrase "humankind in particular" is referring to the fact that humans are made in God's image, with a particular relational function and role to play in the purposes of God, not least in the caring for God's earth. God, in initiating and sustaining this process, is not seen as the divine puppet master, least of all the heavenly engineer who occasionally tinkers with the machinery, the picture conveyed by the proponents of Intelligent Design, but rather the God whose immanent faithfulness in guaranteeing the reproducible properties of matter likewise guarantees their propensity for life, the biological diversity which we as scientists then seek to describe and understand. Creation is about ontology, authorship, the fact that something exists rather than nothing.

Now I am not sure that this kind of theological narrative should commit Christians to any particular theory of progression in evolutionary biology, except in a rather weak sense discussed further below. We do not need to deny that prokaryotes such as bacteria remain the most abundant and in many ways most successful independent life-forms that have remained essentially the same for billions of years. We do not need to believe that every evolutionary lineage inevitably tends to develop in the direction of greater complexity. Many do. Many do not. Some degenerate and lose more complex items, like the eyes that get lost when animals go and live in caves, no longer necessary for their ecological niche. We certainly do not need to believe in any kind of *élan vital* that is impelling life-forms upwards and onwards to some future higher state. Instead we can happily accept that in evolutionary terms we are indeed one small twig on the evolutionary bush, and given that our lineage likely passed through one or more genetic bottlenecks of around twelve thousand interbreeding individuals or less, realize that we might so easily not have been here. We can realize all of these things, and many more, but I am not so sure that they really make any difference to the theological narrative that I have outlined.

Instead what we notice is that our twig does have some rather special properties. The evolutionary process has delivered beings with a kit of attributes that do in fact render the theological account both feasible and coherent. These beings have big frontal lobes that facilitate cogitation and moral decision-making. They have a basic moral tool kit, which is quite possibly inherited in the same way that the neuronal machinery for learning languages is inherited. Their language, grammar, and memory together generate the continuity and development of culture and of relationships. They have a religious sense, a cognitive bias perhaps to believe in God's agency. Many of these characteristics appear numerous times in evolutionary history—communication, culture, music, creativity, all in their various forms keep popping up all over the place in different evolutionary lineages, but it is in the particular package of these attributes within the human twig that we find located the possibility of a freely chosen relationship with God and real moral responsibilities. We just do not take animals or plants to court. And religion, to the best of our knowledge so far, does seem to be a unique property of this little twig. This is what, as a matter of fact, evolution has delivered, and certainly we can call this *progressive* in comparison with the far more numerous prokaryotes, but equally the recognition of the special qualities of our twig need not make us deny the intrinsic value of all the other twigs on this wonderful bush. It is just that we actually *care* about the extinction of the other twigs, whereas they do not.

There is, however, a weaker sense in which we can label the evolutionary process as *progressive*. Given the theological framework already introduced, that God fulfils his intentions and purposes through the evolutionary process, just as he does through the created order taken as a whole, we can, if we so wish, label the process as *progressive* because it does, as a matter of fact, take us from a to b, from non-living matter to living matter through to thinking, feeling, morally choosing human beings. Of course. How could God's world not fulfil God's purposes? But by engaging in such a definitional exercise, we are, ipso facto, rather removing the language of *progression* from the ways in which it has generally been used in an evolutionary context over the past two centuries. Questions of progress and purpose cannot be discussed in vacuo, as it were, as if one were starting the discussion with a blank slate. Whether it be Lamarck's great escalator of life, or Spencer's philosophy of the evolution of just about everything, with Chambers following in his train, or Huxley's conviction that evolution represented the inexorable working out of natural laws, or de Chardin's grand evolutionary narrative in which the whole process is heading towards the "omega point," there seems to be no particular good reason why

Christian theology should feel a need to incorporate biological evolution within such narratives of progress. In a contingent universe, God can bring about his intentions and purposes, which include the creation of humanity, any way that he chooses, and so there seems to be no good reason why it should occur by one way more than another. The contingency of the world entails that it would be unwise for Christians to hook their wagon to the latest progressionist narrative to emerge from evolutionary theory.

But if theology does not commit us to any particular theory of progression, except in the very general terms that I have outlined, then it does act as a brake to a particular flowering of progressionism—the idea that humankind itself is progressing morally and even inevitably towards some future better state. This previously popular view was of course largely killed off by the horrors of the twentieth century, although surprisingly the idea still crops up even today. Be that as it may, Christian theology simply points out, using the fall narrative as a resource, that general moral progress is not an intrinsic capacity of human being on the earth, and that outside of redemption and grace there is little hope of genuine progress.

Is There Purpose in Evolution?[1]

Now what about the other "*p*-word" hovering alongside progression—*purpose*? It is quite possible to be an ardent progressionist in evolution, without believing that the process taken as a whole has any purpose in any ultimate sense. This is clearly the position of Richard Dawkins when he writes, perhaps on a rainy day in Oxford: "The universe we observe has precisely the properties we should expect if there is, at bottom, no design, no purpose, no evil and no good, nothing but blind pitiless indifference" (Dawkins and Ward 1995, 133). The philosopher Daniel Dennett agrees—Dennett asks whether the complexity of biological diversity can "really be the outcome of nothing but a cascade of algorithmic processes feeding on chance? And if so, who designed that cascade?" Dennett answers his own rhetorical question by saying: "Nobody. It is itself the product of a blind, algorithmic process. Evolution is not a process that was designed to produce us" (Dennett 1995, 59).

The discussion here is not then about the question of progression as such, but about the question of purpose, the word that Dawkins uses. And here we do have to mark a parting of the ways. For clearly the idea of purpose is implicit in the kind of theological outline that I have introduced. The idea is that indeed the evolutionary process is fulfilling God's creative intentions and purposes. But on the other hand I rather agree with Dawkins and Dennett that if you look at the evolutionary process as an atheist and simply through the window of biology, then there is nothing there that *forces* upon you a narrative of ultimate purpose. Without the revelation of God in Christ there is no framework, no matrix, within which to place evolutionary history, to provide it with an overall purpose and coherence.

---

[1] A much expanded treatment of this question may be found in Alexander (2018).

So does that leave advances in our understanding of evolution completely divorced from our theological understanding of purpose? Not completely. We certainly cannot in my view *derive* theology from the evolutionary process itself, though some have tried to do just that, but I do think that our current understanding of biology renders less plausible the suggestion that evolutionary history *necessarily* lacks any plan or purpose. Let me give seven examples of what I have in mind.

First, and most obviously, evolution taken as a whole is not a chance process. This at least is where atheists and theologians can sing from the same song sheet. As Dawkins writes in *The Blind Watchmaker*: "One of my tasks will be to destroy this eagerly believed myth that Darwinism is a theory of 'chance'" (Dawkins 1986, xi). Of course the process incorporates chance in the generation of genomic variation, of course there are stochastic events leading to mass extinctions, but the winnowing effect of natural selection ensures that in the finely tuned interaction between chance and necessity it is necessity which wins in the end. So evolution is a highly organized and in many ways highly conservative process.

There is a certain irony in the reflection that the secular Thomas Henry Huxley was suspicious of the role of chance in generating variant phenotypes of organisms upon which natural selection then acted. For Huxley, chance sounded like an opening for God's special creation, whereas he wanted to see evolution as emerging out of natural scientific laws. The irony arises from the fact that in his day Huxley resisted the idea of chance, because he thought that it had theological overtones, whereas creationists today resist the idea of chance because they think that it has atheistic overtones. Often people interpret essentially the same data in quite different ways depending on their political, economic, and cultural contexts. In any event, had Huxley been alive today he would most likely have been pleased with the tendency in contemporary evolutionary theory to highlight the law-like behaviour of the trends observed within the evolutionary process.

Secondly, stand back and look at the 3.8 billion years of evolution as a whole, and the striking increase in biological complexity is obvious. For the first 2.5 billion years of life on earth, living things only rarely got bigger than 1 mm across, about the size of a pinhead. There were no birds, no flowers, no animals on land, no fish in the sea, but at the genetic and cellular level there was considerable development and diversification, with the generation of many of the genes and biochemical systems that were later used to such effect to build the bigger, more interesting living things that we see all around us today. At the same time the oxygen levels in the atmosphere increased to the point at which more complex life-forms could be sustained.

It is not until the advent of multicellular life from around perhaps 1.2 billion years ago that living organisms start to get bigger, although even then they were generally on a scale of millimetres rather than centimetres. Only in the so-called Cambrian explosion during the period 505–525 million years ago do we find sponges and algae growing up to 5–10 cm across, and the size of animals began to increase dramatically from that time onwards, until today we have creatures like ourselves with our brains with 10^{11} neurons with their 10^{14} synaptic connections or more, the most complex known entities in the universe.

As Sean Carroll from the University of Wisconsin-Madison remarks in a *Nature* review:

> Life's contingent history could be viewed as an argument against any direction or pattern in the course of evolution or the shape of life. But it is obvious that larger and more complex life forms have evolved from simple unicellular ancestors and that various innovations were necessary for the evolution of new means of living. (Carroll 2001, 1102)

Third, underlying biological complexity are networking principles that are turning out to be fewer and simpler than they might have been. Networking principles refer to those organizational systems that are used in all living organisms. Just as similar traffic control systems are used in all the world's cities, because there is only a limited array of methods that can be used for organizing traffic, so all cells in all organisms display a limited number of organizational motifs. Given that in every cell, complex networks of interactions occur between thousands of metabolites, proteins, and DNA, this is quite surprising. As Uri Alon from the Weizmann Institute comments:

> Biological networks seem to be built, to a good approximation, from only a few types of patterns called network motifs.... The same small set of network motifs, discovered in bacteria, have been found in gene-regulation networks across diverse organisms, including plants and animals. Evolution seems to have "rediscovered" the same motifs again and again in different systems. (Alon 2007, 497)

We may link this to what Sean Carroll has called "deep homology." This means that if you look at complex organs in animals such as limbs and eyes, in many cases it is possible to track back their evolutionary histories to see how such structures arose by the modification of pre-existing genetic regulatory circuits, established very early in animal development.

Fourthly, the very limited array of protein structures used by living organisms compared to the astronomically huge number of possible structures is also very striking. Proteins are made up of a specified sequence of 20 different amino acids and a single protein may contain hundreds of amino acids. Yet if you look at all the known proteins in the world, and their structural motifs, based on the three-dimensional structures of more than 100,000 proteins that have now been published, you find that the great majority can be assigned to just a few thousand domain "families." In other words, all living things are united not only by having the same genetic code, but also by possessing an elegant and highly restricted set of protein structures.

Recent findings also suggest that proteins can only evolve along certain quite restricted pathways because of the internal constraints built into their own structures. For example, a research group from Harvard published a paper entitled "Darwinian evolution can follow only very few mutational paths to fitter proteins" (Weinreich et al. 2006). They studied an enzyme called β-lactamase which breaks down antibiotics such as penicillin. Provided that bacteria have versions of this enzyme that are functioning efficiently, they grow quite happily in media containing antibiotic. If a gene is under natural selection then it needs to evolve in small incremental steps, each step increasing the fitness of the organism or at least not decreasing it. There are five amino acids needed at five key positions in the sequence of amino acids that make up the β-lactamase enzyme that enable it to function well enough to enable the bacteria to grow in antibiotic. So by random events you could imagine the gene evolving to this state through five mutations that might occur in any order; in principle there could be $5 \times 4 \times 3 \times 2 = 120$ different mutational pathways to achieve the goal of optimal enzyme efficiency. But in practice the Harvard researchers found that 102 of these pathways are barred because they decrease the fitness of the bacteria, i.e., their ability to flourish in the presence of antibiotic, and of the remaining 18 trajectories only a very few were really favoured. It is intriguing to read the authors' conclusion of their paper reporting this work: "We conclude that much protein evolution will be similarly constrained. This implies that the protein tape of life may be largely reproducible and even predictable."

The idea of *fitness landscapes* can be quite useful for envisaging how evolution occurs at the molecular level. These traditionally represent topographical pictures of the adaptation of different populations to local ecological niches, visualised in the same way that three-dimensional models can be used to give a good idea of mountainous areas like the Swiss Alps. The peaks represent those areas of optimal fitness at which a population is well adapted to its particular environment.

The concept of fitness landscapes can also be applied to enzyme structure and function. Again and again it turns out that the evolutionary pathways to arrive at a particular function of a particular enzyme are remarkably constrained. In other words, there are only a few ways to arrive at a particular protein function because only some genetic mutations will get you there and not others. It is as if an evolutionary path is laid out in front of the gene encoding the enzyme, and the genetic dice keeps being thrown until the enzyme structure is generated that optimises fitness for its particular function. This is no random process, each step along the way being preserved by benefits to the organism that uses the enzyme. In a recent review on this approach to investigating the evolution of protein functions, the authors conclude: "That only a few paths are favored also implies that evolution might be more reproducible than is commonly perceived, or even be predictable" (Poelwijk et al. 2007, 386).

Overall it appears that around 98% of all the amino acids in all proteins cannot change because of the striking decrease in fitness of the organism that would result (Povolotskaya and Kondrashov 2010). This means that the genes that encode these amino acids cannot change either, at least not by mutations that change the amino acid sequence. This might sound like a recipe for a static protein world. In practice this is not the case: proteins do evolve, but they just do so really slowly and cautiously. For example, if other random mutations occur in the same protein, then the constraint on the 98% of "frozen" amino acids is lifted somewhat. It is unlikely that the evolutionary search engine has yet completed its job of searching the complete repertoire of protein "design space," but it has come a long way in 3.8 billion years, and the present "snapshot" that we have certainly points to a highly constrained molecular world. In practice what this means is that if a random jumble of amino acid sequences is generated, the vast majority (indeed an astronomically huge number) will have no function at all, and it is up to the evolutionary search engine to find the tiny number that have functions useful for life.

Fifthly, as with proteins, so with genes, there are underlying biological principles that constrain the location and type of gene evolution. The "raw material" for evolution is provided by "random" mutations, gene flow, and the genetic recombination that occurs during the generation of the germ-line cells. But note that *randomness* here means only that genetic variation occurs without the needs of the organism in mind. By contrast the genetic variation that leads to evolution is not random in the sense that any kind of variation in any kind of gene will do. In reality there are so-called hotspot genes, those that are far more likely than others to play key roles in evolutionary change, such as a gene that delights in the name *shavenbaby* found in the *Drosophila* fruit fly. Such genes act as "input/output genes," encoding key switching proteins that integrate whole sets of information that are then mediated to downstream effectors. The *shavenbaby* gene regulates the existence and distribution of fine trichomes or cellular hairs on the surface of the larvae of *Drosophila*, so that mutations in *shavenbaby* lead to a lack of trichomes—hence the name (Stern and Orgogozo 2008, 2009).

It is genes such as *shavenbaby*, hotspot genes, that render evolution possible because they regulate an integrated programme of events, in this case converting cells into hair-making cells. The mutations that occur are in the regulatory sequences of this gene that control how much of the protein is actually made. So far about 350 of these kinds of hotspot genes have been identified in plants and animals. As the authors of a recent review entitled "Is Genetic Evolution Predictable?" comment: "Recent observations indicate that all genes are not equal in the eyes of evolution. Evolutionarily relevant mutations tend to accumulate in hotspot genes and at specific positions within genes. Genetic evolution is constrained by gene function, the structure of genetic networks, and population biology. The genetic basis of evolution may be predictable to some extent" (Stern and Orgogozo 2009, 746).

Sixthly, there is the remarkable phenomenon of convergence, the repeated evolution in independent biological lineages of the same biochemical pathway, or organ or structure, to which writers such as Dawkins and the Cambridge palaeobiologist Simon Conway Morris have drawn repeated attention. In his fine book *Life's Solution – Inevitable Humans in a Lonely Universe*, Professor Conway Morris brings together hundreds of examples of convergence in evolutionary history (Conway Morris 2003). For example, the convergence of mimicry of insects and spiders to an ant morphology has evolved at least 70 times independently. The technique of retaining the egg in the mother prior to a live birth is thought to have evolved separately about a hundred times amongst lizards and snakes alone. Compound and camera eyes taken together have evolved more than 20 different times during the course of evolution. If you live in a planet of light and darkness, then you need eyes—so that is precisely what will emerge as the adaptive requirement arises. The hedgehog tenrecs of Madagascar were long thought to be close relatives of "true" hedgehogs, because their respective morphologies are so similar, but it is now realized that they belong to two quite separate evolutionary lineages and have converged independently upon the same adaptive solutions, complete with spikes. Hundreds of other examples of evolutionary convergence may be found at this website: www.mapoflife.org.

Evolutionary convergence at the phenotypic level does not mean that a complete set of new genes evolves separately each time to build, for example, an eye. Far from it. Genomes contain genes that may be switched off, ready for use at some future time as required, or genes that presently have quite different functions, which can be pressed into service. There are many examples of genes that encode moonlighting proteins—proteins that carry out quite different tasks depending on whether they are inside the cell or outside, on the particular tissue in which they are located, or even on which specific location they occupy inside a cell (Jeffery 1999).

There are many examples illustrating the way in which convergence to generate similar adaptations operates at the molecular level. Echolocation is the method that mammals such as bats, porpoises, and dolphins use for hearing. It involves sending out pings of sound that bounce off objects and are then received back and analysed—an animal sonar system, used not just to detect the presence of objects, but to locate and identify prey. The brain works out how long it takes for the ping of sound to come back, and so how far away the object is. Bats can detect the presence of a tiny crawling insect or even a human hair, and can recognize each other's "voices."

A special protein called prestin is key to this sophisticated high-speed process (Jones 2010). The gene that encodes this protein is unique to mammals and has evolved independently several times since mammals split off from the birds in evolutionary history more than 100 million years ago. Prestin is found in the outer hair cells of the inner ear of the mammalian cochlea, a fluid-filled chamber. As the fluid is compressed by the sound waves the ear receives, so the sensory hairs surrounding the chamber move very slightly and convert their movements into nerve impulses via thousands of "hair cells." The outer hair cells that serve as an amplifier in the inner ear refine the sensitivity and frequency selectivity of the mechanical vibrations of the cochlea.

The specialized prestin found in echolocating mammals provides a much faster system for converting air pressure waves into nerve impulses than the prestin found in mammals (like us) that do not use echolocation. The convergent story became apparent when it was discovered that the prestin gene has accumulated many of the same mutational changes in bats, porpoises, and dolphins, changes that are essential for prestin to perform its unique functions (Jones 2010). Similar changes have occurred in unrelated lineages of different bats. Genetic evidence suggests that these changes have undergone natural selection. In other words, here is an adaptation that is of great advantage to the animal that has it, so animals carrying this particular set of mutations in the prestin gene are more likely to reproduce and spread the beneficial gene around an interbreeding population. The particular advantage may well be the necessity to hear very high frequencies, far above the ability of the human ear to hear. The advantage of possessing this specialised piece of echolocation equipment has helped shape the evolution of the prestin gene such that it has converged on the same adaptive solution independently on multiple occasions.

In a commentary on Gould's idea of contingency, Professor Conway Morris writes that

> it is now widely thought that the history of life is little more than a contingent muddle punctuated by disastrous mass extinctions that in spelling the doom of one group so open the doors of opportunity to some other mob of lucky-chancers.... Rerun the tape of the history of life ... and the end result will be an utterly different biosphere. Most notably there will be nothing remotely like a human.... Yet, what we know of evolution suggests the exact reverse: convergence is ubiquitous and the constraints of life make the emergence of the various biological properties [e.g., intelligence] very probable, if not inevitable. (Conway Morris 2003, 283)

So indeed the rolling of the genetic dice is a wonderful way of generating both novelty and diversity, but at the same time it appears to be restrained by necessity to a relatively limited number of living entities that can flourish in particular ecological niches. If you live in a universe with this kind of physics and chemistry, and on a planet with these particular properties, then the biological diversity that we do in fact observe is what is most likely to emerge. Evolution is a search engine for exploring design space. Biological diversity is definitely not a case of "anything can happen." Only some things can happen, not in a deterministic way, but in a highly constrained way.

The seventh example that highlights the highly organized nature of the evolutionary process relates to the emergence of the human mind. Only a very ordered and constrained process could produce something as elegant and complex as the human mind. Personhood developed in evolution with increased self-awareness and what we now call a theory of mind, in turn dependent upon the rapid increase in brain size that has taken place in the hominin lineage. In this understanding, mind is an emergent phenomenon, meaning that something with totally new properties has emerged which cannot be adequately understood or described using the language of biology.

Only two million years ago did hominin brain size begin to seriously surpass that of our nearest living cousin, the chimpanzee, which has a brain volume of about 400 cubic centimetres. Bipedality does not appear to be the critical factor that has driven this rapid cultural evolution, for certainly the hominins (most likely *Australopithecus afarensis*) who left their fossilised footprints in the volcanic ash at Laetoli in Tanzania, 3.5 million years ago, were bipedal, but the first stone tools do not start appearing until 2.6 million years ago, so bipedalism and tool use may be necessary but not sufficient to explain the rapid brain evolution that occurred later. A more likely explanation is the increasing complexity of hominin social life over the past two million years, in which the need for cooperativity in hunting and other social activities gave significant evolutionary advantages to those with larger brains.

Increasing brain size was characterized by an increase in the number of "orders of intentionality" (Dunbar 2004). The idea of orders of intentionality comes from the theory of mind, the ability of our own minds to realise that there are other minds that think like ours and that have intentions and purposes that may be similar or even quite different to ours. We take this "mind reading" completely for granted but it is in fact a crucial aspect of our identity as humans. To engage in communal religious beliefs, for example, several different orders of intentionality are required, in fact four and perhaps as many as five. In an example given by Robin Dunbar (2004), with each level of intentionality italicized and numbered: I *suppose* (1) that you *think* (2) that I *believe* (3) that there is a God who *intends* (4) to influence our futures because he understands our *desires* (5). Dunbar speculates that fourth order intentionality would not have appeared until about 500,000 years ago, about the time of the emergence of archaic *H. sapiens*, with fifth order intentionality appearing with anatomically modern humans around 200,000 years ago, perhaps along with language.

A comment from Martin Nowak, professor at Harvard University, is striking in this context. Nowak has published much recently on the mathematics of game theory in the evolution of social cooperation. Nowak comments (in 2009): "My position is very simple. Evolution has led to a human brain that can gain access to a Platonic world of forms and ideas."

Might it be scientifically feasible to link up all these seven examples (and more) in some larger theory that might show how the constraints imposed upon matter by the laws of physics and chemistry in a planet with these particular properties lead inevitably to the kind of biological systems that we in fact observe? In principle there seems to be no good reason why such a "grand theory" might not eventually prove possible, although in practice we are presently very far from such a scenario. But recent advances mean that asking the question does not seem quite as silly as it might have done only a few decades ago.

Even with our present rather limited understanding, if we reflect on just these seven examples of order and constraint in evolutionary history, it is clear that far from looking stochastic and random, evolution looks highly organised and constrained—predictable to some extent, perhaps even with inevitable outcomes. Note that I am not suggesting that if we read the evolutionary narrative just as biology that *therefore* evolutionary history per se displays some ultimate purpose, but rather a more modest claim: that these kinds of data—and many more—render the claim that evolutionary history is necessarily a purposeless history less plausible. In science it is often data that count *against* a theory that are the most powerful, as Karl Popper was fond of reminding us, and it is the biological data in this case which count against the idea that evolutionary history is a purposeless random walk without rhyme or reason. In human experience, narratives that are highly ordered and constrained are not normally without some kind of purpose.

Of course there are other types of argument that have been mounted against the idea that evolutionary history is consistent with a God who has intentions and purposes in bringing such a history into being. For example, it took too long, or it was too wasteful, or it involved so much death and suffering. The first two objections are rather trivial; the third is weighty, but space does not allow me to address these further questions here. I have addressed them elsewhere (Alexander 2008, 2011), and the problem of animal suffering in evolution has likewise been addressed extensively by others (Southgate 2008; Murray 2008).

We can conclude with the kind of comments that we frequently make in the discussion section of our scientific papers—the data are *consistent with* our favourite model, whatever that happens to be at the time. In like manner this highly organized and constrained evolutionary history is *consistent with* the theological claim that there is a God who has intentions and purposes for the world in general and for us in particular. Evolutionary history fits comfortably within the overall matrix of a theistic universe in which God wills that carbon-based intelligent life-forms emerge (us) who have the ability to respond freely (or not) to his love for us. And the fact that we are sitting on a very small twig in the great bush of evolutionary history should act as a reminder that we are only here by God's grace, and also that his grace extends to our future prospects as well.

References

Alexander, Denis. 2008. *Creation or Evolution: Do We Have to Choose?* Oxford: Monarch Press.
———. 2011. *The Language of Genetics: An Introduction*. Philadelphia: Templeton Foundation Press.
———. 2018. *Is There Purpose in Biology?* Oxford: Monarch Press.
Alon, Uri. 2007. Simplicity in Biology. *Nature* 446 (7135): 497.
Augustine of Hippo, and Marcus Dods. 2009. *The City of God*. Peabody: Hendrickson Publishers.
Bacon, Francis, and G.C. Moore Smith. 1900. *New Atlantis*. Cambridge: At the University Press.
Berry, Andrew. 2013. Alfred Russel Wallace: Evolution's Red-Hot Radical. *Nature* 496: 162–164.
Carroll, Sean. 2001. Chance and Necessity: The Evolution of Morphological Complexity and Diversity. *Nature* 409 (6823): 1102–1109.
Conway Morris, Simon. 2003. *Life's Solution: Inevitable Humans in a Lonely Universe*. Cambridge: Cambridge University Press.
Dawkins, Richard. 1986. *The Blind Watchmaker*. Harlow: Longman.
———. 1997. Human Chauvinism: Review of *Full House* by Stephen Jay Gould (New York: Harmony Books, 1996; also published as *Life's Grandeur* by Jonathan Cape, London). *Evolution* 51 (3): 1015–1020.
Dawkins, Richard, and Lalla Ward. 1995. *River Out of Eden: A Darwinian View of Life*. London: Weidenfeld & Nicolson.
Dawkins, Richard, and Yan Wong. 2016. *The Ancestor's Tale: A Pilgrimage to the Dawn of Life*. 2nd ed. London: Weidenfeld & Nicolson.
Delbourgo, James, and Nicholas Dew. 2008. Introduction: The Far Side of the Ocean. In *Science and Empire in the Atlantic World*, ed. James Delbourgo and Nicholas Dew, 1–28. New York: Routledge.
Dennett, Daniel C. 1995. *Darwin's Dangerous Idea: Evolution and the Meanings of Life*. London: Allen Lane.
Dunbar, Robin. 2004. *The Human Story: A New History of Mankind's Evolution*. London: Faber and Faber.
Gould, Stephen J. 1988. On Replacing the Idea of Progress with an Operational Notion of Directionality. In *Evolutionary Progress*, ed. M.H. Nitecki, 319–338. Chicago: University of Chicago Press.
———. 1991. *Wonderful Life: The Burgess Shale and the Nature of History*. New York: W.W. Norton & Co.
———. 2002. *The Structure of Evolutionary Theory*. Cambridge, MA/London: The Belknap Press.
Harrison, Peter. 2007. *The Fall of Man and the Foundations of Science*. Cambridge: Cambridge University Press.
Hooykaas, Reyer. 1977. *Religion and the Rise of Modern Science*. Edinburgh: Scottish Academic Press.

Huxley, Julian. 1942. *Evolution: The Modern Synthesis*. London: Allen and Unwin.
Jeffery, Constance J. 1999. Moonlighting Proteins. *Trends in Biochemical Science* 24 (1): 8–11.
Jones, Gareth. 2010. Molecular Evolution: Gene Convergence in Echolocating Mammals. *Current Biology* 20 (2): R62–R64.
Murray, Michael J. 2008. *Nature Red in Tooth and Claw: Theism and the Problem of Animal Suffering*. Oxford: Oxford University Press.
Poelwijk, F.J., D.J. Kiviet, D.M. Weinreich, and S.J. Tans. 2007. Empirical Fitness Landscapes Reveal Accessible Evolutionary Paths. *Nature* 445 (7126): 383–386.
Povolotskaya, Inna S., and Fyodor A. Kondrashov. 2010. Sequence Space and the Ongoing Expansion of the Protein Universe. *Nature* 465 (7300): 922–926.
Robertson, John M., ed. 2011. *De dignitate et augmentis scientiarum*. London: Routledge.
Roe, Shirley. 2010. Biology, Atheism, and Politics in Eighteenth-Century France. In *Biology and Ideology: From Descartes to Dawkins*, ed. Denis Alexander and Ronald Numbers, 36–60. Chicago: University of Chicago Press.
Ruse, Michael. 2010. Evolution and the Idea of Social Progress. In *Biology and Ideology: From Descartes to Dawkins*, ed. Denis Alexander and Ronald Numbers, 247–275. Chicago: University of Chicago Press.
Shapin, Steven. 1998. *The Scientific Revolution*. Chicago: University of Chicago Press.
Southgate, Christopher. 2008. *The Groaning of Creation: God, Evolution, and the Problem of Evil*. Louisville/London: Westminster John Knox Press.
Spencer, Herbert. 1851. *Social Statics: Or, the Conditions Essential to Human Happiness Specified, and the First of Them Developed*. London: John Chapman.
Spencer, Stephen. 2014. *Race and Ethnicity: Culture, Identity and Representation*. Abingdon/New York: Routledge.
Stern, David L., and Virginie Orgogozo. 2008. The Loci of Evolution: How Predictable Is Genetic Evolution? *Evolution* 62 (9): 2155–2177.
———. 2009. Is Genetic Evolution Predictable? *Science* 323 (5915): 746–751.
Weinreich, Daniel M., Nigel F. Delaney, Mark A. Depristo, and Daniel L. Hartl. 2006. Darwinian Evolution Can Follow Only Very Few Mutational Paths to Fitter Proteins. *Science* 312 (5770): 111–114.

Is There a Created Order for Cosmic Evolution in the Philosophy of Herman Dooyeweerd?

Jitse M. van der Meer

Abstract The Christian doctrine of creation entails among other things that order characterizes the cosmos and that the Creator is not subject to this order. In that sense lawful order marks the boundary between Creator and creation. Herman Dooyeweerd generalizes the idea of law as boundary to develop an ontology of created kinds of order. Boundaries demarcate distinct ways in which phenomena operate, limiting causal interactions between them. For instance, physical things cannot produce living things, prohibiting evolutionary emergence. This was in line with the speculative nature of evolutionary thought in the early twentieth century, but not today. To accommodate evolutionary emergence I propose revisions in Dooyeweerd's ideas of law, causation, and time. I replace his idea of law as a given static boundary with a model for the evolution of boundaries. This allows for causal continuity and discontinuity between phenomena with different kinds of order—a conundrum in emergence theory. Theoretically, this combines the possibility of causal explanation of evolutionary emergence with a critique of ontological reductionism. Further, I relocate causal power from intangible laws existing outside of time to the dispositions of concrete things in time and interpret them as dynamic manifestations in creation of God's ordering activity. This amounts to a rejection of Dooyeweerd's theory of time. I justify this effort by what I value in his ontology. Phenomena have many ontologically irreducible dimensions, and their understanding requires many irreducible explanations. More specifically, I value his interpretation of the ontological relations of living things with their physical infrastructure.

Keywords Evolutionary emergence · Ontology · Creation order · Herman Dooyeweerd · Law · Causation · Time · Divine agency

---

J. M. van der Meer (✉)
The Pascal Centre for Advanced Studies in Faith and Science, Redeemer University College, Ancaster, ON, Canada
e-mail: jmvdm@redeemer.ca

Introduction

The Christian doctrine of creation entails among other things that order characterizes the cosmos and that the Creator is not subject to this order. In that sense lawful order marks the boundary between Creator and creation. The Dutch philosopher Herman Dooyeweerd uses the idea of law as boundary to develop an ontology of created order. In this contribution I evaluate two elements in this ontology, namely, *law* and *causation*. Following the biblical metaphor depicting God as a king issuing decrees, Dooyeweerd (1969, 1:99, 101, 108) understands *law* as divine decree or creation ordinance. I evaluate this notion of created order for two reasons. First, a metaphor depicts a reality. What reality does the metaphor of creation ordinance refer to? Traditionally, it has been taken to refer to lawful order. Do laws exist outside of time as Dooyeweerd appears to believe and does God act according to them? Or does God act in the dispositions of creatures and do we capture parts of this in law statements? This concerns the relation between creation and temporal becoming.

Second, is Dooyeweerd's understanding of lawful order as applied to temporal becoming acceptable? He postulates that entities in the world display different kinds of lawful order.[1] Evolutionary emergence described in these terms would be the emergence *over time* of entities displaying different kinds of order.[2] Organisms operate according to laws that are different in kind from those that govern the matter of which they consist. Persons perform mental operations according to laws that are different in kind from laws governing their bodies. But Dooyeweerd (1959, 126–129; 1969, 3:94–95) argues that entities of one kind cannot evolve into entities of a different kind with a different lawful order because this would involve a causal relationship that erases their difference in kind.

Dooyeweerd's prohibition on evolutionary emergence is understandable. He developed his philosophy in the 1930s when evolutionary emergence as a metaphysics was speculative and evolution as a biological theory had little empirical support. Today, however, there is plenty of support at least for the biological theory of evolution. So, his contemporary interpreter Jacob Klapwijk proposes a way of incorporating evolutionary emergence into this philosophy while preserving differences between kinds of order. But he keeps Dooyeweerd's prohibition on causal evolutionary relations between entities displaying different kinds of order. As a result, Klapwijk must accept that evolutionary emergence is not explicable. In this chapter I try to avoid that consequence.

---

[1] Dooyeweerd's *kinds of order* are best understood as natural kinds even though he does not characterize them that way. The terms *aspect*, *mode*, and *dimension* are synonyms for *kinds of order*.

[2] I use the term *to emerge* in the generic sense of "to appear" or "to produce," not in the technical sense in the literature on emergence. Thus, when A produces B, B emerges from A. I use *evolutionary emergence* as a shorthand for *diachronic evolutionary emergence* as distinct from *diachronic embryonic emergence*. For a discussion of theories of emergence, see O'Connor and Wong (2015) and Sartenaer (2016).

The common ground for this project includes that the cosmos is created by and dependent upon God. This—among other things—is what marks Dooyeweerd's philosophy as Christian. Further, with Dooyeweerd, and Augustine, I hold that time is a creature. This distinction between the eternity of the Creator and the temporality of the creation raises the question of how the eternal Creator can act in time to create a diversity of kinds of order. Dooyeweerd (1959, 115–119; 1940b, 216) and Klapwijk (2008, 34) suggest that divine creative activity is supra-temporal. It *precedes* time in some sense that humans as creatures *in time* cannot imagine.[3] He postulates the existence of a super-temporal realm also referred to as a created eternity. It serves as the location of laws by which Dooyeweerd accounts for divine action in temporal reality. It also functions as the location from which a knower can see the unity in the diversity of kinds of order—a need created by his philosophy of theorizing.

My preliminary thesis is that Dooyeweerd conceives of law and causality in ways that forbid evolutionary emergence. Instead, I will explore whether his static notion of order can be replaced with a dynamic notion that locates created order in the dispositions of concrete entities interpreted as God's ordering activity. I define *evolutionary emergence* as the production of new concrete entities over time. I take evolutionary emergence to be a causal process because "change is always preceded by causal *powers* and *dispositions*" (Koons and Pickavance 2015, 45).[4] When I use the term *cause* without further specification I am referring to an unanalysed concrete relation of productive determination that has many so-called modal aspects.[5] Entities are new by virtue of displaying powers and a kind of lawful order absent in their precursors. Thus, both entities and laws are new in the ontological sense and are ontologically irreducible to their precursors.

I focus on the emergence of living things from non-living things and of different kinds of living things from ancestors. My thesis entails that his concepts of lawfulness and causality lead Dooyeweerd to reject such evolutionary emergence. My assessment of this rejection focuses in part on *causal* evolutionary emergence because that is the target of Dooyeweerd's critique. The following three sections explain Dooyeweerd's philosophy of theorizing, law, and causality so that the question whether his ontology can account for emergence can be made more specific. Then I explain why Dooyeweerd accepts non-evolutionary causal emergence and rejects evolutionary emergence, and assess his ontological, epistemological, empirical, and religious objections to evolutionary emergence.

Next, I consider hypothetically how living things might emerge over time from physical whole-part relations without erasing their difference in lawful order. This requires a revision of his philosophy of theorizing as well as of his concepts of law

---

[3] Dooyeweerd uses *supra-temporal* and *super-temporal* as synonyms.

[4] Non-causal forms of determination could be considered. Whether emergence could be non-causal depends on whether one has a deterministic or indeterministic interpretation of quantum physics. This issue falls outside the scope of this chapter.

[5] The terms *entity*, *order*, and *aspect* (or *modal aspect*) are explained in the introduction by Gerrit Glas and Jeroen de Ridder.

and causality. Together they entail a rejection of created eternity as the location of law. In the final section, I conclude that created order is located in the orderly functioning of concrete entities that are endowed with the power to change. Evolutionary emergence is the successive manifestation of different kinds of powers interpreted as divine action in the world.[6]

Why the effort? The physical and chemical processes incorporated in the models for evolutionary emergence to be discussed have not been shown to produce living things. Moreover, the revisions to be suggested in Dooyeweerd's ontology are far-reaching. Nevertheless, two reasons justify this project. First, experimentally speaking, the question of evolutionary emergence is open. On the one hand, the in vitro production of self-reproducing and self-correcting proteins seems promising (Saghatelian et al. 2001; Leman et al. 2004). On the other hand, polymer chemistry offers empirical reasons for doubting the possibility of cross-boundary emergence (Vollmert 1983; Thaxton et al. 1984). In my view emergent evolution remains a metaphysical research program for now. The goal of finding experimental support and the indirect empirical evidence for the emergence of life and mind justify attempts at developing theoretical scenarios that direct attention to hitherto neglected possibilities.

Second, the revisions are intended to preserve the idea that there exists a plurality of different kinds of order and that their interrelations are organized according to a meta-order (called "cosmonomic time"). These features provide a strong framework for a critique of ontological reductionism. It also offers an interpretation of the ontological relations of living things with their physical infrastructure. This may suggest an empirical response to a conundrum in emergence theory, namely, how there can be both ontological continuity and discontinuity between different "levels" (Dooyeweerd 1959, 126; Klapwijk 2008, 290; Sartenaer 2016). Finally, relations between Christian faith and science are best mediated philosophically. The proposed revisions also aim at making Dooyeweerd's philosophy suitable for mediation between Christian faith and evolutionary emergence.

The Structure of Theorizing According to Dooyeweerd

This section summarizes those elements of Dooyeweerd's philosophy of theorizing that explain his view of lawfulness and causation. I simplify how his notions of abstraction and transcendence explain why he thinks one can have a *philosophy* of evolutionary emergence, but not a causal explanation in a scientific theory.[7]

---

[6] In my view, divine action constitutes the power of creatures to act; see Jaeger (2012, 53–57). The implications of my anti-realism regarding laws for rationality, morality, and religiosity fall outside the scope of this chapter.

[7] For a critique of Dooyeweerd's theory of knowledge, see Dengerink (1977), Strauss (1984), and Hart (1985, 143–166). In addition, contemporary studies in the history and philosophy of science invalidate Dooyeweerd's separation of intuitive and theoretical knowledge, the details of which are beyond the scope of this chapter.

Abstraction

According to Dooyeweerd everyday experience gives rise to pre-theoretical knowledge, which is unanalysed, intuitive, and holistic. Theorizing occurs by analysing entities that are experienced as unities with many aspects. Comparison reveals similarities and differences in the way entities function. Complete abstraction isolates one of these ways and characterizes it by a modally defined concept.[8] Since most entities function in many different ways, abstraction creates as many modally defined concepts as there are aspects in the entity. Concepts are logically discontinuous when they refer to modally different ways in which an entity functions (Dooyeweerd 1969, 1:30, 2:74). Thus, theories in which these concepts operate are irreducible. This is not surprising since theory reduction is often not possible even within a mode. The result of abstraction is theoretical knowledge. Since it focuses on one aspect of a phenomenon it is partial knowledge as distinct from pre-theoretical knowledge, which is holistic. This raises questions about Dooyeweerd's views of law and causation. He considers laws as abstractions, yet attributes agency to them (see sections below on law as well as on time and the laws of nature). Analysis of causation produces modal causal concepts that refer to aspects of actual causes, not to real causes.

Transcendence

Theorizing aims at grasping how different modal aspects with their laws and causal concepts are unified. Unification requires the self to transcend the theoretical diversity of aspects produced by analysis and synthesize them (Dooyeweerd 1969, 2:548). Taking that synthetic stance is to take a philosophical viewpoint with respect to phenomena (ibid., 1:82–89, 2:548). Therefore, in theoretical analysis the unity of the modal aspects of a phenomenon can only be a philosophical presupposition—that is, a transcendental idea. Its choice involves adopting a religious perspective on philosophy (ibid., 1:55, 57, 79). This requires the self to transcend "cosmic time" (Dooyeweerd 1940b, 179, 181; 1969, 1:24, 28, 31, 2:298–299, 3:783).

Cosmic Time

The Creator has eternal existence. Creatures have temporal existence in cosmic time. The term *cosmic* refers to the mutual coherence between the kinds of order on display in an entity. Such coherence is seen in the sequential arrangement of the

---

[8] "Theoretical abstraction of the modal aspects from cosmic time is necessary for a theoretical insight into the modal diversity of meaning as such" (Dooyeweerd 1969, 2:40).

kinds of order. For instance, physical order is a necessary but insufficient condition for biotic order. Accordingly, different kinds of order are arranged in a succession of necessary but insufficient conditions revealing a hierarchical architecture (Dooyeweerd 1969, 1:28–29, 3:78–79). Coherence is also seen in subject-object relations illustrated below with the production of DNA. Thus, cosmic time has a law-side which is a meta-order for different kinds of order. The transcendental idea of cosmic temporal order or cosmic time refers to this meta-order (ibid., 1:174). Dooyeweerd locates it in created eternity between the eternity of the Creator and the temporality of creatures.[9]

Cosmic time also has a subject-side, or factual side, because there are entities subject to this meta-order.[10] On the factual side, cosmic time is displayed as different kinds of duration according to the kind of order (ibid., 1:28). For instance, in physical things duration is displayed as movement, in living things as birth, maturation, and death. Thus, cosmic time is an idea that straddles created eternity (law) and created temporality (fact). Only humans can transcend cosmic time (Dooyeweerd 1940a, 181; 1969, 1:24, 31, 2:535) because they are created with this capability (Dooyeweerd 1940a, 182). In this super-temporal existence the human self functions religiously (Dooyeweerd 1935–1936, 1:37; 1939, 5; 1969, 1:31). From there the self determines the transcendent conditions for philosophical knowledge (Dooyeweerd 1969, 1:5, 2:548).

In what follows I focus on two implications of abstraction. First, Dooyeweerd treats law as an abstraction that has causal agency and exists in created eternity. Second, abstraction of causes leads to anti-realism about causes.

Law

The epistemology of Dooyeweerd implies that law is an abstraction. As such it has no causal power. But Dooyeweerd claims that type law forbids evolutionary emergence.[11] So, is there something more to the status of law than its abstract character that allows him to assert that type law has the causal power to forbid emergent evolution?

To answer this question I will describe the ontological status of law. In *De wijsbegeerte der wetsidee* (1935–1936) as well as in *A New Critique of Theoretical Thought* (1969) Dooyeweerd appears to hold two different notions of law: law as plan of God and law as something existing in created eternity. He affirms that lawfulness is in God's plan (Dooyeweerd 1935–1936, 1:70, 145, 2:491; 1969, 2:559), that this plan originates in the will of the Creator (Dooyeweerd 1935–1936, 1:145;

---

[9] Dooyeweerd (1939, 4–5) distinguishes between created eternity (*aevum*), the uncreated eternity of God, and temporality. See also Dooyeweerd (1969, 1:31, 2:53, 3:65, 88; 1936, 68–69).

[10] I use *subject-side* and *factual side* as synonyms.

[11] For the difference between modal law and type law, see the introduction by Gerrit Glas and Jeroen de Ridder.

1969, 1:174), that it was established before the foundation of the world (Dooyeweerd 1935–1936, 1:70, 2:491; 1969, 2:559), and that this plan is beyond human understanding (Dooyeweerd 1969, 1:174, 2:559). This he refers to as creation order. He rejects a Christian Platonic interpretation of creation order as *universalia ante rem*. That is, he rejects the notion of ideas in God's mind in the sense of Platonic ideas transformed into ideas of divine reason, because he considers the latter as absolutizations of human reason (Dooyeweerd 1935–1936, 2:491; 1959, 147; 1969, 2:559; see also 1969, 2:551).

As for law existing in created eternity, Dooyeweerd emphasizes first that laws occupy no space and have no duration because they were created to be the law for space and duration. Laws are independent from the process of becoming in time (Dooyeweerd 1959, 115–116). He envisions that laws exist in a super-temporal realm of created eternity where they bridge the gap between the eternal and the temporal.[12] Dooyeweerd (1959, 115; 1966, 4) writes,

> Of the creation alone may it be said, according to Gen. 2:1, that it is *completed*. This can never be said of the genetical process in the temporal order. For this process is still going on; individual men, animals, plants, etc. are formed, and this is not a *temporal* continuation of God's work of creation, but only a *consequence*, within the order of time, of the completed creation.

Here, the "creation alone" is completed outside of time while "the genetical process of becoming" occurs in time (Dooyeweerd 1959, 114–116).[13] This distinction implies that laws exist before their subjects do. Laws are pre-existing.[14] Second, Dooyeweerd (1966, 4) refers to Gen. 2:1: "Thus the heavens and the earth were completed in all their vast array" (NIV). That is, the completed creation includes diversity. But the completed creation excludes the diversity of individual creatures because they are still in "the genetical process of becoming." Therefore, the diversity of the completed creation refers to the diversity of laws for a diversity of subjects. Consider, further, that laws are created (Dooyeweerd 1940b, 195–196; 1959, 116; 1969, 1:101, 108, 2:52). Since laws were created outside of time, they are located in created eternity. This applies to all kinds of law because their creation is complete.[15] But laws are not Platonic forms (Dooyeweerd 1969, 2:559; 1959, 147). Thus, between 1935 and 1969 Dooyeweerd is a realist with respect to law. His ontic notion of law appears to refer to something existing in created eternity. In the section on causality confined by type law, I will show that he attributes causal power to

---

[12] See n. 9 above.

[13] See also Dooyeweerd (1969, 3:65): "The modal aspects of reality find their deeper identity in the central religious sphere alone. But temporal things are perishable." The "central religious sphere" is where the self transcends cosmic time and exists in created eternity.

[14] Klapwijk (2008, 178) makes this implication explicit. He characterizes evolutionary emergence as a "pre-structured and level-transcending disclosure," thereby assuming the pre-existence of different kinds of lawful order to be disclosed. See also Klapwijk (2012, 73).

[15] A more extensive argument can be given for the claim that law exists in created eternity, but that falls outside the scope of this chapter.

this "something." In the section on evolutionary emergence, I interpret his perspective on law in terms of divine action.

Causality

Anti-realism

Dooyeweerd's notion of structural causality refers to the effects entities, phenomena, and processes have on each other in everyday experience. Structural causes have many aspects (Dooyeweerd 1969, 2:41, 3:40, 63, 159, 160). Abstraction isolates as many different *modal causal concepts* as there are different kinds of lawful order the structural cause displays (ibid., 3:62). For instance, Dooyeweerd distinguishes, among other things, between mechanical, biological, and psychological concepts of cause (ibid., 2:40).

A modal causal concept does not refer to a real process (ibid., 2:41, 3:63) because it is a theoretical abstraction (ibid., 2:41), and abstractions have no causal effects (ibid., 3:40). Thus, Dooyeweerd is an anti-realist with respect to modal causal concepts in scientific theories. Only structural causes have effects because they exist in the real world (ibid., 3:63, 64). But structural causes cannot be grasped theoretically in a scientific concept nor function in a scientific explanation because they transcend theoretical thought. They can only be handled as transcendental ideas; i.e., as causality in the transcendental sense of the unity of its modal aspects. Causality in this transcendental sense is a philosophical presupposition because this requires a synthesis of the modal aspects of structural causation (see section on transcendence; Dooyeweerd 1969, 2:40, 3:159; see also 3:40–41). Therefore, the status of causal accounts for evolutionary emergence in Dooyeweerd is that of a philosophical presupposition.

Abstraction creates logical discontinuity between causal concepts of modally different kinds of order. Therefore, the causal concepts and theories about the biotic kind of order cannot be reduced to the physical kind of order. Thus, there can be no causal explanation for the evolutionary emergence of living things from physical things.

Ontological Continuity and Discontinuity

Below I will explore how his ontology might be revised to include causal evolutionary emergence. In preparation I note here that Dooyeweerd's exclusion of causal evolutionary emergence does not imply the exclusion of all ontological continuity. So how does he think ontological discontinuity and continuity can co-exist?

According to Dooyeweerd, things function as subject or object.[16] Thus there are subject-subject relations and subject-object relations. Both exemplify ontological continuity. Subject-subject relations include relations among physical subjects such as the gravitational attraction between planets. Relations among biotic subjects include those between plants. Relations between physical and biotic subjects include those between plants and nutrients. Subject-object relations illustrate how ontological continuity and discontinuity can co-exist. For instance, when molecular nutrients enter a cell, causal interaction ensues between biotic and physical subjects. When the cell produces DNA from these nutrients, DNA production cannot be explained in terms of the nutrients considered as physical subjects. Nor can it be explained in terms of biotic law because DNA is not subject to biotic law—it is not alive. Dooyeweerd would consider DNA to be an object of biotic production by a biotic subject. Such a causal subject-object relation is a relation between a physical entity—DNA—that remains subject to physical law and a biotic subject—the cell. A subject under biotic law (the cell) manipulates a subject under physical law (nutrients). In that sense there are causal relations between the biotic aspect and the physical aspect. But physical law is not violated in the process. In that sense there are no causal relations between the biotic aspect and the physical aspect. This is how I take Dooyeweerd's claim that there are ontologically different kinds of order between which there are no causal relations. Thus, causal continuity and discontinuity taken in different senses co-exist.

While subject-object relations illustrate causal relations between the physical and the biotic aspect, they cannot be the mechanism of causal evolutionary emergence because they presuppose what needs to be explained, namely, the existence of a biotic subject. In the example above, the existence of the subject capable of producing DNA must be presupposed to explain the production and operation of the molecular pathways that make DNA synthesis possible. By contrast, accounts of evolutionary emergence must explain the emergence of new subjects that make subject-object relations possible. Below ("Revision 3"), I propose how the co-existence of ontological continuity and discontinuity can evolve.

According to Dooyeweerd such an account is impossible. It would require subject-subject interaction characterized by structural causation. That interaction would have to cross the modal boundary between physical and biotic order. But Dooyeweerd and Klapwijk hold that different modal kinds of order are ontologically irreducible. Evolutionary emergence would involve structural causes which are multimodal. The highest kind of order in which physical entities function as structural causes is physical order. Thus physical causes lack a biotic aspect. But to claim that physical structural causes can produce living things is to claim that they have a biotic aspect. This would not only be self-contradictory but also deny the ontic difference between physical and biotic order.[17] That is the ontological reason why Dooyeweerd thinks causal evolutionary emergence is not possible and why

---

[16] For these distinctions, see the introduction by Gerrit Glas and Jeroen de Ridder in this volume.

[17] Dooyeweerd (1969, 2:36–49) formalized the need to avoid erasing the ontic difference between kinds of order in his principle of excluded antinomy.

Klapwijk envisions evolutionary emergence as a non-causal and continuous process (Dooyeweerd 1969, 3:61–65, 158–160; Klapwijk 2008, 120, 209 [non-causal], 208 [continuous]).[18] This implication does not apply to causal non-evolutionary emergence such as embryogenesis because it does not cross modal boundaries. The following section shows that causal evolutionary emergence is prohibited by type law as conceived by Dooyeweerd.

Causality Confined by Type Law

Both Dooyeweerd and Klapwijk reject a role for causality in the evolutionary emergence of new entities. Dooyeweerd also rejects evolutionary emergence, but Klapwijk accepts it. This section describes how Dooyeweerd interprets embryonic development, showing that he accepts causal diachronic emergence that is non-evolutionary. Then follow case studies on speciation and emergent evolution, showing why Dooyeweerd rejects causal evolutionary emergence. His reasons are evaluated below.

From Fertilized Egg to Embryo

Dooyeweerd (1969, 3:645) holds that a germ-cell

> develops … into a being of a pre-determined structural type…. This individuality-structure is that of a plant, or an animal; only the human germ-cell lacks a radical-typical limitation and refers to the mystery of the spiritual centre of human existence, which transcends all temporal structures. In other words, the qualifying function of a non-human germ-cell is entirely dependent on the individuality-structure of the being destined to develop from it genetically. Only the germ-cell of a plant is *biotically* qualified according to its radical type.[19]

An *individuality structure* is a law for a kind of entity—type law for short. He interprets the development of a fertilized animal egg into a mature animal as pre-determined by a type law that qualifies its complete life cycle as actively subject to sensitive law.[20] By contrast, the development of a plant is pre-determined by a different type law that qualifies the life cycle of a plant. In different contexts,

---

[18] I fail to understand how Klapwijk (2008, 209) envisions that evolution can be both non-causal and continuous or how time can be "a universal process of disclosure of all that is enclosed in created reality."

[19] The "only" in this citation is intended to differentiate biological subjects (fertilized plant eggs) from sensitive subjects (fertilized animal eggs) and humans. The term *qualifying function* is explained in the introduction by Gerrit Glas and Jeroen de Ridder.

[20] For the difference between being actively and passively subject to law, see the introduction by Gerrit Glas and Jeroen de Ridder.

Dooyeweerd (1969, 1:105, 3:78–79) refers to the lawful development of events on the subject-side as the "realization" or "actualization" of the type law.

Type laws limit the potential of individual things to change:

> Everything that has real existence, has many more potentialities than are actualized. Potentiality itself resides in the factual subject-side: its *principle*, on the contrary, in the cosmonomic-side of time. The factual subject-side is always connected with individuality (actual as well as potential).... But it remains bound to its structural laws, which determine its margin or latitude of possibilities. (Dooyeweerd 1969, 1:105)

Such structural laws "are not changeable in time, since they determine the inner nature of perishable factual things, events and social relationships functioning within their transcendental cadre" (ibid., 2:557, 3:83, 94).

In conclusion, the concrete causes that operate on the subject-side during animal embryogenesis are actualizations of pre-determined causal relations included in the type law because these causes are part of its life cycle. The type law limits the potential for change of an entity to the possibilities given within that law. Thus, animal embryogenesis is characterized by diachronic emergence, but within the limits of sensitive type law. This is confirmed by Dooyeweerd's (1969, 3:749–778) description of the causal role of organizers and regulators in embryonic development. Dooyeweerd is referring to the ability of transplanted embryonic donor cells to cause a second embryo in a host embryo. He considers organizers as material objects produced by transplanted cells (ibid., 3:723, 754). Therefore, their causal effect does not cross modal boundaries in the diachronical sense that development does not begin with a physical entity from which biological and sensitive entities emerge in succession. This is not evolutionary emergence.

Thus, from Dooyeweerd's perspective there is within diachronic emergence a distinction between non-evolutionary and evolutionary emergence. Embryonic development is an example of non-evolutionary diachronic emergence because it does not cross modal boundaries, but there is emergence over time. Evolutionary development crosses such boundaries and exemplifies evolutionary diachronic emergence. Henceforth I will refer to the latter as *evolutionary emergence*.

What Dooyeweerd means by *temporality* becomes clear in his discussion of human development from zygote to adult. He asserts that the capacities for feeling, reasoning, cultural formation, and language develop in that sequence. This sequence is a factual manifestation in temporal reality of

> a process of actualization of potentialities already present in the structural principle of human bodily existence. In the *temporal order* of the modal aspects there cannot be a *real* succession. We could only show that this order has a temporal character because it is necessarily related to a genetic process of realization which reveals successive phases of actualization of the different modal aspects in accordance with this order. But these successive phases of realization are bound to structures of individuality which exceed the boundaries of the modal aspects. As such, these typical total structures have no real duration, since they belong to the law-side of cosmic time. But the individual things, processes, *etc.* in which they are realized, do have it. (Dooyeweerd 1969, 3:78–79)

Here Dooyeweerd describes the law-side and the subject-side (factual side) of human development. As in animal development, human development is a "genetic" process pre-determined by a type law. On the subject-side there is a "real succession" in that feeling, reasoning, cultural formation, and language develop in that chronological sequence. He interprets this succession as "successive phases of actualization" of a pre-existent succession of different kinds of modal law. The "temporal order" of the succession of kinds of modal law does not refer to chronological succession because there is no "real duration" on the law-side. Rather, it refers to a timeless ontological order of necessary but insufficient conditions. He infers this order from the chronological succession of developmental phases on the subject-side because the two are correlated. This analysis confirms that laws are located in created eternity.

From One Biological Species to the Next

My first test showing that Dooyeweerd rejects causal evolutionary emergence in the factual sense uses the development of new biological species. Speciation would cross modal boundaries in the evolution of humans from animals. Dooyeweerd considers biological variability and its extent. On the one hand, he keeps the possibility of factual variability open:

> Obviously, one may not exclude a priori the *possibility* that many of the currently known types of species considered as type laws have in fact realized themselves by means of a more or less gradual structural transformation of groups of individuals whose ancestors displayed a different type of species although this possibility cannot be verified scientifically.[21]

Here, Dooyeweerd acknowledges that a group of organisms may have produced new species and higher taxons such as the genus, the family, the order, the class, and the phylum, but that this cannot be established scientifically. This suggests that, hypothetically, Dooyeweerd allows that a species may change to such an extent that it falls under another type law. But he immediately excludes such variability:

> In the final analysis, common ancestry does not decide whether descendants belong to the same type of species. Rather, the standard for measuring the type of species that has been realized must be the type law for the species. The latter has the constancy characteristic for a law. The question of the factual descent of the members of a species is to a certain extent indifferent with respect to type law for the species. One can assert that the type law for a species implies the possibility of the production of fertile offspring, but only within the limits of the same type law. Irrespective of how the species concept derived from this type law will be defined further in a new systematics, it will remain bound to the postulate of species constancy in the sense of the constancy of the type law for the species. (Dooyeweerd 1959, 146–147; see also 1969, 1:105)

---

[21] "Natuurlijk mag niet aprioristisch de *mogelijkheid* worden uitgesloten dat vele der thans als zodanig bekende soort-typen als ordeningstypen zich feitelijk hebben gerealiseerd langs de weg van een geleidelijke of meer sprongsgewijze structurele omvorming van groepen van individuen in wier voorouders een ander soorttype tot openbaring kwam, ook al is die mogelijkheid niet wetenschappelijk te verifiëren" (Dooyeweerd 1959, 146; see also 153).

Dooyeweerd approves of the species concept of Ray and Linnaeus (Dooyeweerd 1959, 145). He would have established the type law for a species empirically in the way Linnaeus classified plants and animals on the basis of the structure of reproductive organs, not on the basis of common ancestry. He also accepts the independence of biological systematics from phylogeny (ibid.). At the time of Dooyeweerd's publication Linnaean taxons did not match the evolutionary taxons. Today classification is designed to identify ancestor-descendant relationships. But common ancestry still does not decide which offspring belongs to one species (or higher taxons) because the descendants of common ancestors often comprise several different biological species or higher taxons. Moreover, if by the *common ancestors* Dooyeweerd means to refer to a parental couple and their immediate offspring, there are other organisms than this offspring that also belong to that same species. In sum, common ancestry is a necessary but insufficient condition for offspring to belong to the same species. A sufficient criterion is that two individuals have fertile offspring. Dooyeweerd accepts this criterion with the proviso that variability is limited to within the boundaries of the type law for the species.

From Molecules to Man

My second example showing that Dooyeweerd rejects causal evolutionary emergence in the factual sense is that of the evolution from molecules to man. In the "genetic" processes occurring in factual reality, modal laws are actualized in a temporal succession. For instance, "organic life could only develop after the realization of an inorganic world adapted to its needs" (Dooyeweerd 1969, 3:78–79, 86–87). But this is because the inorganic world is a necessary but insufficient condition for the existence of organic life. Dooyeweerd's "temporal succession" is a succession of necessary and insufficient conditions on the law-side, not of evolutionary duration on the subject-side. Dooyeweerd rejects a causal-genetic explanation of organic life as the product of the interaction of physical-chemical factors on the subject-side because it ignores the ontological irreducibility of different kinds of lawful order. Moreover, it violates the theoretical irreducibility of modal aspects the denial of which produces theoretical antinomies (Dooyeweerd 1969, 3:94–95; 1959, 127). Klapwijk agrees by pointing out that causal emergence equates physical and biotic causes and that this contradicts the claim that a new kind of causality has emerged (Klapwijk 2008, 143, 164, 182, 185). He therefore rejects causal emergence *at a time* as in the synchronous emergence of mind from brain (ibid., 142) as well as causal emergence *over time*, i.e., evolutionary emergence (ibid., 182, 185; 2011, 23).

When Dooyeweerd speaks of time as duration or process on the subject-side he speaks about modal as opposed to evolutionary duration. That is, he treats time in its modally restricted sense such as the biological order of birth, maturing, adulthood, aging, and dying. He does not refer to time as the evolutionary duration of a diachronic process that produces entities with new kinds of active properties over time as in the emergence of organisms from non-living things (Dooyeweerd 1969, 1:28). Instead he envisions emergence on the factual side to be a non-causal, non-genetic succession of actualizations of unchanging type laws (Dooyeweerd 1959, 126–129, 132). These type laws are ordered in a series of necessary but insufficient conditions on the law-side such that entities qualified by an ontologically succeeding modal aspect cannot exist until after entities qualified by the ontologically preceding modal aspects have been actualized.

Conclusion

In sum, according to Dooyeweerd, embryonic development and thus the entire life cycle of a type of animal is pre-determined within the limits of an unchangeable type law. Within those limits "a process of actualization of potentialities already present in the structural principle of human bodily existence" unfolds dynamically. But no modal boundaries are erased nor antinomies created.

Regarding biological evolution, Dooyeweerd (1969, 3:83, 94) assumes that a modally qualified type has sub-types, each qualified by its own type law. Presumably, type laws for sub-types also prohibit causal evolutionary emergence so that an evolution of biological species is not possible.

As for evolutionary emergence, physical phenomena are actively subject to a type law that is qualified by the physical kind of order. A type law limits the potential for change of an entity to the possibilities given within that law (Dooyeweerd 1969, 3:83). Thus the causal powers of a given entity cannot exceed those specified by its type law. For instance, physically qualified causes as defined by Dooyeweerd lack a biological aspect and, therefore, a physical process cannot produce living things.[22] This excludes causal evolutionary emergence—that is, the production over time of entities with new kinds of active properties on the factual side of reality.

Evaluation of Dooyeweerd's Objections to Causal Evolutionary Emergence

Different Kinds of Order Are Ontologically Irreducible

We have encountered two ontological arguments against *causal* evolutionary emergence. First, Dooyeweerd and Klapwijk hold that the causal influence of entities, events, and processes does not cross modal boundaries because this would erase an ontological difference between kinds of order. Second, according to Dooyeweerd,

---

[22] Analogously, Bunge (1963, 203–204) asserts that causality renders genuine novelty impossible, but he asserts this about efficient causation (33). As explained, Dooyeweerd holds a wider notion of physical causation.

but not Klapwijk, type laws limit the potential for change of an entity to the possibilities given within that law.

In my view these two arguments originate in part in their view of causation. If entities actively operating according to physical law could cause entities operating according to biological law, then the causal difference between these two orders is erased. But a causally continuous ontology of evolutionary emergence that equalizes kinds of order and a causally discontinuous ontology that differentiates kinds of order cannot be true at the same time. This is the assumption behind the principle of excluded antinomy (Dooyeweerd 1969, 2:36–48). Klapwijk agrees that the notion of causal evolutionary emergence implies self-contradiction. But he does not share Dooyeweerd's notion of type law. Rather, he grounds the contradiction directly in the ontological irreducibility of different kinds of order because they are given (Klapwijk 2008, 209–210). Klapwijk accepts evolutionary emergence and avoids self-contradiction by declaring it non-causal and thus closed to causal explanation (ibid., 143, 182, 185; Geertsema 2011, 56). Given the assumptions of Dooyeweerd and Klapwijk, we are left with what appears to be a valid reason to reject the existence of causal powers that would allow entities with one kind of order to produce entities with a different kind of order. This reason is the theoretical erasure of ontologically different kinds of order. I will suggest that entities actively operating according to physical law can cause entities operating according to biological law without erasing their ontological difference.

Further, I agree with Klapwijk (2008, 248–253) that Dooyeweerd holds an essentialist notion of type law as ontological boundary and boundary concept that prohibits the evolutionary emergence of concrete realities with new kinds of order. Dooyeweerd is not a Christian Platonist because he rejects the notion of laws as eternal ideas in divine reason. Rather, he holds that "before the foundation of the world this order was present in God's plan" and this plan is beyond human understanding (Dooyeweerd 1969, 2:559; 1959, 147). Neither is he an Aristotelian essentialist with respect to lawful order because he rejects the notion of substance (Dooyeweerd 1950, 84; 1959, 147; 1969, 2:11). Nevertheless, Dooyeweerd is an essentialist in that physical entities do not produce biological entities because type laws prohibit evolutionary emergence.[23] He is also an essentialist with respect to biological species because the type law for a biological species is constant (Dooyeweerd 1959, 146–147; Klapwijk 2008, 248–252). Dooyeweerd does not justify his essentialism, but it likely was intended to avoid antinomy and to be consistent with empirical evidence at the time. His essentialist concept of type law prohibits the crossing of modal boundaries (to avoid antinomy) as well as the transformation of sub-types qualified by the same type law. The latter includes the evolution of biological species for which Dooyeweerd (1959, 129; 1969, 3:95) felt there was no empirical evidence.

---

[23] Dooyeweerd (1940b, 199, 219). See also Dooyeweerd (1969, 2:557): "structural *types of laws*, which, just as the structural *modi of laws*, ... are not changeable in time"; Dooyeweerd (1969, 3:83): "structures of individuality" are "invariable"; and Dooyeweerd (1969, 3:97): geno-type as a structure of individuality has a "constant identity."

Yet, essences may be required in accounts of biological evolution. Sober (1980, 354, 455) takes essentialism as the claim that there are hidden structures including causal mechanisms which unite diverse individuals into a natural kind. Thus, the essence of a natural kind can be a causal mechanism which works on each member of the natural kind, making it the kind that it is. Causal mechanisms functioning as essences have been identified in individuals (Walsh 2006), species (Boyd 1991, 1999), and populations (Sober 1980). Thus essences can be identified at every level of biological organization.[24] Therefore, Dooyeweerd's type law can have an essentialist interpretation and be consistent with biological evolution—namely, by describing a mechanism that causes changes of type. This is the approach I will propose later on in "Revision 2."

Time and the Laws of Nature

I concluded that Dooyeweerd holds an ontic notion of lawful order that refers to something existing in created eternity with or without actual subjects. Further, type law has the causal power to prohibit evolutionary emergence. But it remains unclear how a type law can have causal power or whether Dooyeweerd attributes causal power to laws more generally. Sometimes he describes laws as agents. They "rule" (Dooyeweerd 1969, 3:80), "guarantee," "guide," and "determine" (ibid., 1:105). Other descriptions suggest that laws are implemented. There is a "genetic" process of realization or actualization of law (ibid., 3:78–79) in which laws are "realized" or "actualized" (ibid., 1:105). But he does not make explicit that implementation involves divine action.

Nor does Dooyeweerd explain what his concept of law refers to that exists outside of time and can act as an agent in time. He did not envision law as a mind-independent abstract object existing in created eternity.[25] First, he rejected the reality of Platonic forms, which are a kind of abstract object. Second, he rejected the view that knowledge is restricted to what can be constructed mentally. Attributing causal power to law conceived that way would amount to a hypostatization or reification of a concept against which he has argued extensively. Particularly relevant is his characterization of the concept of substance in Leibniz as "the hypostasis of the modern functional concept of law. The functional coherence between variant phenomena, construed by thought, becomes the 'invariant,' the substance of reality" (Dooyeweerd

---

[24] Additional arguments and references in Boulter (2012). Further, linking essentialism with stasis is historically incorrect: see Wilkins (2013).

[25] There is no agreement on a definition of *abstract object*. Abstract objects can be concepts obtained by abstraction or mind-independent objects. The concept of abstract object used here is explained in Cheyne (2001) and Rosen (2014). Likewise, Zylstra (2004) and Stafleu (2008) are realists with respect to laws, although Stafleu does not locate laws in a transcendental realm outside of space and time.

1969, 1:202–203).[26] Finally, abstract objects do not have the causal powers he attributes to them (Cheyne 2001; Copan and Craig 2004, 181–182). Thus the effect attributed to type law cannot be implemented. Therefore, an alternative interpretation is called for. I assume that Dooyeweerd is speaking metaphorically when referring to law as an agent and is referring to God acting according to law. This is consistent with Dooyeweerd's language of the law being actualized and realized. Thus, on divine action Dooyeweerd is best understood in terms of his early emphasis on divine plan and providence (Henderson 1994, 149–150). Therefore, without rejecting the notion of created eternity, I am rejecting its use as a location for law. Below ("Revision 1") I introduce an alternative that works without his notion of law.

Theoretical Erasure of Modal Difference Contradicts Their Ontological Difference

Type law and irreducibility are the ontological reasons why Dooyeweerd rejects the possibility of a *theory* of causal evolutionary emergence of entities with different qualifying functions. From the perspective of Dooyeweerd and Klapwijk causal evolutionary emergence erases the ontological difference between kinds of order because they pre-exist. They claim that it is self-contradictory to maintain both their difference in kind and the causal power to erase such differences. From my perspective, however, self-contradiction is not a problem. Causal evolutionary emergence does not erase ontological differences between kinds of order because they do not pre-exist. Rather, physical entities produce biological ones which have new kinds of properties and display new kinds of order including new kinds of causation. Modal boundaries can be distinguished *after* they have emerged. Therefore, causal evolutionary emergence cannot erase modal boundaries.

No Theoretical God's Eye Point of View

From Dooyeweerd's ontological perspective, could the evolution of the entire cosmos be considered as the realization of a single comprehensive type law under which a diversity of types could unfold? This is not possible, he argues, because a particular type can only be known in distinction from other types. Since we are considering the entire cosmos as a single type there are no other types and the entire cosmos cannot be identified as a type. Accordingly, a single type law encompassing the entire architecture of different kinds of order in the universe is not possible

---

[26] See also Dooyeweerd (1969, 1:44, 92–93, 107, 122, 136, 143, 231, 297, 384, 416, 449, 475, 2:26, 56, 84–85, 386–389, 435, 560–565, 569–572, 578, 583, 3:4, 27, 166–167, 218, 243–244, 246).

(Dooyeweerd 1969, 3:630–632). Moreover, from Dooyeweerd's epistemological perspective a single type law cannot be a presupposition derived from pre-theoretical experience since there is no pre-theoretical experience of evolutionary emergence of the entire cosmos. Hence, once again Dooyeweerd has no room for a *theory* of evolutionary emergence.

Empirical Objection

So far no physical causes have been discovered that could produce living things. But what if experiment were to show that living things can emerge causally from physical things? Dooyeweerd (1959, 129) claims that he was prepared to sacrifice his philosophical theory about the irreducibility of the biological aspect if "living protein" could be produced by physical-chemical means. But this sacrifice would never have to be made because for Dooyeweerd, questions about the causal relationship between entities in different modal aspects are philosophical questions by definition:

> Every evolutionary hypothesis about this point that tries to offer a physical-chemical explanation of the origin [of life] moves from the domain of the natural sciences into that of a philosophical vision of the origin of our world. This philosophy erases the modal boundaries between the physical aspect and that of organic life. (Dooyeweerd 1959, 129)

If the origin of life consisted of the emergence of a lump of protoplasm without any cellular structure, the protoplasm, according to Dooyeweerd (ibid., 141), "would appear within a highest individuality structure (phylum), namely that of the unicellular animals or plants, and then it could not count as the ancestor of the successive series of phyla."

Thus a lump of protoplasm could not be considered a physical stage developing into a biological stage because the type law for the lump prescribes that it is a biological thing. Something similar happens with embryonic development. An animal zygote could never be considered a biological stage preceding a sensitive stage because its type law prescribes that it is a sensitive thing. In sum, entities fall into either one or the other modal kind with no transition. On evolutionary emergence Dooyeweerd's philosophy is closed to correction by empirical evidence despite his declared intention.

What if speciation had crossed the modal boundary between animal and human? Dooyeweerd limits species variability to within the boundaries of the type law for the species. This is an a priori limitation on the evolution of species. It is at variance both with his willingness to consider that "types of species considered as type laws have in fact realized themselves by means of a more or less gradual structural transformation of groups of individuals" (Dooyeweerd 1959, 146; see also 153) and with his empirical approach to ontology.[27] Thus, with Geertsema (2011, 51) and Klapwijk

---

[27] This conclusion disagrees with Geertsema (2011, 60).

(2008, 253–254) I take the essentialist notion of type law as the rationale for Dooyeweerd's rejection of the evolution of species. But essentialism and empirical evidence no longer justify this rejection.

A Religious Argument Against Causal Evolutionary Emergence

For Dooyeweerd a causal emergence of living things from physical things conceived naturalistically amounts to an absolutization of the physical aspect of creation. This violates the law as boundary between creation and Creator (Dooyeweerd 1959, 127).[28] It attributes the divine status of Creator to creatures functioning in that originating physical mode. This is unacceptable to him because it would be a form of idolatry (Dooyeweerd 1969, 1:99–104). In my view the possibility of causal relations between entities qualified by different modes does not entail a divine status for one or the other mode. Dooyeweerd rightly observes that philosophical naturalism and its implied philosophical reductionism can function as a pseudoreligion. But naturalism as methodological simplification is not a pseudoreligion.

Evolutionary Emergence

The greatest challenge for a theory of evolutionary emergence is to discover how an ontologically continuous process of causal emergence can produce a world displaying ontological and logical discontinuity (Sartenaer 2013). This section addresses that question for the emergence of living entities from physical entities by proposing three revisions of Dooyeweerdian ontology.

Revision 1: Closing the Distance of the Creator from Creation

Dooyeweerd's theory of time creates two problems. First, the early Dooyeweerd (1926, 69) equates the notion of law with that of divine providence in order to mark divine authority over creation. Further, the notion of divine providence and plan warranted unity in a diversity of laws (Henderson 1994, 149–163).[29] This early notion of divine providence as law derives from the metaphor of a king issuing decrees. It distances the Creator-king from his creature-servants. Later, Dooyeweerd relocates law outside of space and time but still within creation (ibid., 150, 152).

---

[28] Dooyeweerd (1969, 3:763) characterizes Woltereck's theory of the emergence of life from physico-chemical constellations as "an overstraining of the modal aspect of biotic development in its *subject*-side."

[29] For the references to the original Dutch sources in Dooyeweerd, see Henderson (1994, 150–152).

This further increases the distance. Second, this metaphor also creates tension between the temporal and the eternal in God's action in the world and detracts from its unity.[30] Others have noticed this tension when he describes the spiritual development of individuals and societies and fails to keep divine action above time separated from divine action in time. This difficulty has led to the conclusion—which I share—that his theory of time is unnecessary (Dooyeweerd 1969, 1:32–33; Brüggemann-Kruyff 1982, 55). In my view it is equally problematic to distinguish between an eternal and temporal aspect of divine action in nature. Specifically, it is unclear what laws located outside of time and space are and how they can have effects in time. Moreover, given that different modal time durations are unified in concrete things and that this unity is governed by the type law for a concrete thing, there is no need for this unity to also be controlled by cosmic time.[31]

These two problems disappear when the law metaphor is interpreted in terms of the reality it depicts (Soskice 1985, 97ff., chap. 7, chap. 8). This reality is the divine action which brings about the orderly behaviour of concrete things in nature. First, the distance between Creator and creatures is eliminated by locating the different kinds of order in the created dispositions of concrete things. This obviates the need for laws existing above and beyond this concrete behaviour outside of space and time. Law statements suffice. Cosmic time as the law for coherence of different kinds of order is replaced by the evolutionary emergence of entities with dispositions that cohere even though they display different kinds of order. Their coherence is provided by the initial and boundary conditions for existence provided by a given kind of order for the next order. The early Dooyeweerd suggested something similar for his static system, namely, that this coherence could be provided from within the different kinds of order and their different qualifications.[32]

I locate causal power in the dispositions of concrete things and interpret them as dynamic manifestations in creation of God's ordering activity. That is, divine action is manifest in (among other things) the ways things function and in how different functions cohere to establish identity. This avoids both the causal impotence of law conceived as existing outside of space and time and the tension within divine action between the temporal and the eternal. This interpretation honours Dooyeweerd's intent with the notion of law, which was to acknowledge that the eternal Creator can act in time. It is close to how Dooyeweerd equates law with divine providence. In this way one can envision that new concrete entities emerge in time and display new powers or dispositions that can be described as a different kind of order.

---

[30] For critique of the idea of cosmic time, see Hart (1973); Dooyeweerd (1969, 1:32–33); Brüggemann-Kruyff (1981, 1982); Steen (1983); Olthuis (1985, 21–40); and McIntire (1985, 81–117).

[31] Van Woudenberg (1992, 170) also observes that Dooyeweerd accounts for unity of particular kinds of order in an entity in terms of type law as well as cosmic time.

[32] "De kosmische eenheid der tijdsorden bouwt zich op uit de onderscheiden gekwalificeerde tijdsorden, doch is niet zelve een tijdsorde" (Dooyeweerd [1928, 121n86], quoted in Verburg [1989, 114]).

Revision 2: Widening Physical Causation to Include Evolutionary Emergence

Dooyeweerd's ontological notion of type law prohibits evolutionary emergence by specifying a type of causation that cannot cross modal boundaries. If we were to replace this notion with a law statement that describes evolutionary emergence, it would have to satisfy at least four requirements.

First, the law statement must attribute to concrete entities a notion of physical causation such that physical entities can produce living ones and living entities can produce sensitive ones. I use this concept of causation in the generic sense of *lawful way of becoming* or *productive determination*.[33] I take causal evolutionary emergence to be determined probabilistically in lawful ways by the external and the internal conditions of the subject in question with each condition determined in turn by other types of determination, such as mechanical or teleological determination. The question is where to look for conditions with the power to generate biotic entities.

One may look for a rare *combination* of *external* physical causes and conditions. The probabilistic character of such subject-subject combination would constitute the initial conditions for evolutionary emergence (Fig. 1). Boundary conditions act by preventing some combinations and selecting other combinations leading to biotic entities. I borrow these roles for initial and boundary conditions in evolutionary emergence from hierarchy theory which has resources for developing emergent evolution into a research program reviewed in the next section.

The generation of biotic entities may also involve unknown *internal changes of state* of a physical entity. Internal conditions, I assume, include lawful internal relations among properties of matter that produce the powers to produce living things. Internal conditions also include so-called uncaused quantum phenomena which produce spontaneous events in lawful ways[34] with the proviso that quantum phenomena may be produced by hidden deterministic variables (Bohm 1952a, b; Bohm and Hiley 1993, chap. 15). Depending on internal conditions, the causal powers of an entity, event, or process may be released or suppressed.

Finally, a living thing may be generated by a specific constellation of causes and conditions both internal and external either in parallel or in succession. An external cause may trigger an internal change of state and generate biotic entities. The constellation as a whole is not necessary for the effect because other constellations may have the same effect. Neither is a single condition sufficient for the effect to occur because other conditions may also be required (Mackie 1974, 62; Salmon 1998, 22–23).

---

[33] The term *productive determination* is from Bunge (1963, 108–116, 194). For a discussion of theories of causation, see Dowe (2008).

[34] Spontaneous events are sometimes called non-causal in that they are not produced by external (efficient) causes (Bunge 1963, 49, 181–182).

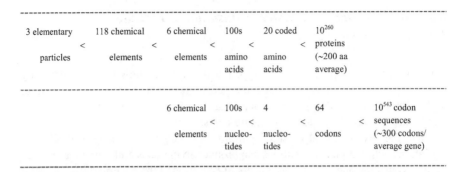

Fig. 1 Power of random combination and selection

In sum, the law statement must describe a notion of causation that involves both "self-determination and extrinsic determination, in which external causes are conceived as unchainers of inner processes rather than as agents moulding a passive lump of clay" (Bunge 1963, 197). The causal mechanisms by which physical entities might generate biological ones may be considered as the essence of a natural kind referred to above.

The second requirement is that a statement of type law must describe evolutionary history. Type laws describe the identity of types of things such as types of elements in chemistry, types of biological species, or types of symmetry of body plans in biology. If chemical elements, biological species, and body plans are in part the result of evolutionary history, statements of their respective type laws must describe this history.

Third, the notion of type law may appear to be a contradiction between the requirement that a law is a generalization and the fact that no two individual organisms have the same identity. If type law is too closely connected with individual organisms, their diversity will make generalization impossible and the notion of type law as generalization loses its meaning. This is why I suggest a revised statement of type law. It should describe a mechanism for the *generation* of diversity of a particular type. For instance, it could describe the difference in identity between humans and chimpanzees as the proximate result of the generation of differences during embryonic development and as the distal result of the evolution of genomic differences during evolutionary development. In each case the type law would describe a class of individuals sharing a given identity. To this class the definition of Stafleu (2002, 2) applies: "This book takes a natural character [type] to be *a cluster of natural laws determining a class of individuals and sometimes an ensemble of possible variations.* Individuals may be things, plants or animals, events or processes, including numbers, spatial figures and signals."

Fourth and finally, the law statement must describe the dispositions of concrete entities without referring to a transcendental realm outside of space and time as a location of a succession of laws.

Revision 3: Evolutionary Emergence Without Erasing Modal Boundaries

The third revision proposes how the ontological discontinuity Dooyeweerd sees between physical and living things may have emerged in a causally continuous evolutionary process. I envision this process as beginning with the partial causal isolation of physical entities from each other followed by the differentiation of some physical entities into living things. Salthe and Deacon propose specific models suggesting how this may have happened.

Partial Causal Isolation

I have two suggestions for the development of partial causal isolation between entities of any kind. The first suggestion concerns a step in preparation for the transition from physical to biological order, namely, the development of partial causal isolation between physical entities. Salthe (1985, 2002, 2006) starts with a growing physical entity such as might have existed at the start of the Big Bang. Due to size increase it becomes physically unstable and divides into parts as a growing soap bubble forms smaller compartments within itself. Each part in turn divides into still smaller parts.

The possibility of causal interaction—whether internal, between parts within a whole, or external, between entities—depends on their relative physical scale (difference in size, mass, and rate of change).[35] Entities of similar size and mass complete their cycle of characteristic behaviour in similar time intervals and can, therefore, interact in unlimited ways, generating many combinations. For instance, random combination and selection of entities at a given level produces an expanded number of entities one level up (Fig. 1, left to right). The three different elementary particles—electrons, neutrons, and protons—combined in different numbers yield 118 different chemical elements. The six chemical elements found in organisms (CHONSP) combined in different numbers yield hundreds of amino acids. The 20 amino acids coded in DNA yield ten to the power 260 proteins, assuming 200 amino acids as the average length of a protein.

Likewise, as long as the whole and its parts are relatively similar in size and mass they continue to interact. But as subdivision continues, a whole is subdivided into parts each of which is in turn subdivided into smaller parts. In the end, size and mass may differ by many orders of magnitude. The largest wholes will then change at rates far below the smallest parts physically contained in them. As a result, the smallest parts will have completed a cycle of change before they can be influenced by a change in the large whole in which they are contained. The large whole can no longer participate in generating combinations of the smallest parts. It can only limit

---

[35] In a different context Dooyeweerd (1969, 3:79) describes somewhat similar differences in duration between entities but does not discuss causal isolation as its implication.

the number and kind of combinations that continue to be produced by random interaction among the smallest parts at the lower level.

Causal isolation can be partial because it depends on the relative difference in size and mass between entities. Only wholes that have grown past a threshold of size and mass cease to interact causally with the smallest parts. Below this threshold causal interaction and combination continue. That is, wholes continue to interact with medium-size parts also contained in them. This applies to internal causal interaction between a whole and its parts as well as to external causal interaction between parts within a whole. This suggests how causal evolutionary emergence may have produced entities in which ontologically discontinuous kinds of order co-exist with causally continuous subject-object relations. Complete causal isolation would allow ontological differentiation in kinds of order between entities as detailed below. Partial causal isolation between a differentiating whole and its parts would allow subject-object relations to develop between them in which the whole would use the parts analogous to how a cell uses its DNA. The entity in which subject-object relations are developing is simultaneously evolving into a living entity. Finally, before isolation, the interaction between the parts is described as internal to the whole. Afterwards we only have the external interaction between what used to be parts that depended on the whole but are now causally independent entities. In this way the process of evolutionary emergence may be determined by internal and external conditions.

My second suggestion of how a temporally continuous process could produce discontinuous kinds of order applies to biological evolution. It involves the notion of thresholds of complexity characteristic for cybernetic control systems in organisms. A temporally continuous process of biological evolution may produce discontinuous natural kinds of organisms when a threshold of complexity is crossed. My strongly simplified example is the endosymbiont theory of the evolution of eukaryotic cells.[36] All eukaryotic cells have mitochondria. Many also have chloroplasts. According to this theory, eukaryotes evolved when a heterotrophic cell acquired the property of ingesting autotrophic cells looking like mitochondria. In turn, this eukaryotic cell acquired the capacity for photosynthesis when it acquired the property of ingesting cyanobacteria which look like chloroplasts. In each event a threshold of complexity was crossed and a new type of cell emerged. The eukaryotic cell emerged first and ultimately gave rise to the animal kingdom. The photosynthetic cell evolved next and produced the plant kingdom. According to Dooyeweerd these kingdoms display different natural kinds of order. Assuming for the sake of argument that he is correct, this means that two new natural kinds of order emerged. The two suggested scenarios for the causal evolution of a new kind of order from a base order do not erase the difference between them.

---

[36] Details and evidence may be found in introductory biology textbooks such as http://www.biology-pages.info/E/Endosymbiosis.html (accessed January 9, 2017).

Differentiation

Causal isolation between physical parts and a physical whole is complete when a large entity ceases to participate as a cause in generating combinations of small entities. It still participates, but only as a boundary that limits the number and kind of combinations that continue to be produced by random interaction among small entities at the lower level. When a whole has a concrete property by which it controls its parts in the case of a whole-part configuration, this property is a boundary condition. Such a property has a constant magnitude with a specific value. One or more properties that function as boundary conditions constitute a boundary. Thus, as a result of the completion of causal isolation, a boundary has emerged that selects which possible combinations of parts can be realized. But this is not yet a boundary that functions as an ontological discontinuity between physical and biological order. That requires the existing whole-part relation to develop into an encaptic relation between biotically and physically qualified wholes.

What is an *encaptic relation*? Dooyeweerd defines a *whole-part relation* as a relation in which the parts are qualified by the same function as the whole and cannot function without the whole. For instance, a plant cell is a part of a plant because both are biotically qualified—i.e., both are alive. By contrast, in an *encaptic relation* entities have mutually independent qualifying functions. Their qualifying functions may be different, as in a chromosome, which is not a part of a plant cell because it is not biotically qualified (it is not alive). Rather, a chromosome is encapsulated in the cell. Or the qualifying functions may be the same as in a symbiotic relation between animals, which are sensitively qualified (based on Dooyeweerd 1969, 3:125n1, 637, 648, 695). Thus the emergence of an encaptic entity is preceded by the emergence of the spatial whole-part relation. This is consistent with the sequence of kinds of order in cosmic time according to which the spatial whole-part relation arises first in physically qualified things and is repeated in modified form in entities qualified by later modes. Repetition of the spatial whole-part relation in a form unique for each mode of existence would explain a particular kind of encaptic relation referred to by Dooyeweerd as an "encaptic whole." In it the relation between an encapsulated whole and an encapsulating whole coincides with a whole-part relation (Dooyeweerd 1969, 3:702–703).

We were asking how an ontological boundary between physical and biotic order could develop. This would require that the existing physical whole-part relation develop into an encaptic relation between biotically and physically qualified wholes. This requires that the physical parts differentiate into biological wholes. Deacon (2006a, b) proposes that this transition occurred stepwise. In each step, the random interaction of the smaller physical entities produces an extremely large number of combinations among which, occasionally, there is a rare autocatalytic combination that, while still physically qualified, has the potential of contributing to the evolutionary emergence of a biologically qualified entity (Fig. 1). Deacon (2006a, b) postulates an iteration of such random combination—some autocatalytic—producing increasingly complex physical entities until a combination emerges that operates as a living entity.[37] In this living entity continuity of ontological (physical) order co-exists with discontinuity between physical and biotic order. In Dooyeweerdian terms, we would have a biologically qualified entity that encapsulates physically qualified entities—for instance, chemical compounds that can enter into equilibrium reactions of the kind found in metabolic pathways. This biologically qualified encaptic entity selects new combinations of lower-level entities. Thereby it acquires new properties so that it can repeat this entire process and, over time, enter into new combinations, producing an ever expanding number of new entities at higher levels.

In sum, when the relative size difference between a physical whole and some of its physical parts passes a threshold of mass and size, they enter into complete causal isolation. This sets the stage for the physical parts to differentiate into living entities. These differentiating physical parts have physical parts themselves. If the latter do not pass that threshold, they will continue in direct causal interaction with the differentiating physical entity. They will end up in a biotic whole where they can become included in causal subject-object relations. Thus, ontological discontinuity between the physical and biotic kinds of order can co-exist with the ontological continuity of subject-object relations.

The overall architecture of the universe corresponds to that expected as the result of the different kinds of order that emerge. In the models of Salthe (2006) and Deacon (2006a, b), biological entities emerge from physical entities. The biological entities are located between large and small physical entities in a hierarchy of physical wholes and parts. From a physical perspective there is a hierarchy with three levels of physical constitution: the large-scale physical environment which contains the medium-scale biological entities which in turn contain the small-scale physical constituents. From a functional perspective there are two modes of functioning: the physical and the biological. They constitute a specification hierarchy (van der Meer 2011).

Discussion

The Direction of Causation

In the literature on hierarchy theory, the direction of causation is usually defined in terms of the physical whole-part relation. In that context, downward causation is the effect of a physical whole on its physical parts, and vice versa for upward causation. I refer to this as *whole-part causation* or *part-whole causation*, as the case may be. If, however, causal direction is defined in terms of Dooyeweerd's modal scale, downward causation is the effect of an entity on another entity with a lower qualification. Upward causation is the effect of an entity on another entity with a higher qualification. I refer to them as *downward* or *upward causation*, respectively.[38]

---

[37] For an experimental example of an autocatalytic process that breaks the symmetry between left-handed and right-handed molecules, see Viedma (2005).

[38] The argument that downward causation is incoherent applies only to synchronic emergence (Kim 2006). A solution for this incoherence has been proposed by Wilson (2015). Since this chapter deals with diachronic emergence, the incoherence does not apply (Welshon 2002).

In my model for causal evolutionary emergence, causation follows a specific sequence. First, a large physical whole limits the random variation of its smaller physical parts. This effect is neither upward nor downward because it is physically qualified. It is physical whole-part causation. The growing size difference eventually blocks causal interaction between the whole and its smallest parts. Second, the smallest physical parts combine in many configurations (Fig. 1), some of which differentiate over time into biological wholes. This is an upward effect from physical entities to biological entities, resulting in a biological entity that encapsulates physical entities. The causal block makes it possible for living things to maintain a different kind of order than that of the physical things from which they emerged. Third, the higher-level encapsulating entity acquires new properties over time which enable it to enter into further interactions and produce rare entities still qualified by the same mode. For instance, a single cell may acquire the ability to synthesize DNA which encodes the property of contacting other cells and forming a colony. The new properties have the same qualifying function as the encapsulating entity, but their production requires downward causation because the cell produces DNA from physical building blocks. As I suggested, the causal interactions between entities displaying different kinds of order do not erase their difference. Finally, upward causation would involve subsequent interactions between encapsulating entities that produced an entity qualified by a higher function.

Unpredictability

The number and kinds of next-generation combinations of elements in this model cannot be predicted from knowledge of the combining elements because the combinations are not numerical but depend on the characteristics of the elements (Fig. 1). Probability theory requires, among other things, that a characteristic used for prediction be homogeneous across space. This does not apply to our model of emergent evolution because the characteristics of the combining elements that are relevant for emergent evolution, whether physical or biological, are heterogeneous (Deacon 2006a, b).

There are further ontological reasons why physical theories specifically cannot always predict physical phenomena. For instance, in chaotic phenomena interaction between physical parts is deterministic. Yet one cannot predict which combinations will emerge because there are forks in the causal path. Finally, quantum phenomena are not determined and this blocks prediction.

Nor can physical theories predict biological phenomena, because the causality of a living whole is of a different kind than the causality of its physical parts. There is no *logical* connection between the interaction of the physical parts and the behaviour of the living whole. You cannot *derive* descriptions of a living thing from descriptions of its physical precursors. Thus, the law statements about the living thing are irreducible to those about the physical precursors. The latter are necessary, but insufficient for the former (van der Meer 2011).

The rarity of this transitional process can be compared with the rarity of, for instance, a succession of mutations each of which by itself has no effect while the entire succession produces cancer. The product rule of probability shows that the entire series of low probability events that produce cancer is extremely rare. In our model the probability of each combination is not known. But it is very low, given the many possibilities. Therefore, the product of the probabilities of the entire series of combinations is extremely low.

Excluded Antinomy

The principle of excluded antinomy depends on the assumption that ontological causal continuity erases the logical differences between kinds of order. This principle would be invalidated if, as I have proposed, causally continuous processes would produce causally discontinuous kinds of order rather than erase their difference. These causally discontinuous kinds of order would also be logically distinct. Thus, one could use self-contradiction as an indicator that one is dealing with different kinds of order that have emerged by evolution, as is widely recognized by those unfamiliar with Dooyeweerd's diagnosis of ontological antinomy.[39] Moreover, this would eliminate the inadequacy of unspecified notions of novelty as an indicator of the evolutionary emergence of new kinds of order. But it would not do so as an indicator of evolutionary emergence *within* a kind of order.

Nominalism

Dooyeweerd opposes attempts at reducing God's plan for the cosmos to a product of God's will, as in nominalism, or to a manifestation of God's reason, as in the rationalism of Christian Neoplatonism and Christian Aristotelianism (Dooyeweerd 1969, 3:559). He views both nominalism and rationalism as forms of idolatry with human faculties which ought not to be incorporated into our understanding of God's plan because it is beyond human understanding. One can agree without disqualifying God's will and reason or creating a false dilemma between them. My proposal avoids nominalism by locating the different kinds of lawful behaviour in the causal mechanisms that underlie the dispositions of concrete entities, events, and

---

[39] Different forms of this argument have been raised by various authors—including Charles Darwin, in a letter to W. Graham, July 3, 1881 (in Francis Darwin [1897] 2005, 1:285). See, for instance, Haldane (1932, 209); Woltereck (1940, sect. 176); Straus (1963, 298–304); Grene (1974, 42, 44); and Polanyi (1966, 38).

processes. These mechanisms can be viewed as essences of natural kinds. I do not see the dispositions as mental constructs, but as ontological manifestations of divine action, without grounding them in God's will or reason.

Divine Action

Finally, my proposal is not a form of non-reductive physicalism. I interpret the dispositions of concrete entities involved in the evolutionary emergence of different kinds of lawful order as the manifestation of God's ordering activity. In this view God is neither at the bottom of an emerging hierarchy controlling the world from the bottom up in a non-interventionist way (Pollard 1958; Russell 1988, 362; 2008, 181–183) nor at the top controlling the world from the top down (Polkinghorne 1998, chap. 3; 2000, chaps. 6 and 7). Rather, God is present at every level as Klapwijk (2008, 209) suggests. In this way the idea of creation order continues to be crucial in a Christian perspective on the cosmos.[40]

References

Bohm, David. 1952a. A Suggested Interpretation of the Quantum Theory in Terms of "Hidden Variables" I. *Physical Review* 85: 166–179.

———. 1952b. A Suggested Interpretation of the Quantum Theory in Terms of "Hidden Variables" II. *Physical Review* 85: 180–193.

Bohm, David, and Basil J. Hiley. 1993. *The Undivided Universe: An Ontological Interpretation of Quantum Theory*. Abingdon/New York: Routledge.

Boulter, Stephen J. 2012. Can Evolutionary Biology do Without Aristotelian Essentialism? *Royal Institute of Philosophy Supplement* 70: 83–103.

Boyd, Richard N. 1991. Realism, Anti-Foundationalism and the Enthusiasm for Natural Kinds. *Philosophical Studies: An International Journal for Philosophy in the Analytic Tradition* 61 (1 and 2): 127–148.

———. 1999. Kinds, Complexity and Multiple Realization: Comments on Millikan's "Historical Kinds and the Special Sciences." *Philosophical Studies: An International Journal for Philosophy in the Analytic Tradition* 95 (1 and 2): 67–98.

Brüggemann-Kruyff, Atie Th. 1981. Tijd als omsluiting, tijd als ontsluiting (I). *Philosophia Reformata* 46 (2): 119–163.

———. 1982. Tijd als omsluiting, tijd als ontsluiting (II). *Philosophia Reformata* 47 (1): 41–68.

Bunge, Mario. 1963. *Causality*. Cleveland: World Publishing Company.

Cheyne, Colin. 2001. *Knowledge, Cause, and Abstract Objects: Causal Objections to Platonism*. Dordrecht: Kluwer Academic Publishers.

Copan, Paul, and William Lane Craig. 2004. *Creation out of Nothing: A Biblical, Philosophical, and Scientific Exploration*. Grand Rapids/Leicester: Baker Academic/Apollos.

Darwin, Francis, ed. (1897) 2005. *The Life and Letters of Charles Darwin*. 2 vols. Boston: Elibron.

---

[40] I thank Roy Clouser, Harry Cook, Henk Geertsema, Gerrit Glas, Dick Stafleu, Richard van Holst, Uko Zylstra, and two anonymous reviewers for comments on an earlier draft. Failures are mine.

Deacon, Terrence W. 2006a. Emergence: The Hole at the Wheel's Hub. In *The Re-Emergence of Emergence*, ed. Philip Clayton and Paul Davies, 111–150. Oxford: Oxford University Press.

———. 2006b. Reciprocal Linkage Between Self-Organizing Processes is Sufficient for Self-Reproduction and Evolvability. *Biological Theory* 1 (2): 136–149.

Dengerink, Jan D. 1977. Ontisch of/en intentioneel? Een bijdrage tot de discussie inzake de aard en structuur van het wetenschappelijk denken binnen de reformatorische wijsbegeerte. *Philosophia Reformata* 42: 13–49.

Dooyeweerd, Herman. 1926. Calvinisme contra neo-Kantianisme. *Tijdschrift voor wijsbegeerte* 20: 29–74.

———. 1928. Het juridisch causaliteitsprobleem in 't licht der wetsidee. *Antirevolutionaire Staatkunde* 2: 21–124.

———. 1935–1936. *De wijsbegeerte der wetsidee*. 3 vols. Amsterdam: H.J. Paris.

———. 1936. Het tijdsprobleem en zijn antinomieën op het immanentiestandpunt. *Philosophia Reformata* 1: 65–83.

———. 1939. Het tijdsprobleem en zijn antinomieën op het immanentiestandpunt II. *Philosophia Reformata* 4 (1): 1–28.

———. 1940a. Het tijdsprobleem in de wijsbegeerte der wetsidee I. *Philosophia Reformata* 5: 160–182. English translation: The Problem of Time in the Philosophy of the Law-Idea. Trans. J. Glenn Friesen. https://jgfriesen.files.wordpress.com/2016/12/tijdsprobleem.pdf. Accessed 13 Jan 2017.

———. 1940b. Het tijdsprobleem in de wijsbegeerte der wetsidee II. *Philosophia Reformata* 5: 193–234. English translation: The Problem of Time in the Philosophy of the Law-Idea. Trans. J. Glenn Friesen. https://jgfriesen.files.wordpress.com/2016/12/tijdsprobleem.pdf. Accessed 13 Jan 2017.

———. 1950. Het substantiebegrip in de moderne natuur-philosophie en de theorie van het enkaptisch structuurgeheel. *Philosophia Reformata* 15: 66–139.

———. 1959. Schepping en evolutie. *Philosophia Reformata* 24: 113–159.

———. 1966. *Creation and Evolution. Translation of Dooyeweerd (1959) by Adrian Kooymans.* http://www.reformationalpublishingproject.com/pdf_books/Scanned_Books_PDF/CreationandEvolution.pdf. Accessed 10 July 2017.

———. 1969. *A New Critique of Theoretical Thought*. 4 vols. Philadelphia: The Presbyterian and Reformed Publishing Company.

Dowe, Phil. 2008. Causal Processes. In *The Stanford Encyclopedia of Philosophy*, ed. Edward N. Zalta. http://plato.stanford.edu/archives/fall2008/entries/causation-process/. Accessed 13 Jan 2017.

Geertsema, Henk. 2011. Emergent Evolution? Klapwijk and Dooyeweerd. *Philosophia Reformata* 76: 50–76.

Grene, Marjorie. 1974. Biology and the Problem of Levels of Reality. In *The Understanding of Nature. Essays in the Philosophy of Biology*, Boston Studies in the Philosophy of Science 23, 35–52. Dordrecht/Boston: Reidel Publishing Company. First published in 1967 in The New Scholasticism 41: 94–123.

Haldane, John B.S. 1932. *Possible Worlds and Other Essays*. London: Chatto and Windus.

Hart, Hendrik. 1973. Problems of Time. An Essay. *Philosophia Reformata* 38 (1): 30–42.

———. 1985. Dooyeweerd's Gegenstand Theory of Theory. In *The Legacy of Herman Dooyeweerd: Reflections on Critical Philosophy in the Christian Tradition*, ed. C.T. McIntire, 143–166. Lanham: University Press of America.

Henderson, Roger D. 1994. *Illuminating Law: The Construction of Herman Dooyeweerd's Philosophy 1918–1928*. Amsterdam: Buijten & Schipperheijn.

Jaeger, Lydia. 2012. *What the Heavens Declare: Science in the Light of Creation*. Eugene: Cascade Books.

Kim, Jaegwon. 2006. Emergence: Core Ideas and Issues. *Synthese* 151: 547–559.

Klapwijk, Jacob. 2008. *Purpose in the Living World? Creation and Emergent Evolution*. Cambridge: Cambridge University Press.

———. 2011. Creation Belief and the Paradigm of Emergent Evolution. *Philosophia Reformata* 76: 11–31.
———. 2012. Nothing in Evolutionary Theory Makes Sense Except in the Light of Creation. *Philosophia Reformata* 77: 57–77.
Koons, Robert C., and Timothy H. Pickavance. 2015. *Metaphysics: The Fundamentals*. Oxford: Wiley Blackwell.
Leman, L., L. Orgel, and M.R. Ghadiri. 2004. Carbonyl Sulfide-Mediated Prebiotic Formation of Peptides. *Science* 306: 283–286.
Mackie, John L. 1974. *The Cement of the Universe*. Oxford: Oxford University Press.
McIntire, C.T. 1985. Dooyeweerd's Philosophy of History. In *The Legacy of Herman Dooyeweerd: Reflections on Critical Philosophy in the Christian Tradition*, ed. C.T. McIntire, 81–117. Lanham: University Press of America.
O'Connor, Timothy, and Hong Yu Wong. 2015. Emergent Properties. In *The Stanford Encyclopedia of Philosophy*, ed. Edward N. Zalta. http://plato.stanford.edu/archives/sum2015/entries/properties-emergent/. Accessed 13 Jan 2017.
Olthuis, James H. 1985. Dooyeweerd on Religion and Faith. In *The Legacy of Herman Dooyeweerd: Reflections on Critical Philosophy in the Christian Tradition*, ed. C.T. McIntire, 21–40. Lanham: University Press of America.
Polanyi, Michael. 1966. *The Tacit Dimension*. Garden City: Doubleday & Company, Inc.
Polkinghorne, John. 1998. *Belief in God in an Age of Science*. New Haven/London: Yale University Press.
———. 2000. *Faith, Science and Understanding*. New Haven/London: Yale University Press.
Pollard, William G. 1958. *Chance and Providence*. New York: Charles Scribner's Sons.
Rosen, Gideon. 2014. Abstract Objects. In *The Stanford Encyclopedia of Philosophy*, ed. Edward N. Zalta. http://plato.stanford.edu/archives/fall2014/entries/abstract-objects. Accessed 13 Jan 2017.
Russell, Robert John. 1988. Quantum Physics in Philosophical and Theological Perspective. In *Physics, Philosophy and Theology: A Common Quest for Understanding*, ed. Robert John Russell, William R. Stoeger, and George V. Coyne, 343–374. Vatican City State/Notre Dame: Vatican Observatory/University of Notre Dame Press.
———. 2008. *Cosmology: From Alpha to Omega*. Minneapolis: Fortress Press.
Saghatelian, Alan, Yohei Yokobayashi, Kathy Soltani, and M. Reza Ghadiri. 2001. A Chiroselective Peptide Replicator. *Nature* 409: 797–801.
Salmon, Wesley C. 1998. *Causality and Explanation*. New York/Oxford: Oxford University Press.
Salthe, Stanley N. 1985. *Evolving Hierarchical Systems*. New York: Columbia University Press.
———. 2002. Summary of the Principles of Hierarchy Theory. *General Systems Bulletin* 31: 13–17.
———. 2006. Two Frameworks for Complexity Generation in Biological Systems. In *Evolution of Complexity*, ed. Carlos Gershenson and Tom Lenaerts, 99–104. Bloomington: Indiana University Press. http://www.nbi.dk/natphil/salthe/A-life_Conf_paper_Word.pdf. Accessed 13 Jan 2017.
Sartenaer, Olivier. 2013. Neither Metaphysical Dichotomy nor Pure Identity: Clarifying the Emergentist Creed. *Studies in History and Philosophy of Biological and Biomedical Sciences* 44 (3): 365–373.
———. 2016. Sixteen Years Later: Making Sense of Emergence (Again). *Journal for General Philosophy of Science/Zeitschrift für Allgemeine Wissenschaftstheorie* 47 (1): 79–103.
Sober, Elliott. 1980. Evolution, Population Thinking, and Essentialism. *Philosophy of Science* 47 (3): 350–383.
Soskice, Janet. 1985. *Metaphor and Religious Language*. Oxford: Oxford University Press.
Stafleu, Marinus D. 2002. *Een wereld vol relaties. Karakter en zin van natuurlijke dingen en processen*. Amsterdam: Buijten & Schipperheijn. English translation: A World Full of Relations: Character and Meaning of Natural Things and Events. Trans. M.D. Stafleu. https://www.scribd.com/document/29057727/M-D-Stafleu-A-World-Full-of-Relations. Accessed 13 Jan 2017.

———. 2008. Time and History in the Philosophy of the Cosmonomic Idea. *Philosophia Reformata* 73: 154–169.
Steen, Peter J. 1983. *The Structure of Herman Dooyeweerd's Thought*. Toronto: Wedge.
Straus, Erwin. 1963. *The Primary World of Senses*. New York: Free Press of Glencoe.
Strauss, Danie. 1984. An Analysis of the Structure of Analysis. *Philosophia Reformata* 49 (1): 35–56.
Thaxton, Charles B., Walter L. Bradley, and Roger L. Olsen. 1984. *The Mystery of Life's Origin: Reassessing Current Theories*. New York: Philosophical Library.
van der Meer, Jitse M. 2011. Stratified Cosmic Order: Distinguishing Parts, Wholes, and Levels of Organization. In *Science and Faith Within Reason: Reality, Creation, Life and Design*, ed. Jaume Navarro, 145–164. Farnham/Burlington: Ashgate.
Van Woudenberg, René. 1992. *Gelovend denken*. Amsterdam: Buijten & Schipperheijn.
Verburg, Marcel E. 1989. *Herman Dooyeweerd: Leven en werk van een Nederlands christenwijsgeer*. Baarn: Ten Have.
Viedma, Cristobal. 2005. Chiral Symmetry Breaking During Crystallization: Complete Chiral Purity Induced by Nonlinear Autocatalysis and Recycling. *Physical Review Letters* 94, 065504: 1–4. http://eprints.ucm.es/16555/1/e065504.pdf. Accessed 13 Jan 2017.
Vollmert, Bruno. 1983. *Polykondensation in Natur und Technik*. Karlsruhe: E. Vollmert-Verlag.
Walsh, Denis. 2006. Evolutionary Essentialism. *British Journal for the Philosophy of Science* 57: 425–448.
Welshon, Rex. 2002. Emergence, Supervenience, and Realization. *Philosophical Studies* 108: 39–51.
Wilkins, John S. 2013. Essentialism in Biology. In *The Philosophy of Biology*, ed. Kostas Kampourakis, 395–419. Dordrecht: Springer.
Wilson, Jessica. 2015. Metaphysical Emergence: Weak and Strong. In *Metaphysics in Contemporary Physics*, Poznań Studies in the Philosophy of the Sciences and the Humanities 103, ed. Tomasz Bigaj and Christian Wüthrich, 251–306. Amsterdam/New York: Rodopi.
Woltereck, Richard. 1940. *Ontologie des Lebendigen*. Stuttgart: F. Enke.
Zylstra, Uko. 2004. Intelligent-Design Theory: An Argument for Biotic Laws. *Zygon* 39 (1): 175–191.

Creation Order and the Sciences of the Person

Gerrit Glas

Abstract In this chapter, conceptions of order in neuroscience and psychology are compared and assessed from the perspective of a philosophy of creation order with its strong view on laws as necessitating principles, or conditions. Three questions guide the discussion: (1) Does it make a difference for the sciences of the person to maintain a strong notion of law? (2) Can the apparent tension between the creation order view and evolutionary accounts of lawfulness and order be diminished by employing the concept of emergence? (3) Can the concept of emergence be made compatible with a strong concept of law? I explore whether, and if so, to what extent, Herman Dooyeweerd's reformational philosophy—representative of the creation order approach—can accommodate evolutionary accounts of lawfulness. After a critical analysis of the emergence approaches of Philip Clayton and Evan Thompson, I conclude that the term *emergence* is used ambiguously, referring both to some sort of causal activity and to more abstract principles of self-organization. With respect to the three guiding questions, it is concluded that (1) a strong notion of law (or lawfulness) can play an important role in the struggle against reductionism in the sciences of the mind and the brain; (2) accommodation between the creation order view and evolutionary accounts is possible to a certain extent and the concept of emergence might play a role in this accommodation; (3) *emergence* should in that case be interpreted as a boundary concept.

Keywords Emergence · Laws · Reformational philosophy · Creation order · Self-organization · Structural principle · Evolutionary theory · Herman Dooyeweerd · Evan Thompson · Philip Clayton

---

The original version of this chapter was revised.
An erratum to this chapter can be found at https://doi.org/10.1007/978-3-319-70881-2_16

G. Glas (✉)
Department of Philosophy, Vrije Universiteit Amsterdam, Amsterdam, The Netherlands
e-mail: g.glas@vu.nl

Introduction

The aim of this chapter is to discuss the notion of creation order and its future in the context of the sciences of the person, especially neuroscience and psychology. As described in the introductory chapter, the concept of creation order is strongly associated with the idea of a lawful order, an order which is both preexistent (i.e., functions from the very beginning of the universe) and ordained by God in the act of creation. Divine ordaining brings the notion of law that is implied in the concept of creation order close to what in philosophy of science has become known as the *necessity view* on laws. According to the necessity view, the relation between the events (or entities) that are identified by a law holds with necessity, which means that it holds under all circumstances, in all possible worlds (Armstrong 1985; Lange 2009; Roberts 2008). Laws, in other words, specify the nomic conditions under which certain relations between things hold and sequences of events occur. In the creation order view the nomic nature of these relations/sequences is founded in the will of God. I will call the notion of law that is implied in the term *creation order* "strong" because of its association with both the preexistence and the necessity of laws.

This strong notion of law seems at odds with conceptions of order in neuroscience and psychology, which tend to be more pragmatic and patchy (or *dappled*, to borrow a term used by Cartwright [1999]). The majority of these conceptions is based on, or at least related to, evolutionary theory (Cartwright 1992; Kendler 2012). What comes closest to laws and the idea of a lawful order in these disciplines are such concepts as function (McLaughlin 2001), pattern (Sober 2000), mechanism (Bechtel 2008; Craver 2007), model (Shaffner 1993), system (Thompson 2007), and property cluster (Kendler et al. 2011). Mechanisms, functions, patterns, and the like, are typically conceived as having emerged from a contingent process of evolution. They are, in other words, not preexistent, nor necessary, let alone backed up by the will of a divine being at the beginning of the universe.

There is a lot of discussion in philosophy of biology about the conceptual status of these terms (Rosenberg and McShea 2008; Davies 2001; Kitcher 1998; Godfrey-Smith 1998). So much is clear that these functions, mechanisms, and patterns are, generally speaking, considered not to refer to necessitating powers but to relations between properties or entities or events—relations with a sufficient degree of regularity and explanatory relevance. These terms have an explanatory function and help to order and interpret the empirical material.

The discrepancies between the creation order view and the pragmatic instrumentalism with respect to lawfulness that is characteristic for neuroscience and psychology lead to three mutually related questions that will play a role in the background of this chapter. These questions will pop up regularly and require our attention, though not necessarily in each section or in the order in which I present them now. One question is whether it makes a difference for the sciences of the person to maintain a strong notion of law—strong in the sense that laws are considered to be preexistent and to function with necessity. A second question is whether the tension between the creation order view and evolutionary accounts of lawfulness and order, insofar as relevant for neuroscience and psychology, can be diminished by employing the concept of emergence. More specifically, and thirdly, there is the question whether the concept of emergence is, or can be made, compatible with a strong concept of law. If it is (or can be), the tension mentioned earlier is dissolved. If it is not (or cannot be), would that imply that the concept of law should be revised? Or should the notion of emergence be given up? Or would it be better to stick to a pre-evolutionary worldview?

We will begin with Dooyeweerd in the following section. I will focus on how his systematic philosophy deals with evolutionary theory. In the analysis of his position the three background questions will recur. Next, the attention will shift toward what thinkers of other traditions have said about emergence; I will mainly discuss Philip Clayton's *Mind and Emergence* and Evan Thompson's *Mind in Life*. Finally, I will demonstrate how far the accommodation reaches between a revised Dooyeweerdian account of creation order and the concept of emergence.

Dooyeweerd, Creation Order, and Evolutionary Theory

Creation and Temporal Becoming; Worldview and Theory

Dooyeweerd is well known for his strong concept of law and of creation order. He combined a creation order view on reality with a positive attitude towards the empirical sciences, biology included. He considered evolutionary theories as compatible with a creation view, because—in his opinion—there exists a fundamental distinction between *creation*—a divine act that in itself is not accessible for the human mind—and *temporal becoming*—the process of unfolding of all creational structures in the course of billions of years, a process that is open for all sorts of empirical investigation. Creation lies beyond the bounds of human understanding, whereas temporal becoming refers to processes that are in principle accessible for empirical investigation.

Alongside this fundamental distinction there is another one that was supportive for a positive account of evolutionary theory: the distinction between *evolutionary theory* as a set of empirically testable hypotheses and *evolutionism* as a worldview that transcends the framework of empirical science. Dooyeweerd adopted views based on evolutionary theory while rejecting evolutionism as worldview.

It is questionable, however, whether the distinction between theory and worldview is as clear-cut as Dooyeweerd tried to maintain. It is at the heart of the concept of creation order to presuppose the preexistence of laws, i.e., the existence of laws from the very beginning of the universe, before the unfolding of temporal structures. However, it is at the heart of evolutionary theory—Darwin was aware of this when he wrote the *Origin of Species*—that the order of creation is *contingent*; that is, it could have been otherwise in the most fundamental sense. How is this contingency compatible with a notion of creation order (cf. the first background question)? One has to either give up the idea of the preexistence of a framework of laws founded in God's creative acts or transform the classical account of evolutionary theory into a theory that supposes that, whatever new structure would emerge in the course of evolution, it would always in some way express a deeper, preexistent order.

Dooyeweerd's Solution

In his discussions with biologist Jan Lever and philosophers of biology Herman Driesch and Jan Woltereck, Dooyeweerd tried to find a third way—i.e., a way between either giving up the idea of preexistent laws or requiring that evolutionary processes, in spite of their contingency, respond to creational laws and principles. He did this (a) by making a distinction between *natural law* as it is known in the natural sciences and *law* as a much more abstract, philosophical boundary concept referring to the coherence, diversity, unity, wholeness, and dependence (on an origin) of all temporal reality; and (b) by emphasizing the close "collaboration" and interdependence of law- and subject-side[1] of reality.

The first and I think most important move consisted in making a distinction between *scientific laws*, i.e., laws such as are discovered and described by the sciences, and *law* as a transcendental boundary concept. This distinction allowed him to salvage the notion of creation order by seeing this order as more fundamental than the fallible scientific interpretations of the laws that are maintained by this order. Interpretations of the law may be contingent, but the order itself is not (see also Jaeger 2008, 2010). The order of creation (or law as a philosophical boundary concept) holds for reality, it conditions it; but the way it holds or conditions is conceptually opaque. Law is, according to this strand of thought in Dooyeweerd's thinking, ultimately a religious idea at the boundary of our knowing faculties. It is—as law-idea—the theoretical intuition we have of the coherence, diversity, unity, wholeness, and dependence on an origin of created reality.[2]

It is because of this (transcendental) status of the law-idea that evolutionary theory, with its emphasis on continuity and contingence, cannot be a threat to creation order. In fact, for Dooyeweerd, it would mean a serious logical flaw to assume that the continuity and contingency of evolutionary processes contradict the notion of a preexisting and non-contingent creation order (with its laws). The holding of these laws, the fact that they condition things in the world and that they make things possible, can never itself be explained on the basis of processes that these laws condition or make possible. Laws are always already presupposed in the understanding of the

---

[1] The term *subject-side* refers to reality as "being sub-jected (Latin: *sub iactare*) to the law" and not to the epistemological subject or ego.

[2] Not everyone in the tradition of reformational philosophy has been very happy with this solution. Van Riessen, for instance, gave up the entire transcendental framework and opted for a connection between the philosophical intuition of lawfulness and the biblical notion of wisdom and its relatedness to order.

processes that they are thought to produce. They can, therefore, never be the product of these processes.

From this perspective, *development*, *continuity*, and *emergence* are terms that refer to the factual side (or subject-side) of reality. As descriptive terms they can only be understood on the basis of a (transcendental) pre-understanding of constancy, discreteness, and diversity. There is, in other words, a conceptual gap between law- and subject-side: one can never, on the basis of characteristics of what is under the law, conclude something about the ontological status of the law itself.[3] The continuity and contingence of processes that are subject to the law as boundary concept do not imply that laws themselves are continuous and contingent.

Moving to the second issue, to wit, the collaboration and mutual dependence of the law- and subject-side of reality, it should be noted that this theme was only later developed in Dooyeweerd's thought. The new terminology was clearly meant to avoid a static and fixed view on creation order—as if this order were some cosmic blueprint. Generally speaking, the new emphasis on interaction and interdependence allowed Dooyeweerd to speak about development not only at the subject-side but also at the law-side of reality.[4] However, as far as I know he never applied this idea to the theory of evolution, nor did he address the problems that—obviously—were introduced with this move: the problem of how to combine order with change, and the problem of the arbitrariness that inevitably creeps into the system if one is going to make distinctions between laws that can change and those that cannot.[5]

After Dooyeweerd

Most reformational thinkers after Dooyeweerd have remained faithful to his original position: all laws, including biotic laws, exist prior to the realities in which these laws express themselves. Laws are dormant, so to say, until the relevant evolutionary processes take place by which they are awakened. In the work of Dick Stafleu, for instance, there is full recognition of the continuity of the evolutionary process with its different types of emerging properties and relations (cf. Stafleu, chapter "Properties, Propensities, and Challenges: Emergence in and from the Natural World," this volume). However, this continuity does not detract from the distinctness of biotic and other laws. These laws are not reducible to physical laws.[6] Danie

---

[3] This point has been made time and again by Danie Strauss and other reformational philosophers.

[4] Not so much, however, as to allow evolution (change) of species (Dooyeweerd 1959). For the issue of development at the law-side, see Van der Hoeven (1981).

[5] There have even been speculations that Dooyeweerd never fully developed his anthropology because of the difficulties he encountered in his thinking on evolutionary theory (cf. Glas 2010).

[6] "The continuity of the evolution (the subjective process of becoming) in cosmic time does not imply that biotic laws, characters, and types of subject-subject relations and subject-object relations are reducible to physical or chemical ones" (Stafleu 2002, 7; see also Stafleu 1999).

Strauss (2009, 479ff.) emphasizes the point mentioned earlier that continuity in the process of origination can never explain the discontinuity in structure. When emergent evolutionists want it both ways (newness of structure as a result of a genetic process of becoming), they burden their position with the inner antinomy we discussed a moment ago: laws cannot originate from processes that presuppose them. Continuity can only be understood on the basis of a prior understanding of discontinuity; change only on the basis of constancy.

Recently, however, reformational philosopher Jacob Klapwijk has defended a view on creation and laws that is compatible with an emergent view on evolution. He does so by making a division between two types of laws. Laws are needed in the reformational account of emergent evolution, because the newness of emerging levels of functioning and organization depends on the origination and development of new laws. So, there are laws that exist from the earliest beginning and there are laws that originate and develop during the process of evolution. The division between these two types of laws parallels the (Dooyeweerdian) distinction between modal laws and type laws. Modal laws are laws that hold for modal aspects or modes of functioning (the *how* of things/events) of reality as a whole. These laws are unchangeable and exist from the moment of creation in Klapwijk's view. They are the deepest laws and hold for reality in its entirety. Type laws hold for particular domains or kinds (or classes) of objects. They may be subjected to change. Laws that characterize species, for instance, are type laws; they may therefore change.

How does Klapwijk deal with the inner antinomy we just discussed? Change, according to Klapwijk, is compatible with the idea of creation order by adopting an Augustinian conception of time.

> Time is a process of disclosure of all the potential that God has enclosed in His creation, thus also the disclosure of creaturely orderings. Nothing forces us to conclude that ... emergent domains with their idionomic [i.e., type law-based] regimes are brought about by bottom-up causation on the basis of physical laws.... Why don't we consider these orderings to be realized by time, yes, as successive expressions of what have been the holy intentions of the creator from the very beginnings? (Klapwijk 2008, 209)

Creation order becomes here the potential that God has enclosed in his creation. The successive expressions of this potential may be interpreted as corresponding with the "holy intentions" of the Creator.

But what about the law–subject distinction, which is so fundamental for Dooyeweerd? Does it function with respect to type laws? Klapwijk says about this that idionomy—which is understood as lawfulness in the functioning of particular types or classes of objects—goes hand in hand with self-organization, i.e., the bottom-up process of interaction between subsystems that leads to the emergence of new phenomena. Idionomy and self-organization are two sides of the same coin (Klapwijk 2008, 121). This formulation seems to echo the distinction between law- and subject-side of the later Dooyeweerd. However, *lawfulness* is not the same as *functioning as law*. *Lawfulness* can be interpreted as representing a variant of the regularity view on laws. The regularity view has no place for the holding of laws. When the lawfulness of idionomic structures is interpreted in terms of the regularity view, the distinction between idionomy and self-organization cannot be equated with the law–subject distinction. However, I am not certain about this interpretation, so I have no clear answer to the question whether idionomy and self-organization correspond with law- and subject-side of emergent processes.

At first sight, the difference between Dooyeweerd and Klapwijk seems a matter of degree. For Dooyeweerd, change at the biotic level stops at the level of species (Dooyeweerd 1959). Klapwijk allows change for all type laws. Species definitions therefore don't refer to fixed orderings. However, Klapwijk seems to deviate from Dooyeweerd also in a more substantial sense; not only by being unclear about the relevance of the law–subject distinction for emergent processes, but also with his Augustinian conception of creation order. This Augustinian inspiration offers, so to say, the hermeneutic and theological background for the theory of emergent evolution (Klapwijk 2008, 210; see also Klapwijk 2011).

It may be questioned whether this time-conception is mature enough to respond to legitimate philosophical concerns. One of these concerns is how the concept of germinative principles laid down at the beginning of the universe fits with an evolutionary narrative that highlights contingency, unpredictability, and deep context dependence. Is, in other words, the implicit telocentrism of these germinative principles compatible with the randomness and contingency of evolutionary processes? Another issue is historical and has been raised by Gousmett (2011), who wonders how to take account of Stoic influences on Augustine's time-conception. He refers particularly to speculations about the so-called *logos spermatikos* and about divine rational seeds that were put into creation at the instant of creation. We leave this issue here, because it would lead us too far from our primarily conceptual concerns.

Upshot to Self-Organization

To conclude this section and anticipating a topic that will be discussed more extensively later, I believe part of the lack of clarity that surrounds Klapwijk's notion of emergent evolution results from unresolved questions with respect to the concept of self-organization. *Self-organization* is usually taken to denote factual processes, describing *processes in time*. This use of the term refers to the subject- or factual side of reality, to the causal history, so to say, or to the process by which interaction between parts leads to the formation of new and larger wholes. However, in other contexts the term *self-organization* is used in a different sense. There, it refers to a *principle* that governs the process of the unfolding of structures; a principle that often is cast in the form of mathematical formulations. In empirical accounts on self-organization, it is not always clear which meaning of the term is used.[7] I would like to add that self-organization should be seen as a boundary concept within the

---

[7] Thompson (2007, 39) says something similar about the term *dynamic system*. This term refers on the one hand to actual systems in the world, like the solar system, but on the other hand to the mathematical models of these systems.

sciences, i.e., as an interpretation of what ultimately and in a more fundamental way escapes from scientific understanding.

The problem of conflating the different uses of the term *self-organization* is relevant for the idea of emergent evolution. This can be illustrated by our previous discussion on the (im)possibility of the emergence of new laws. There, we said that the holding of laws, the fact that they make things possible, can never itself be explained on the basis of processes that are made possible and are conditioned by these laws. Laws are always already presupposed in the understanding of the processes that they are thought to produce. They can, therefore, never be the product of these processes.

Something similar holds for self-organization. As a principle it explains and conditions the process of self-organization. However, this works only one way: the process of self-organization does not explain the holding of the principles of self-organization. Therefore, defending the idea of emergent evolution with a logic that is based on the concept of self-organization does not bring us any further with respect to the original question of how to account for lawful principles that guide the process of evolution. Claiming that the evolutionary paradigm is compatible with a strong notion of emergence (i.e., with ontological newness) and that this notion of emergence is compatible with an Augustinian version of the creation order view *because of* the role of self-organization in evolutionary processes is begging the question. The process cannot be the explanation for the holding of the principle: self-organization as a process does not explain the emergence of new (idionomic) structures. *Self-organization* is a mixed philosophical/empirical (nomic/descriptive) concept that offers a philosophical framework for the understanding of emergent phenomena; a framework that guides our scientific imagination but leaves an important philosophical issue unresolved, namely, how to account for its functioning as a principle. Something similar holds for emergence. It is not an explanatory, but primarily a philosophical concept. As an idea, or paradigm, it opens our imagination. But it is also the cloth that conceals our explanatory ignorance.[8]

In what follows it will remain important to distinguish between law-side and factual side and between scientific, philosophical, and worldview-based uses of the overarching key concepts such as emergence, self-organization, law, and order.

---

[8] Henk Geertsema (2011, 63) combines both points (i.e., that laws cannot be the product of a development that itself presupposes their functioning, and that emergence suggests but does not really explain new orderings) at the end of a discussion of the relation between law and emergent evolution in Klapwijk's book, when he says: "The suggestion of an immanent development, which is implied in the term 'emergent evolution' and which is expressed by terms like 'self-transcendence' and 'self-organization,' can hardly be combined with the recognition of a creational order that is the very condition for this development."

Emergence and Creation Order: Are They Compatible in the Sciences of the Person?

In this section I will delve deeper into the concept of emergence and its importance for the sciences of the person. I have chosen two representatives of the emergence paradigm: the philosopher of religion Philip Clayton and the phenomenologist and dynamical systems theorist Evan Thompson. I will discuss their work in an attempt to get a better understanding of the compatibility of the concepts of emergence and creation order.

Clayton on Emergence

Emergence means that "at each level of complexity, new and often surprising qualities emerge that cannot, at least in any straightforward manner, be attributed to known properties of the constituents" (Davies 2006, x). The concept of emergence is therefore—"roughly speaking" (ibid.)—related to part-whole holism, i.e., the idea that increasingly complex interactions between parts unpredictably may give rise to new properties that cannot be accounted for in terms of individual parts. According to most emergence thinkers, the new patterns and regularities that arise are irreducible to lower level processes and exert "causal" influence on those lower levels. This position is known as *strong emergentism*. Adherents of strong emergence maintain that over the course of evolutionary history, genuinely new causal agents or causal processes come into existence. These causal agents or processes are "real" because they exert causal influence on lower levels. This is what is meant with the term *top-down causation* (a term originally coined by Campbell [1974]). Strong emergence is emergence with an ontological claim: the emerging level is ontologically new and irreducible because it exerts a causal influence that cannot be reduced to the systems out of which the emerging level evolves.

Weak emergentists insist that "as new patterns emerge, the fundamental causal processes remain, ultimately, physical" (Clayton 2006, 7). Weak emergence coincides with epistemological emergence: the new structures only appear to us as new and irreducible—they are not really (or: ontologically) novel, but only at the level of description (or: epistemically). Weak emergence therefore coincides with non-reductive physicalism—i.e., the idea that supervenient levels merit description in their own terms and are epistemologically distinct, but do not violate the principle of causal closure of the physical (Clayton 2006, 26–27; Kim 1993).

Clayton on Different Forms of Irreducibility

To be more precise, Clayton distinguishes between explanatory, causal, and ontological irreducibility (Clayton 2006, 310). When weak and strong emergence are mapped onto this threefold distinction, the line should be drawn between explanatory and causal irreducibility. Explanatory irreducibility is compatible with weak emergence, according to Clayton. Causal emergence is, as we have seen, emergence with an ontological claim. Causal and ontological irreducibility should, therefore, be grouped together. They are compatible with strong emergentism. One can simply not imagine emergent levels having causal influence on lower levels without implying ontological independence or irreducibility of the emerging level, according to Clayton.

I have doubts about this. First, it is problematic to say that levels have a causal influence on other levels. Levels are abstractions. The term *level* refers to (ontologically distinct) kinds of properties or modes of being/functioning. Levels should not be reified into (collections of) things that exert influence on other things. Clayton is, of course, not blind to these difficulties. He acknowledges that there are different types of causality. He speaks about family resemblances between different forms of causality. But he also clearly defends a form of interlevel causation. Interlevel causation is needed, he says, in order to break the hegemony of an exclusively (micro) physical approach to causation. We need interlevel causation, furthermore, to preserve the view that newly emerging properties (and relations) are real and exert a real causal influence on subvenient levels. As we will see in our discussion of Thompson, there are other ways of defending ontological irreducibility than by claiming top-down causation. Thompson and Dooyeweerd belong to those philosophers who choose other routes.

Secondly, without going too much into a discussion on causality, it seems reasonable to say that it depends on one's conception of causality whether top-down causation requires strong emergence. Nancey Murphy (1998), for instance, would say that there is only real causality at the lowest, physical level.[9] Like many others, she endorses the thesis of causal closure of the physical. She nevertheless defends a variant of top-down causation by describing how constraints at supervenient levels exert influence on subvenient levels. Contrary to Clayton, she considers such whole-part constraining as a form of causality; a form for which explanatory (and not ontological or causal) irreducibility of the supervenient level is sufficient. Clayton denies the thesis of the causal closure of the physical.[10] He sees whole-part constraining as a non-causal kind of determination and denies that it can replace strong emergence.[11] I am not suggesting that I agree with Murphy. My point is that

---

[9] Murphy endorses a non-reductive physicalist ontology and defends her positive account of downward causation with a variant of whole-part constraining. To maintain her physicalist ontology, she has to allow that higher order constraints can be redefined in terms of lower level boundary conditions, structures, or causal processes (Murphy 2006, 238–242). Her notion of causal closure of the physical boils down to a very weak form of determinism that says no more than that physical events by definition are preceded by physical events.

[10] Sikkema (2005) rightly notes that non-reductive physicalists usually refer to completely outdated (materialistic) conceptions of physics. The twentieth-century history of physics, however, is a history of dematerialization. From a reformational philosophical perspective it could be added that the thesis of the causal closure of the physical is the typical result of a neglect of the distinction between law and subject. Physical laws hold for reality as a whole, but that does not mean that all reality is (only and entirely) physical.

[11] There is some unclarity with respect to Clayton's view on causation. Strong emergentism presupposes not only a realist metaphysics but also a realistic view on causation. But Clayton does not

top-down causation does not guarantee ontological irreducibility, given the fact that top-down causation is also defended by a weak emergentist like Murphy.

Emergence thinking is usually depicted as a third way between reductionism and dualism. The bottom line of emergentism is that with the emergence of each new level the functioning of the system as a whole changes. The change in the functioning of the system is described in terms of causal influence of the whole on its constituent parts. Note that the adjective *causal* is almost redundant in this formulation. It does not add very much to what already is expressed by the substantive *influence*. *Causal* merely suggests that the influence has a direction.

Clayton on Laws

If it is not causality that explains ontological irreducibility, how can ontological irreducibility then be explained? Are laws the answer? Clayton indeed offers a plea for recognition of the importance of laws (Clayton 2004, 53ff.). This plea aims at preservation of the irreducibility of levels. Each level is "defined by the existence of distinct laws and by distinct types of causal activity at that level" (ibid., 52). By functioning according to their own laws, higher levels exert influence on lower levels and thereby on the system as a whole. However, as a rule, these lower levels keep functioning according to their own laws (or types of causal activity). Clayton describes the relationship between the levels as a relationship of (inter)dependence (ibid., 53), each level depending on a lower level in a different way.

I must admit that this is somewhat confusing. Interlevel dependence is not the same as interlevel causation. If interlevel causation is identical with interlevel dependence, then causation is clearly not a process in time. This is indeed what Clayton seems to imply; and maybe better so. Because if he would view top-down causation as a process in time, this would bring his position dangerously close to ontological separation between the levels, i.e., to a form of dualism or pluralism. However, given what Clayton says elsewhere, I take it that causation refers not primarily to a process in time but to immediate top-down influencing on the basis of relationships of dependence between the levels. Thus conceived, top-down causation becomes the reverse of emergence. Emergence as a factual process can be described empirically but not explained. Likewise, top-down influencing can be described empirically but has not yet been theoretically explained.

---

always seem to be consistent in this respect. Speaking about the counterfactual definition of causality, he says, for instance, that when "other factors influence the outcomes of processes in the world in a counterfactual fashion, there is no reason not to speak of them as actual causes" (Clayton 2004, 57). This may be true, but the counterfactual approach to causality depends on a possible-worlds semantics and belongs, therefore, to another brand of metaphysics than Clayton's own realist metaphysics. Counterfactual definitions of causality state that the occurrence of event e depends on the presence of condition c, provided that e is distinct from c: "Where c and e are two distinct *actual* events, e *causally depends* on c if and only if, if c were not to occur e would not occur" (Menzies 2014).

Moreover, the kinds of emergence and top-down causation differ from level to level; no uniform pattern exists, and a discontinuity remains between the levels and their corresponding sciences (Clayton 2004, 54).[12]

Discussion and Evaluation of Clayton's Approach

We have seen that Clayton makes a case for strong emergence, which implies top-down causation. Top-down causation, in turn, requires distinctness of levels, and distinctness of levels requires laws. One of the strengths of Clayton's approach is its empirical orientation. Philosophical theories should at least be consistent with the results of empirical study. He warns against pre-established ontologies that function as frameworks into which empirical findings have to be squeezed. It should be the other way around: philosophers should feel challenged to make sense of the relationships and dependencies that scientists are investigating.

Clayton's approach runs parallel to reformational philosophical approaches as long as the subject under investigation is viewed "from below," i.e., as a picture of the factual side (or subject-side) of reality. In the reformational philosophical view, science ultimately aims at the law-side of reality. It tries to unravel patterns, regularities, structures, and laws that hold for reality and find their place in explanations. Scientists explain in terms of the way patterns and regularities hold for particular types of objects or events. Causality is the result of the particular ways in which laws hold for certain classes of objects. Laws, in other words, explain causality. It is not the other way around; that is, causality does not explain lawfulness—at least, not if lawfulness is understood as an effect of really existing laws. From a reformational philosophical point of view there is one problem with respect to Clayton's account. This concerns the lack of clarity in his conceptualization of the relation between law and causality. Clayton sees how important laws are for the philosophical defense of the distinctness of levels. However, he also tends to identify the functioning of laws with causal activity. This is unclear, because laws do not cause. They hold and thereby support and undergird causal sequences.

The attractiveness of Clayton's approach—its empirical orientation—is also its weakness. Some of the more fundamental issues are not addressed. For instance, do laws emerge or were they already there when the relevant patterns emerged? And what is the ontic status of laws: do they hold for a particular realm of reality, or are they just patterns of regularity that are reconstructed in scientific models? In short, Clayton's approach could have benefited from a clearer distinction between law and causality, i.e., between law-side and factual side of reality.[13]

---

[12] Clayton nevertheless prefers the term *monism* over *ontological pluralism*, because *monism* "better expresses the commitment of science to understand the interrelationship of levels as fully as possible" (Clayton 2004, 54).

[13] Now and then Clayton comes close to recognizing the importance of this distinction. In Kauffman and Clayton (2006), for instance, the authors conclude that the current debate lacks an adequate

With this we return to the three background questions mentioned in the introduction. Regarding the first question—i.e., whether it makes a difference to maintain a strong notion of law—we have to say that laws do matter on Clayton's account. They support the distinctness of ontic levels. These levels function differently because of differences between types of law. The question whether laws are preexistent or emerge in the course of evolution is not answered straightforwardly, but I believe Clayton would choose the latter option, given his inclination to identify laws with types of causal activity. With respect to the second question—i.e., whether the tension between the creation order view and evolutionary accounts can be solved with the help of the concept of emergence—it is safe to say that in Clayton's work the concept of emergence indeed functions as a mediating term between evolutionary theory and creation theology. There are many different forms of emergence, however.

I will close with a few remarks on the third question—i.e., whether emergence is compatible with a strong concept of law. Clayton rightly points at the importance of laws to distinguish between levels. By doing so, he counteracts a tendency to put one-sided emphasis on continuity. However, at the same time he tends to neglect more fundamental questions, for instance, about the ontological status of laws. Clayton's work does not provide an answer to the question whether emergence as a concept is compatible with a strong view on laws (as preexisting and necessary).

This is different in reformational philosophy. This philosophy defends and explains differences between types of causality on the basis of differences between laws. It is not causality which explains the existence and nature of laws, but the other way around. It is a matter of debate where to locate these laws. All factual law formulations are fallible interpretations, functioning within explanatory scientific frameworks. One could call them *scientific laws* and distinguish them from deeper (transcendental) laws. The latter are then conceived as preexistent and necessary. We need them conceptually to make sense of the holding of laws, both the transcendental and the scientific ones. The holding of laws should be distinguished from causal activity that is determined by these laws.

Evan Thompson: Key Concepts

I will now turn to Evan Thompson, and especially to his landmark study *Mind in Life* (2007). I have chosen this book because it is one of the most detailed and groundbreaking studies I know on the principles of self-organization applied to

---

theory of (self-)organization. This theory should serve as a linking pin between the empirical study of continuous development (emergent evolution) and the philosophical discussion on emergence versus reduction, they say. As indicated in the section entitled "Upshot to Self-Organization," self-organization is a hybrid construct in the sense that it fulfills two functions: it refers to the empirical process of increasing organization and differentiation, and it functions as principle by specifying standards for what counts as emergent phenomenon.

psychology and neuroscience.[14] Moreover, it shows how phenomenological accounts of the structure of experience can enrich neuroscientific and psychological investigation of the mind. The book is written, as the author says, as a contribution to the idea of "deep continuity of life and mind." Life and mind share "a core set of formal ... properties," i.e., the self-organizing features of the mind are an enriched version of the self-organizing features of life. The self-producing or *autopoietic* organization of biological life is, in fact, a form of cognition (Thompson 2007, ix).

Central to this conception is a cluster of notions: self-organization, autonomy, circular causality, nonlinear dynamics, complexity, and emergence.

Thompson's concept of self-organization is based on ideas that first were developed by Maturana and Varela (1980) in their book on *autopoiesis* and by Varela (1979) in his work on the autonomy of biological systems. Varela construes autonomy as the result of complementary top-down and bottom-up processes.

Top-down approaches focus on the interaction of the system with other systems and processes in the system that help maintain and mold these interactions. The autonomy of the system appears in this approach as the product of the interaction with other systems—a process also indicated with such terms as *organizational closure, operational closure,* and *structural coupling* (i.e., coupling between systems in which the process of coupling has a structuring effect on the stability and functioning of each system). The key idea here is that the boundaries of the system are co-defined by the interaction with other systems.

The bottom-up approach addresses the energetic and thermodynamic processes that are required for the instantiation of basic autonomy—"the management of the flow of matter and energy ... that ... regulate, modify and control: (i) internal self-constructive processes and (ii) processes of exchange with the environment."[15] Take, for example, the chemical processes in the cell that are involved in the production of ATP (adenosine triphosphate). ATP is a nucleotide which is responsible for the storage and transfer of energy in and between cells. ATP plays a role in self-constructive processes within the cell and in the exchange of molecules and energy between the cell and its surroundings.

From the late eighties, these ideas were applied to cognitive neuroscience and psychology in a book on embodied cognition (Varela et al. 1991) and later in *Mind in Life*. Varela's complementarity of top-down and bottom-up approaches returns in Thompson's own account in which emergence is described as "dynamic co-emergence,"[16] meaning that "a whole not only arises from its parts, but the parts also arise from the whole" (Thompson 2007, 38). Dynamic co-emergence more or less coincides with the concept of circular causality.

---

[14] Other important scholars in the field of biology and self-organization are Stuart Kauffman (1993) and Terrence W. Deacon (2012).

[15] Thompson (2007, 46) quotes here from Ruiz-Mirazo and Moreno (2004, 240).

[16] *Mind in Life* was initially conceived as a follow-up on earlier work by Varela and Thompson, but was severely delayed by the premature death of Varela in 2001. The text that eventually appeared contained the original ideas of Varela but had been recast and rewritten by Thompson.

With respect to its empirical support, Thompson (2007, 39ff.) says that the book relies to a considerable extent on the work of Kelso and some others on brain dynamics (Kelso 1995; Le Van Quyen 2003). Kelso (1995, 257) considers the brain as "fundamentally a pattern forming self-organized system governed by potentially discoverable, nonlinear dynamical laws." This means that "behaviors such as perceiving, intending, acting, learning, and remembering arise as metastable spatiotemporal patterns of brain activity that are themselves produced by cooperative interactions among neural clusters" (ibid.). With this statement Kelso leaves primitive localizationism as well as Cartesian representationalism far behind.

The reference to nonlinear dynamics implies that self-organization is basically interpreted as a mathematical concept. Generally speaking, the behavior of systems over time can be modeled by specifying the values of relevant quantifiable variables of the system in a mathematical model. As long as the behavior of these variables over time is smooth and continuous, the behavior of the system is linear and differentiable.[17] Nonlinear systems lack this predictability and are not differentiable in the ordinary way.[18] The behavior of these systems is analyzed in terms of their behavior in phase space, which is the geometric representation of all possible states of the system. The behavior of the system appears then as the curve or trajectory in phase space. The brain belongs to those nonlinear systems of which the behavior—supposedly—can be described in this way.[19]

Another important concept in this context is the concept of *complexity*. It is important because nonlinearity is not the same as randomness. Complex systems are not unstable but metastable or dynamically instable, indicating that they usually live at the edge of instability where they can "bifurcate" into two (or more) different states in phase space. This dynamic instability is thought to add to the system's flexibility and adaptability (Kelso 1995, 22). Returning to the topic of dynamic co-emergence, Thompson explains (2007, app. B, 420) that in the case of autonomous systems what emerges is simultaneously a "self," or "individual," and a correlative niche: the domain or set of interactions possible for such a system given its organization and concrete structural realization. The system's interaction with the world is such that it helps shape the condition under which it can retain its autonomy.

---

[17] Differential equations are equations in which an unknown function (or dependent variable) is construed as the function of one or more independent variables. Differential equations are further classified according to the order of the highest derivative of the dependent variable with respect to the independent variable appearing in the equation.

[18] Mathematical tools that are used to analyze the behavior of such systems are recurrence plots and Poincaré maps. In the experimentation with nonlinear systems, scientists try to identify "control parameters" and "collective variables" for dynamic patterns (Kelso 1995, 259; Thompson 2007, 419).

[19] Other names for nonlinearity are chaos theory and butterfly effect (the butterfly in the Pacific "causing" a hurricane at the Atlantic).

Thompson on Emergence, Autonomy, and Downward Causation

Thompson's working definition of emergence is as follows:

> Relational holism, N, of interrelated components exhibits an emergent process, E, with emergent properties, P, if and only if:
>
> 1. E is a global process that instantiates P, and arises from the coupling of N's components and the nonlinear dynamics, D, of their local interaction.
>
> 2. E and P have a global-to-local ("downward") determinative influence on the dynamics D of the components of N.
>
> And possibly:
>
> 3. E and P are not exhaustively determined by the intrinsic properties of the components of N, that is, they exhibit "relational holism." (Thompson 2007, 419)

Autonomy can be defined in different ways. There is the topological and morphodynamic formulation we just gave. Another way of characterizing autonomy is by distinguishing between decomposable, nearly decomposable, minimally decomposable, and non-decomposable systems. Nonlinear autonomous systems are either non- or minimally decomposable. A system is minimally decomposable if the components of the system are "less governed by intrinsic factors and more by the system's organization" (ibid.). In non-decomposable systems, the components lose their identity and are no longer separable.[20]

So, emergence occurs in systems that are non-decomposable or minimally decomposable. But it is also the other way around: as a result of emergence, systems are behaving as non-decomposable or minimally decomposable systems.

In earlier chapters, Thompson has described experiments in which large-scale dynamics modulate local neuronal activity by "entraining" or "pulling" the behavior of individual neurons into a particular pattern of global activity. This global-to-local ("downward") influence is not equivalent to top-down control in a sequential hierarchy of processing stages. It therefore differs from top-down causation.

Thompson opposes here a dominant view on brain functioning that conceives neural networks as consisting of a hierarchy of more or less distinct processing levels, from peripheral (lower) to central (higher) areas. Lower and higher are mea-

---

[20] In decomposable systems, there is a hierarchical organization in which each component functions according to its own intrinsic principles, independent of other components. In nearly decomposable systems, although there is some top-down and bottom-up interaction, the causal interactions *within* subsystems are more important in determining component properties than causal interaction *between* subsystems (Thompson 2007, 420). The four degrees of decomposability are interesting from a Dooyeweerdian point of view. In Dooyeweerd's philosophy, there is a threefold distinction between aggregates (comparable with decomposable systems), part-whole structures (non-decomposable), and encaptic structural wholes (top-down and bottom-up "interaction" but in such a form that the relative independence of the constituent parts is retained), which are more or less comparable with nearly decomposable systems. The distinction between nearly decomposable and minimally decomposable is interesting: it seems empirically and conceptually relevant and it is lacking in Dooyeweerd's account.

sured here in terms of synaptic distance from sensory stimulation (and motor output). There is reciprocity between the interconnected brain areas at each level and each level has relative independence. Thompson's problem with this picture is that it owes to the old "sandwich model" of the brain (input - black box - output; the term *sandwich model* was coined by Susan Hurley). Sensory input and motor output are always and at each level connected, and this connection does not run via a hierarchy that has brain processes at its top. Broadly speaking, Thompson's phenomenologically inspired enactivism entails that perception (input) and action (output) are coupled, and that the one cannot be understood without the other. We perceive with our bodies; and it is with our bodies that we perceive—think, for instance, of how spoons, knives, keys, smartphones, and computers become part of our body schema.[21] This occurs not only subconsciously but also with conscious awareness and with conscious reflection.

One of the problems of the old model is that it does not recognize that the brain often functions as a non-decomposable system rather than a hierarchy of nearly decomposable or even decomposable subsystems. Thompson's dynamical systems model poses that the brain functions as "brainweb." This brainweb generates global processes that "subsume their components so that they are no longer clearly separable as components."[22] From an outsider perspective, it is indiscernible in such cases whether the components (i.e., local neuronal activities) emerge from the whole or the whole (i.e., dynamic patterns of large-scale integration) emerges from the components (Thompson 2007, 423). From an insider perspective, this question does not matter; there is just an immanent movement towards emergence that does not require any extra external influence in order to occur.

In a subsequent discussion of *downward causation*, it is—again—remarked that (i) this concept should be interpreted from a dynamical systems perspective; and that (ii) it should not be taken as referring to influence of one (emergent) level on lower levels but as influence of the system as a whole on the parts that are integrated into it.

The first point highlights that downward causation should be taken as a topological and formal notion, indicating the constraints that follow from the fact that the relational properties of parts are both limited and opened up as a consequence of their being integrated into a systematic (global) network. The second point follows from the minimally or non-decomposable nature of large-scale emerging dynamic patterns. The right way to think about how these patterns influence constituent parts is not in terms of top-down causation but system causation. Thompson quotes Searle in this context: "The system, as a system, has causal effects on each element, even though the system is made up of the elements" (Searle 2000, 17; Thompson 2007, 426–427). The term *causation* indicates in this context nothing more than an "organizational constraint of a system with respect to its components." Causality is not an

---

[21] There is a rich literature on this coupling of perception and action: see, for instance, Merleau-Ponty (1945), Gallagher (2005), and Noë (2004, 2009).

[22] Secondly, and more specifically, the old model overlooks the importance of endogenous brain activity as reflected in states of preparation, expectation, emotional tone, and attention, and the influence of this activity on the sensory perception.

external force of a higher and more encompassing level—or structure—acting on a lower and smaller level—or substructure—but an "interconnectedness" and "relatedness among processes" (Thompson 2007, 427).[23] The concept of *relational holism* aims at this state of affairs: it expresses that "holistic relations do not simply influence the parts, but supersede or subsume their independent existence in an irreducibly relational structure" (ibid., 428). Given relational holism, downward causation seems a "misnomer."

Discussion and Evaluation

Thompson's account differs considerably from Clayton's. He thinks, as we just saw, in a different way about top-down causation. Top-down causation is, in fact, system (or whole-part) causation. But system causation is always entangled with bottom-up causation. This entanglement is expressed by the idea of dynamic co-emergence. The interaction between constituents that gives rise to emergent phenomena goes hand in hand with global-to-local (system) influences, and it is in and by this bidirectional interaction that emergent phenomena appear.

Another, subtler difference between Thompson and Clayton concerns Clayton's emphasis on the distinctness of forms of emergence (and causation). The emergence of *level* B out of A is a different kind of emergence than the emergence of C out of B, according to Clayton. Reading Thompson, one gets the impression that all forms of emergence are structurally similar, namely, mathematical. This is what is meant with the term *isomorphism*: an isomorphism exists between system levels of functioning, which can be described in mathematical terms.

> Because dynamic systems theory is concerned with geometrical and topological forms of activity, it possesses an ideality that makes it neutral with respect to the distinction between the physical and the phenomenal, but also applicable to both. Dynamical descriptions can be mapped onto biological systems and shown to be realized in their properties (for example, the collective variable of phase synchrony can be grounded in the electrophysiological properties of neurons), and dynamical descriptions can be mapped onto what Husserl calls eidetic features, the invariable phenomenal forms or structures of experience. (Thompson 2007, 357)

Geometry and topology are branches of mathematics.[24] Without getting too technical, the quotation says that the topological redescription is able to capture dynamic properties of neuronal networks, and that it relates these properties to physical properties of neurons as well as to phenomenal aspects of experience. The isomorphism

---

[23] Thompson (2007, 427) cites and supports the comparison between organizational constraints and Aristotle's notion of formal cause.

[24] *Topology* is the application of set-theory on geometry. It studies properties of geometric forms that remain invariant under transformations such as bending or stretching. These invariants help define what is called a topological space. Important topological properties include convergence, connectedness, and continuity.

between the levels remains epistemic.[25] But Thompson's ambitions with respect to integration are big enough to include even subjective experience in the new science of the mind. Not only does the ideality of the mathematical redescription allow findings from one field to be mapped onto findings in other fields—what in other contexts is indicated with the term *translation*, as in *translational neuroscience*—but Thompson also suggests that this ideality can be expressed in ways specific enough to include (or at least approximate) individuality. The brain signatures are so individual that they may be conceived as correlates of the phenomenal properties of our experience. This is what Thompson means when he suggests that the ideality of the topological description can be mapped on phenomenal forms of experience.

First-person accounts of mental phenomena should, and can, be integrated in the new science of the mind. First-person accounts refer to subjective experience, to the what-it-is-like-for-me aspect of experiences, thoughts, and the like. This first-person perspective is indicated with the term *phenomenology*. Not surprisingly, the new science that attempts to bring neurobiology, phenomenology, and dynamic systems theory together is called *neurophenomenology* (Thompson et al. 2005). It is one of the most empirical variants of the broader program of Varela and others that aims at "naturalizing" phenomenology.

Does Thompson solve the above-discussed problem of the concept of self-organization—i.e., that it refers to both a factual process and a principle that guides this process? I doubt it. Thompson's references to both bottom-up and top-down processes are descriptive and, as we have seen, derived from a variety of sciences—mathematics, biology, neuroscience, and cognitive science. In these descriptions there seems to be no place for the holding of overarching principles that explain emergence. What holds are highly abstract mathematical principles which explain isomorphism, mapping, translation, and the like. But these principles do not offer a theory about the "engine" behind emergent processes. They specify the conditions, but they do not determine the process itself. Isomorphism is a boundary concept that aims at interlevel similarity. Emergence is a different boundary concept: it is a concept that refers to dissimilarity, the evolving of new, more complex structures out of simpler ones.

Reformational Philosophy and the Concept of Emergence: Divergence or Accommodation?

In this final section I will begin with an observation and a lesson. Then, I will continue with a brief exposition of my own position and conclude, finally, with an attempt to answer the three questions that were raised in the introduction.

---

[25] Kelso goes one step further when he suggests an *ontological* (instead of epistemic) isomorphism between the levels (Kelso 1995, 288).

An Observation and a Lesson

What impresses me in the vast empirical and philosophical literature on emergence in neuroscience and psychology is the creativity, ingenuity, and rigor with which scientists try to do justice to the range of phenomena indicated with the term *emergence*. One can only be perplexed by the level of technical sophistication, mathematical refinement, and complexity of their research, especially of neuroscientific research.

Throughout this chapter I have pointed at the lack of conceptual clarity and rigor of some of the sciences' fundamental concepts. How is it possible that the sciences have made so much progress, in spite of these conceptual limitations? Obviously, conceptual inadequacy does not always stand in the way of success and scientific progress. There is, of course, the fundamental inadequacy in the construction of models. By definition, models do not capture the phenomenon under study entirely—they capture only aspects of it. But the inadequacies I am referring to are more practical and widespread, often even well known. To mention only one example, think of the enormous success of stress research in spite of the obvious inadequacy of the concept of stress itself.[26] There is a lesson for philosophers to be drawn from this observation, I think: they should give scientists sufficient space to work even though their concepts may be inadequate, and not prematurely close discussions at the empirical level with philosophical arguments, however sound these arguments may be.[27]

This may sound counterintuitive, yet I think it is wise. Philosophers should be aware of the provisional, heuristic, and hypothetical status of much of what scientists claim to be true. And they should allow scientists to use terms that cannot yet bear the weight of age-old philosophical discussions.

Creation Order and the Dispositional View on Laws

Throughout this chapter, one of my concerns has been the notion of the holding of laws. The epistemic (or descriptive) approach to laws, which considers them as patterns or lawful regularities, seems limited in that it cannot do justice to this element of *holding*. But how can we make sense of this holding if we stick—as I would suggest we do—to the idea that at least some laws (perhaps the most fundamental laws,

---

[26] The term *stress* refers to the stress that is built up in an iron rod when one tries to bend it. Selye (1956), who introduced the concept of stress, hypothesized that, just like there is a point at which the iron rod will snap instead of bend further, there will be a maximum in the level of psychological pressure that an individual can bear. Beyond that point, the individual will "snap" (break down) psychologically. The metaphor is of course very mechanistic and inadequate, but it has given rise to much scientific research.

[27] Having said that, it is of course also important to warn against the sweeping statements of reductionist popscientists and -philosophers who declare that "we are our brains," that humanity can best be seen as a "cosmic accident," and so on.

or laws beyond the grasp of our understanding) hold with necessity? Are laws a matter of divine decree from the beginning of the universe? And if we *think* so (or not), how can we *know*? And if we think we know, what kind of knowledge is it that we think we possess: is it systematic philosophy, or worldview, or theology, or science?

I am inclined to adopt a moderate position with respect to this issue. I endorse the idea of creation order, not as a strictly philosophical concept but as a key concept that is based on a faithful and sensitive understanding of the Scriptures. Creation order should not be thought of as a cosmic blueprint or a metaphysical idea but as an expression of the trustworthiness and orderliness of the world we live in. Thus conceived, creation order is a term with a clearly "worldviewish" status (see Geertsema's chapter "Creation Order in the Light of Redemption (1): Natural Science and Theology" in this volume for a similar approach).

I am also inclined to defend the view that the concept of law implied in this worldview comes closest to what philosophers call the *necessity view*. However, I am also aware that the necessity view might entail some highly abstract implications that are difficult to consider as being implied within the worldview conception of law. An ontology of possible worlds, for instance, is not necessarily implied in the concept of creation order, whereas it is implied in (one of the variants of) the necessity view.

As a philosopher, I am also inclined to agree with Dooyeweerd when he describes law as, fundamentally, a boundary concept for our understanding, and when he distinguishes law as a necessarily presupposed boundary concept from law in the sense of a fallible (scientific, or other) interpretation of the lawful functioning of created reality.

The question is what this means for the concept of emergence. What I am going to say is tentative; it is a suggestion about the direction in which I am looking for an answer, rather than an account that addresses all the tricky topics.

We have seen that, at the background of the discussion on emergence, the contrast between necessity and contingency plays an important role. This especially appeared to be the case when authors tried to accommodate evolutionary theory to the emergence paradigm. Classical accounts of evolutionary theory seem to presuppose a metaphysics of contingency.[28] Emergence is a lucky accident on this account. However, for religious thinkers, this model is at odds with the idea that God has a plan with the cosmos. And it is definitely at odds with a concept of law that emphasizes the preexistence and necessity of laws.

The move I want to make at this point consists of an attempt to disentangle the worldview conception of law, not only from actual discussions in biology and cosmology, but, particularly, from metaphysical speculations that are derived from these discussions. I am not implying that worldview, biology/cosmology, and metaphysics stand apart and should live their lives completely separated from one another. What I am suggesting is that the relation between the worldview account of law and the necessity view as metaphysical position, on the one hand, and between

---

[28] See, however, Denis Alexander's chapter "Order and Emergence in Biological Evolution" in this volume for a different view based on more recent developments in evolutionary theory.

evolutionary theory and the metaphysics of contingency, on the other hand, might be less compelling than some discussants are inclined to suggest. If this is true, it implies that philosophy, science, and worldview have their own internal logic and agenda. These fields are linked, of course. A need exists for mutual attunement, which implies the comparison of concepts and careful probing of their implications and strengths. But this does not lead to a merger between worldview, science, and metaphysics. Elsewhere, I have argued for this position with respect to the relationship between philosophy and worldview (Glas 2011). Just like the sciences, philosophy has its own agenda. It is informed by insights derived from the worldview of the philosopher. However, Christian philosophy is not a merger between Christian worldview and philosophy.

What does this mean? It means that the Christian worldview thinker is simply overstating her case if she demands that every individual product of the process of evolution which happens to manifest lawful properties should have such properties on the basis of a preexistent will of a God whose decrees hold with necessity from the beginning of the universe. All the Christian worldview commits itself to is the idea that lawfulness is more than just regularity, that laws in some way hold, and that the immense order and potential of creation are somehow a reflection of the intentions of a Creator.

The Christian worldview differs from a Deist universe in that it suggests that God, instead of acting like a distant watchmaker, is in an intimate and actual way concerned with the work of his hands. Dooyeweerd's ontology of law- and subject-side working together in the disclosure of spheres of reality can be conceived as a philosophical way to articulate this divine concern with the actual world. It can also be seen as a way to express that the powers that keep creation running should not be located one-sidedly in laws—and even less, I would say, in preexistent laws, as the latter would bring us again close to Deism.

The powers that propel the grand process of opening up the spheres of reality are located in the world itself, in the very interaction and attunement of law- and subject-side. They are in some way wanted by God, according to the Christian worldview. But how exactly this interaction and attunement occurs—i.e., at which moment, to which degree of detail, by decree or by laying down propensities (tendencies, dispositions)—we simply do not know. From a worldview perspective one can understand that a philosophical translation of the idea of divine concern could exist in a vocabulary of law- and subject-side, operating in strict correlation (I would say *attunement*, or *conjunction*). But philosophical translation may sometimes just go a little bit too far. If a necessity interpretation of laws would forbid certain assumptions—for instance, the assumption that development occurs at the law-side of reality (because laws are necessary, exist from the beginning, and in every possible world determine the outcome)—then maybe the worldview conception of law has been philosophically overinterpreted. This, in turn, might indicate that we need other characteristics than preexistence and necessity alone to undergird the idea of a lawful order. One candidate characteristic could be, for instance, the notion of propensity (or tendency, disposition, power) as defended by Lydia Jaeger in her contribution to this volume.

Christian philosophers are, in my view, indeed entitled to endorse the view that there are powers in created reality itself that explain how creational structures are opened up. These powers are not forces-in-themselves but capacities that God in his wisdom has laid down in reality and that are activated under the right circumstances. Propensities do not hold; rather, they come to expression, as capacities. The working of these propensities and capacities is lawful (i.e., their working obeys rules, principles, laws, and so on), and this lawfulness might again be conceived as an expression of the will of God and of the existence of a creation order.

With this latter statement, we transgress the boundary between philosophy and worldview and are back in the realm of worldview. For the phenomenon of emergence, a similar going back and forth obtains. The unpredicted and unpredictable emergence of new structures will always remain somewhat enigmatic. The philosopher and the scientist have to respect the enigma and should be aware of the boundaries of their thinking. The new structures manifest new forms of orderliness and lawfulness; a lawfulness that can partially be captured in scientific terms, mostly in the language of general patterns, boundary conditions, and phase transitions. However, there is also a language beyond science and philosophy. It is a language that suggests an order with even more intriguing qualities, qualities that we can also find in our daily experiences and worldviews: meaning, beauty, and the capacity to imbue feelings of awe and wonder and, perhaps, of being at home in this vast universe (Kauffman 1995).

Emergence: A Boundary Concept

With this in mind I come back to the three issues that were raised in the introduction. The first question was whether it makes a difference for the sciences of the person to maintain a strong notion of law—strong in the sense that laws are considered to be preexistent and necessary. We have seen that strong notions of law indeed make a difference. They are an important antidote against reductionism in the sciences. They are helpful in clarifying why there cannot be such a thing as unified science. In addition, they make us aware of why the notion of top-down causation should be used with caution.

However, we have also seen that the two core characteristics of the strong interpretation of laws—i.e., preexistence and necessity—are associated with certain philosophical assumptions that not every supporter of the creation order view would be happy to adopt in every respect. Many Christian thinkers would hesitate, for instance, to commit themselves to the view that every individual law can be traced back to an individual divine decree at the beginning of the universe. And for good reason: requiring preexistence in such a detailed way could easily amount to a form of Deism.

Christian thinkers would also be inclined to reject the idea that—given the necessity claim—one has to assume that the evolution process can only have one outcome, which is, of course, at odds with the core hypotheses on which evolutionary theory is based. If laws by definition are necessary, and if necessity means that that which is necessary exists in all possible worlds, then evolution, with its emergence of new structures, can only amount to these particular structures—structures that in some way would have already been prefigured (in God's mind, in a cosmic blueprint, or in fundamental principles or forms).

However, there are other options, as we saw. Lawfulness may also partially refer to propensities or powers ("seeds") that are laid down in creation. Such powers *work* but, since they are not laws, they do not *hold*. The lawful functioning of these powers may be a reflection of a deeper, more general kind of holding of a lawful order. But from a philosophical point of view it is impossible to further specify the empirical conditions and circumstances under which this holding occurs. The holding of laws is, in other words, already implied by a general (boundary) concept of cosmic order. My suggestion is that this is all we need to maintain a minimal notion of the holding of laws.

Our second question was whether the (presumed) tension between evolutionary accounts within neuroscience and psychology and a philosophy of creation order can be solved with the concept of emergence. The answer to this question is, yes—but, again, only to a certain degree and primarily at a philosophical level. Emergence is a boundary concept for the sciences. One should presuppose it in order to better understand certain phenomena. As a boundary concept it helps to expand our scientific imagination. Previously in this chapter I have explained that the emergence paradigm is a philosophical way of looking at things; it is not a scientific theory, let alone a theory of everything. As a paradigm, it does not (scientifically) explain evolutionary or ontogenetic development. Instead, it provides a conceptual framework that helps to make sense of certain phenomena and their scientific explanations.

One of the reasons for this restricted view on the concept of emergence is provided by our discussion on emergence and the concept of law. This discussion showed that the concept of emergence has a limitation: laws cannot be presupposed in the explanation of emergence, nor can they be seen as the emergent product of lower level functioning. The first would lead to self-contradiction, the second to the ascription of a form of self-productivity that would lead—if driven to its extreme and translated to a metaphysical level—to emergence as a form of self-origination or self-causation. This is why, by the way, in the context of philosophy of religion, concepts of emergence tend to be connected with a form of panentheism. The creative powers within creation are then conceived as divine seeds that illuminate and guide emergent processes to their ultimate destination. In this chapter, a different view is defended; a view in which emergence functions as a philosophical guiding idea at the boundary of the special sciences.

Thirdly, is the concept of emergence compatible with a strong concept of law? And if not, should we revise our concept of law, or should we give up the notion of emergence, or stick to a pre-evolutionary worldview? In the above it has been shown that in many respects emergentism is compatible with a strong conception of law. The caveats are that emergence should be taken in a primarily heuristic and paradigmatic sense, i.e., as a boundary concept at the background of a research program in a Lakatosian sense. In addition, it should not be driven to its utter metaphysical consequences but taken in a more general sense, with specifications that can easily be expressed in worldview terms. It is difficult to draw the line between emergentism as a heuristic paradigm (the position that is defended here) and emergentism as an implicit ontology that puts emphasis on lawfulness as caused by bottom-up processes (a position that is rejected in this chapter). We should judge each individual case on its own merits.[29] Something similar holds for the strong view. It can best be considered as a philosophical working hypothesis that in the long run turns out to need adaptations in the form of auxiliary hypotheses.

The important lesson we have drawn from the emergence literature, especially in the field of neuroscience, is that we are dealing with complexity; that is, we are dealing with phenomena that are in need of creative concepts, illuminating metaphors, and the establishment of unexpected relations. Scientific modeling of these complex phenomena should not be hampered too early by philosophical considerations.

The notion of creation order proved to be helpful and important in the context of neuroscience and psychology. It is connected with a strong view on laws, a view that emphasizes preexistence and necessity of laws. We have seen that this emphasis is driven by primarily philosophical concerns which are by and large compatible with, but not in every respect necessarily implied by, the wordviewish notion of creation order. This distance between worldview and philosophy offers room for a more relaxed stance toward the philosophical implications of the creation order view.[30]

Something similar holds for the conceptual space between science and philosophy. This space offers ample opportunity for playing with one's conceptual imagination—a play that might profit immensely from Dooyeweerd's systematic philosophy with its refined conceptual tools (e.g., analogies, retrocipations, anticipations, subject- and object-functions, different forms of encapsis, and encaptic interlacements). This philosophy can help disentangle concepts, levels, and problems in the philosophy of the empirical sciences.

Creation order confronts us with boundaries in our understanding. I have defended a position in which the notion of holding was retained and taken in a realist sense—in other words, as belonging to reality. There is a lawfulness of and within reality which reflects the holding of an order that is in some way related to the intentions of a Creator. This position maintains the law–subject distinction; it allows emergence, including the emergence of new orderings; and it makes firms distinctions between science, philosophy, and worldview.

---

[29] This brings my position factually close to the one developed by Jacob Klapwijk. My reservations concern his view on Dooyeweerd (as essentialistic thinker), his reference to Augustine's time-conception, and his expectations with respect to the explanatory power of emergence thinking (his expectations are higher than mine).

[30] It may come as a surprise, but I see the reformational philosophical tradition as representing a prime example of this kind of relaxedness. Dooyeweerd and his followers have never interpreted law and cosmic order as metaphysically necessary states of affairs. On the contrary, some of Dooyeweerd's followers have criticized him for his law concept's tendency toward nominalism (Strauss 2009). This is not the right place, however, to go deeper into this issue.

References

Armstrong, David M. 1985. *What Is a Law of Nature?* Cambridge: Cambridge University Press.
Bechtel, William. 2008. *Mental Mechanisms. Philosophical Perspectives on Cognitive Neuroscience*. New York: Psychology Press.
Campbell, Donald T. 1974. "Downward Causation" in Hierarchical Organized Biological Systems. In *Studies in the Philosophy of Biology*, ed. Francisco J. Ayala and Theodosius Dobzhansky, 179–186. London: Macmillan.
Cartwright, Nancy. 1992. The Reality of Causes in a World of Instrumental Laws. In *The Philosophy of Science*, ed. Richard Boyd, Philip Gasper, and J.D. Trout, 379–386. Cambridge, MA: The MIT Press.
———. 1999. *The Dappled World. A Study of the Boundaries of Science*. Cambridge: Cambridge University Press.
Clayton, Philip. 2004. *Mind and Emergence. From Quantum to Consciousness*. Oxford: Oxford University Press.
———. 2006. Emergence from Quantum Physics to Religion: A Critical Appraisal. In *The Re-Emergence of Emergence. The Emergentist Hypothesis from Science to Religion*, ed. Philip Clayton and Paul Davies, 303–322. Oxford: Oxford University Press.
Craver, Carl F. 2007. *Explaining the Brain. Mechanisms and the Mosaic Unity of Neuroscience*. New York: Oxford University Press.
Davies, Paul Sheldon. 2001. *Norms of Nature. Naturalism and the Nature of Function*. Cambridge, MA: The MIT Press.
Davies, Paul C.W. 2006. Preface to The Re-Emergence of Emergence. *In The Emergentist Hypothesis from Science to Religion*, ed. Philip Clayton and Paul Davies, ix–xiv. Oxford: Oxford University Press.
Deacon, Terrence W. 2012. *Incomplete Nature. How Mind Emerged from Matter*. New York/London: W.W. Norton and Company.
Dooyeweerd, Herman. 1959. Schepping en evolutie. *Philosophia Reformata* 24: 113–159.
Gallagher, Shaun. 2005. *How the Body Shapes the Mind*. Oxford: Oxford University Press.
Geertsema, Henk G. 2011. Emergent Evolution? Klapwijk and Dooyeweerd. *Philosophia Reformata* 76: 50–76.
Glas, Gerrit. 2010. Christian Philosophical Anthropology. A Reformation Perspective. *Philosophia Reformata* 75: 141–189.
———. 2011. What Is Christian Philosophy? *Pro Rege* 40 (1): 1–17.
Godfrey-Smith, Peter. 1998. Functions: Consensus Without Unity. In *The Philosophy of Biology*, ed. David L. Hull and Michael Ruse, 280–292. Oxford: Oxford University Press.
Gousmett, Chris. 2011. Emergent Evolution, Augustine, Intelligent Design, and Miracles. *Philosophia Reformata* 76: 119–137.
Jaeger, Lydia. 2008. The Ideal of Law in Science and Religion. *Science and Christian Belief* 20: 133–146.
———. 2010. The Contingency of Laws of Nature in Science and Theology. *Foundations of Physics* 40: 1611–1624.
Kauffman, Stuart. 1993. *The Origins of Order. Self-Organization and Selection in Evolution*. Oxford: Oxford University Press.
———. 1995. *At Home in the Universe. The Search for the Laws of Self-Organization and Complexity*. New York: Oxford University Press.
Kauffman, Stuart, and Philip Clayton. 2006. On Emergence, Agency, and Organization. *Biology and Philosophy* 21: 501–521.
Kelso, J.A. Scott. 1995. *Dynamic Patterns. The Self-Organization of Brain and Behavior*. Cambridge, MA: The MIT Press.
Kendler, Ken S. 2012. The Dappled Nature of Causes of Psychiatric Illness: Replacing the Organic-Functional/Hardware-Software Dichotomy With Empirically Based Pluralism. *Molecular Psychiatry* 17: 377–388.
Kendler, Ken S., Peter Zachar, and Carl Craver. 2011. What Kinds of Things Are Psychiatric Disorders? *Psychological Medicine* 41: 1143–1150.

Kim, Jaegwon. 1993. *Supervenience and Mind*. Cambridge: Cambridge University Press.
Kitcher, Philip. 1998. Function and Design. In *The Philosophy of Biology*, ed. David L. Hull and Michael Ruse, 258–279. Oxford: Oxford University Press.
Klapwijk, Jacob. 2008. *Purpose in the Living World? Creation and Emergent Evolution*. Cambridge: Cambridge University Press.
———. 2011. Creation Belief and the Paradigm of Emergent Evolution. *Philosophia Reformata* 76: 11–31.
Lange, Marc. 2009. *Laws and Lawmakers. Science, Metaphysics, and the Laws of Nature*. Oxford: Oxford University Press.
Le Van Quyen, Michel. 2003. Disentangling the Dynamic Core: A Research Program for Neurodynamics at the Large-Scale. *Biological Research* 36: 67–88.
Maturana, Humberto, and Francisco J. Varela. 1980. *Autopoiesis and Cognition: The Realization of the Living*. Boston: Reidel.
McLaughlin, Peter. 2001. *What Functions Explain. Functional Explanation and Self-Reproducing Systems*. Cambridge: Cambridge University Press.
Menzies, Peter. 2014. Counterfactual Theories of Causation. In *The Stanford Encyclopedia of Philosophy*, ed. Edward N. Zalta, Spring ed. http://plato.stanford.edu/archives/spr2014/entries/causation-counterfactual/. Accessed 20 July 2016.
Merleau-Ponty, Maurice. 1945. *Phénoménologie de la perception*. Paris: Gallimard.
Murphy, Nancey. 1998. Nonreductive Physicalism: Philosophical Issues. In *Whatever Happened to the Soul? Scientific and Theological Portraits of Human Nature*, ed. Warren S. Brown, Nancey Murphy, and H. Newton Malony, 127–148. Minneapolis: Fortress Press.
———. 2006. *Bodies and Souls, or Spirited Bodies?* Cambridge: Cambridge University Press.
Noë, Alva. 2004. *Action in Perception*. Cambridge, MA: The MIT Press.
———. 2009. *Out of Our Heads. Why You Are Not Your Brain, and Other Lessons from the Biology of Consciousness*. New York: Hill and Wang.
Roberts, John T. 2008. *The Law-Governed Universe*. Oxford: Oxford University Press.
Rosenberg, Alex, and Daniel W. McShea. 2008. *Philosophy of Biology: A Contemporary Introduction*. New York/London: Routledge.
Ruiz-Mirazo, Kepa, and Alvaro Moreno. 2004. Basic Autonomy as a Fundamental Step in the Synthesis of Life. *Artificial Life* 10: 235–259.
Searle, John R. 2000. Consciousness, Free Action and the Brain. *Journal of Consciousness Studies* 7: 3–22.
Selye, Hans. 1956. *The Stress of Life*. New York: McGraw-Hill.
Shaffner, Kenneth F. 1993. *Discovery and Explanation in Biology and Medicine*. Chicago: University of Chicago Press.
Sikkema, Arnold E. 2005. A Physicist's Reformed Critique of Nonreductive Physicalism and Emergence. *Pro Rege* 33 (June): 20–32.
Sober, Elliott. 2000. *Philosophy of Biology*. Boulder/London: Westview Press.
Stafleu, Marinus D. 1999. The Idea of Natural Law. *Philosophia Reformata* 64: 98–104.
———. 2002. Evolution, History and the Individual Character of a Person. *Philosophia Reformata* 67: 3–18.
Strauss, Danie. 2009. *Philosophy: Discipline of the Disciplines*. Grand Rapids: Paideia Press.
Thompson, Evan. 2007. *Mind in Life. Biology, Phenomenology, and the Sciences of Mind*. Cambridge, MA: The Belknap Press of Harvard University Press.
Thompson, Evan, Antoine Lutz, and Diego Cosmelli. 2005. Neurophenomenology: An Introduction for Neurophilosophers. In *Cognition and the Brain: The Philosophy and Neuroscience Movement*, ed. Andrew Brook and Kathleen Akins, 40–97. Cambridge: Cambridge University Press.
Van der Hoeven, Johan. 1981. Wetten en feiten. De "wijsbegeerte der wetsidee" temidden van hedendaagse bezinning op dit thema. In *Wetenschap, wijsheid, filosoferen*, ed. Peter Blokhuis et al., 99–122. Assen: Van Gorcum.
Varela, Francisco J. 1979. *Principles of Biological Autonomy*. New York: Elsevier.
Varela, Francisco J., Evan Thompson, and Eleanor Rosch. 1991. *The Embodied Mind. Cognitive Science and Human Experience*. Cambridge, MA: The MIT Press.

Beyond Emergence: Learning from Dooyeweerdian Anthropology?

Lydia Jaeger

Abstract In dialogue with Gerrit Glas' contribution to this volume, this chapter interacts with two clusters of concepts of emergence, drawing on Dooyeweerdian insights in order to provide a thorough critique of them and to provide an alternative proposal. It starts from the concept of emergence as used today in analytical philosophy of mind, discussing Jaegwon Kim's critical work on emergence, Philip Clayton's emergentist "ontological monism," and non-reductive physicalism. Drawing on Dooyeweerd's modal aspects, I conclude that this type of emergence cannot fulfil the promise of providing a satisfactory non-reductive view. The chapter then discusses a concept of emergence used in the context of phenomenology and developed by a group of philosophers inspired by Francisco Varela. Proponents of this second approach share some concerns with a Dooyeweerdian-inspired critique of analytical emergence, but their explicit stance against creation leads them to develop co-emergence in accordance with Buddhist "emptiness." This chapter then examines Dooyeweerd's refusal of mind–body dualism, linked to his rejection of the concept of substance and "*logos* speculation." Based on the biblical warrant for the role of the divine *Logos* in creation, I conclude, over against Dooyeweerd, that the concept of substance can be redeemed and that a minimal form of dualism is necessary in order to account for the Bible's teaching about humans. Adopting a realist reading of the multidimensionality of human existence uncovered by Dooyeweerd's modal aspects analysis, we arrive at a truly non-reductionist view of human nature, which the two forms of emergence examined here aimed at but could not provide.

Keywords Emergence · *Logos* · Non-reductive physicalism · Substance · Dualism · Mādhyamaka Buddhism · Herman Dooyeweerd · Jaegwon Kim · Philip Clayton · Francisco Varela

---

The original version of this chapter was revised.
An erratum to this chapter can be found at https://doi.org/10.1007/978-3-319-70881-2_16

L. Jaeger (✉)
Institut Biblique de Nogent-sur-Marne, Nogent-sur-Marne, France
e-mail: diretudes@ibnogent.org

© Springer International Publishing AG 2017
G. Glas, J. de Ridder (eds.), *The Future of Creation Order*, New Approaches to the Scientific Study of Religion 3, https://doi.org/10.1007/978-3-319-70881-2_11

On Emergence

Lately, *emergence* has become a popular concept in analytical philosophy of mind. It holds the promise of being a third option in addition to reductionism and dualism, which are both deemed to be debunked positions in anthropology. The ever-increasing wealth of neuroscientific data, showing the intimate dependence of all cognitive functions on the brain,[1] and the standard account of the gradual evolution of human intelligence, leave (according to received wisdom) no hope for any dualist account of human nature. But wholesale reductionism—of which eliminative materialism is the most consequent form—provides us with a very diminished picture of the human person. Any claim that it is possible to build a satisfactory anthropology on this ground is no more than a huge promissory note on future developments of science and philosophy, of which we have no idea whatsoever at the present time. Among the most challenging difficulties for any reductionist account of mind are the "hard problem of consciousness," the intentionality of descriptive thought, and the normative character of reason. Thus, emergence raises the hope of offering a non-reductive anthropology compatible with the latest findings of neuroscience and evolutionary biology.

Despite its attractions, however, the concept of emergence faces serious problems. It is far from clear that advocates of emergence have been able to answer these challenges in a satisfactory manner. Let me name some of them.

1. There are not one, but many concepts of emergence, even inside philosophy of mind done in the analytical style. Gerrit Glas mentions weak (or epistemological) and strong (or ontological) emergence. One can also distinguish between global and local emergence (Kim 1993, 68ff., 79–91).[2] Thus it is necessary in each case to explore the weaknesses and strengths of the concept in use. Jaegwon Kim has provided groundbreaking work in this regard, and his conclusions are rather pessimistic with regard to the hopes invested in emergence (Kim 1999, 2006).
2. The standard analytical account of strong emergence makes crucial use of the supervenience relation. *Supervenience* is thought to be the key concept which allows the emergent level to be dependent on the base level without being reducible to it. But it has proven to be awfully difficult to combine these two aspects. Based on his careful and detailed analyses, Kim observes:

---

[1] It bears mention that not all neuroscientific research points in the direction of monism: despite ever more precise observations of the dependence of the mind on brain functions, there is an emerging body of anecdotal evidence related to so-called near-death experiences. They may suggest an operation of the mind during time intervals where no brain function can be measured (cf. Beauregard and O'Leary 2007, 153–166). Several scientific studies are currently being undertaken in order to evaluate the solidity of the evidence reported. For a critical presentation of the phenomenology and neurology of near-death experiences, cf. Blanke and Dieguez (2009).

[2] For an overview of different notions of emergence, see also Clayton (2004, chaps. 1 and 2) and O'Connor and Wong (2009).

The main difficulty has been this: if a relation is weak enough to be nonreductive, it tends to be too weak to serve as a dependence relation; conversely, when a relation is strong enough to give us dependence, it tends to be too strong—strong enough to imply reductibility.³ (Kim 1999, 276)

3. Emergence is a popular catchword, and the underlying intuition has wide appeal. But if emergence is more than wishful thinking, one needs some positive account of how the emergent level emerges from the base level. It is fair to say that emergentist accounts have overall not gone very far in providing a clear, positive description of emergence. It is particularly urgent for any viable emergentist picture to explain how downward causation is possible, without falling prey to causal overdetermination.⁴
4. It seems that the lack of positive content will not be overcome by pushing further the technicalities of emergentist accounts, but that the lack of positive content is, so to speak, a congenital disorder. It is directly linked to the claim that emergence is non-reductive. Remember what this implies: emergence without reduction means that we cannot provide any finite translation between base and emergent properties. But in analogy to Leibniz's analysis of contingent events, "there is still a perfect description 'at the far edge of infinity.'⁵ The supervenience [and by analogy the emergence] claim then still entails only that there is, so to speak, a reduction for God or for the angels, just not for finite beings like us"⁶ (Van Fraassen 2004, 474).

Given that no finite description of the emergence relation will ever be available (as long as it is non-reductive), the Dutch-born philosopher of science Bas van Fraassen asks, in the context of supervenience of the human person on the physical properties of the body: "What are the benefits of believing in such a relation of persons to physical objects? The mere assurance of consistency? Cold comfort! Add to this that no such ideal 'physicalist' language exists, or is likely ever to be had…. Why play these games?" (ibid.). At best, emergence provides a convenient metaphor for the conviction that the human mind is "embodied," but it is the hard work of filling in the metaphor that really matters, as Glas observes: "[Emergence] is not an explanatory, but primarily a philosophical concept. As an idea, or paradigm, it opens our imagination. But it is also the cloth that conceals our explanatory ignorance" (Glas, chapter "Creation Order and the Sciences of the Person," this volume).

---

³Cf. my analysis of the notion of supervenience in David Lewis' philosophy in Jaeger (2007, 152–165).
⁴A point made by Kim (2006, 557). Cf. my critique of non-reductive physicalism in Jaeger (2012a, 295–312).
⁵Expression inspired by Pascal ([1670] 1976, 114, pensée no. 233).
⁶Bas van Fraassen proposed the comparison with Leibniz at a conference on February 10, 2004, at the CRÉA, Paris.

The Crucial Importance of the Base

Emergence is meant to provide an alternative to reductionism and dualism. Unlike dualism, it relies on a unified base level from which higher levels emerge. Unlike reductionism, it considers that higher levels cannot be disposed of in a complete description of the system. The "flavour" of an emergentist account depends crucially on the choice of the base. Under the influence of neuroscience and evolutionary biology, virtually all emergentist accounts rely on a physicalist base. A classic definition, provided by C.H. el-Hani and A.M. Pereira, mentions *ontological physicalism* as one ingredient of emergence: "All that exists in the space-time world are the basic particles recognized by physics and their aggregates" (Clayton [2004, 4], relying on El-Hani and Pereira [2000, 133]). In the *Stanford Encyclopedia of Philosophy*, one reads: "Ontological emergentists see the physical world as entirely constituted by physical structures, simple or composite" (O'Connor and Wong 2009). In contemporary philosophy of mind, an influential emergentist position has even been dubbed, revealingly, non-reductive *physicalism*: "*The physicalist thesis is that as we go up the hierarchy of increasingly complex organisms, all of the other capacities once attributed to the soul will also turn out to be products of complex organization, rather than properties of a non-material entity*" (Murphy 2000, 57; italics in original). In contrast, Philip Clayton explicitly rejects ontological physicalism and offers *ontological monism* instead:

> Reality is ultimately composed of one basic kind of stuff.... The one "stuff" apparently takes forms for which the explanations of physics, and thus the ontology of physics ... are not adequate. We should not assume that the entities postulated by physics complete the inventory of what exists. (Clayton 2004, 4)

As has often been noted, it is remarkably difficult, if not impossible, to give a positive empirically testable content to the claim of physicalism. Nobody would want to claim that present-day physics reveals the whole truth about the world. Thus physicalism must be about the completeness of some future, perfected physics. But who knows today what this perfected physics will look like? Therefore, as Van Fraassen puts it, physicalism "is not identifiable with a theory about what there is, but only with an attitude or cluster of attitudes. These attitudes include strong deference to science in matters of opinion about what there is, and the inclination to accept (approximative) completeness claims for science as actually constituted at any given time" (Van Fraassen 1996, 170).[7]

If physicalism is a difficult notion to pin down, Clayton's ontological monism is even more so. In fact, as it stands, even Descartes could have accepted it; it suffices to define the "basic kind of stuff" as all that is created. But if Clayton's definition does not exclude Cartesian dualism, this definition cannot really accommodate emergentist intuitions. Thus it does not provide an alternative to the standard account of emergence which considers that physics completely describes the most basic level.

---

[7]To be precise, Van Fraassen offers this diagnosis for materialism, but it easily transfers to physicalism.

Even if we leave aside the difficulty of how to precisely define physicalism, the conviction that the basic elements of reality are precisely those which (a conveniently completed) physics includes is, in my diagnosis, the fatal flaw which prevents emergence from offering a sufficiently robust antireductionist view. Despite better intentions, emergence will never truly be non-reductive as long as it sets out with a physicalist base. All the difficulties of emergence we have encountered—how to strike the right balance between dependence on the base level and non-reductivity of the emergent level, how to describe positively the emergence relation, and how to avoid causal overdetermination in downward causation—are consequences of the refusal to break wholeheartedly with such a limited picture of the world.

Let us examine how this insight is confirmed in the specific case of non-reductive physicalism. Holding to a complete physical description of the base level, philosophers of mind who adopt this position also consider that rational thought is possible and even exerts a real influence on the world (typically through top-down causation). As attractive as this position may be, respecting both the physical image of the world and the avoidance of reductionism, it can be retained only if we have an idea of the way in which the complete physical description at the microscopic level can coexist with mental top-down causality. For it is not enough to propose two postulates, even if both are desirable, if we have not shown that they are compatible. The non-reductive physicalist is therefore faced with the delicate task of providing details of the relationship between cerebral states and mental states to show that a complete physical description of the brain is indeed possible without having to give up the existence of the mind. Clearly neither the relationship of identity nor the relationship of causality provides a satisfactory account. If mental states are identical or directly caused by cerebral states, they are at best epiphenomena, which means that no top-down causality can exist.

Non-reductive physicalists consider that *supervenience* provides the appropriate relationship between brain states and mental states, which allows for both a complete physical description and for freedom of thought. The fundamental idea of supervenience is easy to grasp: "No difference of one kind without a difference of another kind" (Kim 1993, 155). Nancey Murphy proposes the following definition:

> Property S is supervenient on property B if and only if something instantiates S in virtue of (as a non-causal consequence of) its instantiating B under circumstance c. (Murphy 1998, 134)

But this definition, despite its technical allure, is no more precise than the simple slogan "no difference of one kind without a difference of another kind." The key point is the relationship between the basic properties B (cerebral states, in this case) and the properties S that supervene (mental states). Designating it by the vague expression "in virtue of" hardly gets us anywhere, and the same is true of the negative statement that it is a non-causal relationship. Thus, non-reductive physicalism does not overcome the congenital vagueness of emergentist accounts which was noted above. At best, it is a promise that requires further and more precise work to show whether it is a tenable position; at worst, it is incoherent.

Moving Beyond Emergence: Dooyeweerd's Modal Aspects

At this point, we have seen that the cluster of concepts of emergence as used in contemporary analytical philosophy of mind does not deliver on the promise of providing a truly non-reductive view. In fact, despite its best intentions, it has not made a wholehearted break from the dogma of physicalism. To put it in Dooyeweerdian terms: it is still caught in the antinomy of nature and freedom, characteristic of autonomous thought in modern times (Dooyeweerd 1975, 36–37, 45–51; see also the below section on Dooyeweerd's rejection of substance dualism).

How does one move forward from this admission of failure? Another of Herman Dooyeweerd's insights can help us at this point, to wit, his concept of modal aspects. Dooyeweerd defines *modal aspects* as "special viewpoints under which the different branches of empirical science examine the empirical world." He takes the ego to be "a supra-temporal, central unity," but human experience "is refracted in the order of time into a rich diversity of modi, or modalities of meaning, just as sunlight is refracted by a prism in a rich diversity of colors" (Dooyeweerd 1975, 7–8). Dooyeweerd lists the following 15 modal aspects: quantitative, spatial, kinematic, physical, biotic, sensory, logical, historical, linguistic, social, economic, aesthetic, justitial, ethical, and fiduciary. They are arranged in a hierarchy of modes of experience. Depending on the context, one of these modal aspects becomes predominant, although the others will never be completely absent (cf. Clouser 1996, 83–86). For the modal aspects are abstractions arising from the distinct methodologies of particular sciences, such that every object already exists in the totality of these spheres. In certain spheres, it has only passive capacities, while in others it has both active and passive capacities. Take a stone, for example. It can move and be moved: as far as kinetics is concerned, it has active and passive capacities. But in terms of linguistics, it only has passive capacities, since a stone cannot speak; it can, however, be spoken of. Similarly, it has passive economic capacities because it can be a currency for trade—i.e., it can be considered a "precious" stone. In this sense, a stone exists in all the spheres, even if only passively.

Emergentism starts from a physical description of reality, into which one tries to fit the higher levels, and, in particular, human cognitive functions. Instead, Dooyeweerd's analysis allows us to start from an honest acknowledgement of the multiple dimensions of reality. In fact, there is not (and never was) a purely physical world from which higher levels of organisation have emerged. Therefore, it should not come as a surprise that an account which singles out one facet of reality (i.e., whatever is accessible to physics) is doomed to fail when we want to use it for understanding other aspects of reality.

Emergence Without Foundations

Evan Thompson and other philosophers inspired by Chilean-born biologist Francisco Varela also point out that any emergentist account starting from physics fails to do justice to human experience. Despite their use of the term *emergence*, they work inside a framework significantly divergent from that of analytical philosophy of mind. The starting point is the concept of *autopoiesis*—literally meaning "self-creation"—developed by Humberto Maturana and Francisco Varela in order to describe the self-sustaining of living beings. The broader framework is provided by Husserlian phenomenology (ultimately going back to Kant). Thinkers in this group are convinced that first-person experience is primitive: "Francisco [Varela]'s insight was that no purely third-person, theoretical proposal or model would suffice to overcome" the "conceptual gap between subjective experience and the brain." Thus, "experience is ... irreducible" (Thompson 2004, 383). But it would be wrong to read this statement simply in the sense of emergentism or dualism: the irreducibility of experience in this approach is of a much more fundamental kind than in any traditional account in philosophy of mind. Whether reductionist, dualistic, or emergentist, they all share the quest for a third-person account of consciousness, but "life can only be known by life" (Thompson [2004, 393], quoting Weber and Varela [2002, 110]).

Philosophers in this group still use the concept of emergence, but in a very different sense from what we have studied so far: there is no question of starting with a base level and deriving emergent levels from this foundation. In fact, they consider that the difficulties of standard analytical accounts of emergence outlined above cannot be overcome without a radical change of paradigm. *Autopoiesis* thus goes beyond the ordinary ideas of emergence and self-organization and holds that in certain circumstances the elements are only constituted by their inclusion in the larger system. A cell membrane is an example: although the membrane is a constitutive part of the cell, it does not exist without the cell; only by their inclusion in the cell do the molecules forming the membrane acquire their function as a membrane. Similarly, neurons do not exist apart from the brain (Gregersen 1998, 335–337). On the level of fundamental particles, the statistics applying to quantum systems[8] show that it is not possible to distinguish individual particles in a quantum system containing multiple identical particles (cf. Bitbol [2007] for further arguments from quantum field theory).

The account which philosophers inspired by *autopoiesis* offer is self-consciously anti-foundational (Thompson 2004, 391):

> Emergence is present when there is no way to analyze a system into pre-existing parts and resultant whole.... Part and whole are completely interdependent: an emergent whole is produced by a continuous interaction of its parts, but these parts cannot be characterized independently from the whole.

---

[8] The Fermi-Dirac statistics apply to fermions (e.g., electrons) and the Bose-Einstein statistics to bosons (e.g., photons).

Thus it would be more correct to speak of co-emergence than of emergence: it is not just that the whole emerges from pre-existing parts, but these parts cannot be understood without reference to the whole. This is not a minor change in wording, as Michel Bitbol, another philosopher of science linked to the same network, explains.[9] In his paper "Ontology, Matter and Emergence," Bitbol emphatically acknowledges the failure of ontological (or strong) emergence in the sense of its ordinary use in analytic philosophy of mind. He diagnoses as the root of all puzzles to which emergence gives rise the reification of both the base and the emergent level, with a resulting asymmetry between them. Instead,

> the overall process of which we partake by our actions and cognitive relations has no fundamental level on which everything else rests. It has no absolute fundamental level and no absolute emergent level either, but it has co-emergent order. According to Wittgenstein's beautiful metaphor: "One might almost say that these foundation-walls are carried by the whole house." (Bitbol [2007], quoting Wittgenstein [1974, 33–34])

Such an analysis can be applied to a variety of contexts: the articulation between the social and the mental level, between the mental and the biological level, and between the biological and the physical level. But even inside physics, co-emergence takes place: as we have seen, the very idea of individual particles having distinctive properties is very problematic in quantum mechanics, so that there is no straightforward atomistic base level from which to start.

This co-emergentist account directly resonates with Buddhist non-ontological thought: the Middle Way of Mādhyamaka Buddhism, systematized by Nāgārjuna in second-century India, is an important source of inspiration for this view on emergence (Bitbol 2007, 2010; cf. Varela et al. 1991).[10] The reification of base and emergent levels is countered by the "emptiness" of both of them, where *emptiness* is "the Buddhist technical term for the lack of independent existence, inherent existence, or essence in things" (Garfield 1994, 219). Instead of considering that fundamental particles exist prior to and independently of emerging levels, these philosophers advocate a relational, interdependent view of reality, in which it is impossible to conceive of individual entities without taking into account the whole universe. Whereas analytical emergentist accounts argue for the "real" existence of emergent properties by trying to show their causal powers (via top-down causation), Varela's friends concur with the critique in the first chapter of Nāgārjuna's *Mûlamadhyamakakârikâ* of properties having causal powers in themselves. Instead, Nāgārjuna makes use of a non-substantial concept of causality that relies on conditions and conditioned events which "are both empty of inherent existence; … yet they are co-dependently arising and are in turn connected similarly with other events or phenomena" (Bitbol 2007, 304). Thus "the whole process [of emergence] is *groundless throughout*.… Not emergence of large scale absolute properties out of small scale absolute properties, but *co-relative* emergence of phenomena. These

---

[9] All have links to the Mind and Life Institute, which started from discussions between Francisco Varela and the Dalai Lama and explores the relationship between Buddhism and science.

[10] For an introduction to the Middle Way, see Garfield (1995).

phenomena, in turn, are to be construed as relative to a certain experimental context" (Bitbol 2007, 304–305).

This "groundless" view of emergence which admits of no truly fundamental level is explicitly offered as an alternative to a view of reality based on creation, as it no longer needs any foundation or grounding of reality:

> The concept of co-origination, or non-foundational origin, ... is currently to a large extent in competition with the opposing paradigm of foundational origin, and causes problems for its metaphysical correlate, the creation-creator pairing....
>
> One has yet to implement the alternative schema of anti-foundationalism: no ultimate basis for a reduction to the lowest level of organization, nor ontologically autonomous emergent properties at higher levels of organization, but a co-production of one by another....
>
> This way of thinking is spreading; it has many areas of application and has met with resounding success. Slowly but surely, it is relegating its foundationalist opponent, and with it the metaphor of creation, to methodological and cultural history.[11] (Bitbol 2004, 28–30)

As the Buddhist view of the world has no place for the transcendent grounding of reality in the Creator God, so this view of co-emergence has no need for a grounding of higher levels in a base level.

The Heart as the Centre of Knowledge

No reformational philosopher would want to follow the anti-creationist interpretation of the concept of co-emergence. Nevertheless, there are significant points of contact between the anti-foundationalist view of co-emergence and the Dooyeweerdian corrective of the standard account of emergence I described earlier. Having both developed in dialogue with Husserlian phenomenology, they take human experience as irreducible, a reality to be reckoned with, prior to any theorizing. Both arrive at a similar diagnosis of the radical evil of emergence as it is ordinarily understood: its singling out of one aspect of reality as basic. And one may also recall Dooyeweerd's rejection of the concept of substance, which is akin to the conviction of co-emergentists that it is an unnecessary, or even noxious, reification of phenomena. In his own words, "the metaphysical concept of substance ever rests upon the hypostatization of theoretical abstractions" (Dooyeweerd 1953, 203; cf. the below section on Dooyeweerd's rejection of substance dualism).

Such common character traits lead one to hope that the work provided by Thompson, Bitbol, and others can also prove fertile for reformational philosophers. Remember Gerrit Glas' praise of Evan Thompson's "landmark study" *Mind in Life* (2007) as being "one of the most detailed and groundbreaking studies... on the principles of self-organization applied to psychology and neuroscience" (Glas, chapter "Creation Order and the Sciences of the Person," this volume). The anti-creationist stance of co-emergentist authors should nevertheless cause us to stop and

---

[11] Translation by Jonathan Vaughan.

think. It will certainly be impossible at a very deep level to integrate their work with a view of reality which seriously takes creation into account. In fact, may I dare to suggest that some of the similarities observed between non-ontological co-emergence and Dooyeweerdian ideas point to an idealistic bent in Dooyeweerd's system of thought itself, becoming apparent in the strategic role he gives to the self in knowledge? This idealistic trend in Dooyeweerd's thinking is intimately linked to his critique of the theory of the *logos*. In this section, therefore, we will examine how Dooyeweerd grounds knowledge in the self, before turning in the next section to his critique of what he calls *logos* speculation. The final two sections will then indicate how to redeem *logos* epistemology from Dooyeweerd's critique and provide some hints as to ways in which Dooyeweerd's ideas might be revised, in order to arrive at a truly satisfactory view of multidimensional reality as based on creation.

The debt which Dooyeweerd owes to Kantian-type idealism has often been emphasized. Despite his severe critique of Kant for falling prey to the antinomy of nature and freedom, Dooyeweerd not only uses some of Kant's terminology (most prominently the term *transcendental*), but he also takes human, pre-theoretical experience as the anchor point of knowledge, an emphasis shared with Kant-inspired strands of philosophy such as Husserlian phenomenology. The central role of the human subject shows itself in Dooyeweerd's account of theoretical thought. Just as Kant had asked how natural science is possible, Dooyeweerd wants to know how theoretical thought is possible. He distinguishes between naïve pre-theoretical experience, which establishes an immediate contact with reality, and theoretical thought (science, in the broad sense of *Wissenschaft*). Naïve experience is integral or holistic in nature (Dooyeweerd 1948, 32–33). In contrast, theoretical thought applies the logical aspect of thought to the concrete reality of naïve experience and thus "produces an *antithetical relation* in which the *logical aspect of our thought* is opposed to *non-logical aspects of reality*." Dooyeweerd calls this antithetical relation "*gegenstand*-relation" (ibid., 29–30). The German term *Gegenstand* is used for an object of theoretical thought, apprehended according to one of the modal aspects which is chosen for investigation (see the above section entitled "Moving Beyond Emergence: Dooyeweerd's Modal Aspects"). As an abstraction, it is to be sharply distinguished from empirical reality: "the *Gegenstand* is always the product of a theoretical abstraction by which a non-logical aspect of reality is opposed to the logical aspect of our thought" (ibid., 51).

Given the antithetical relation involved in theoretical thought, the problem of the unity of the different modal aspects emerges: how can we guarantee the coherence of the different perspectives, each corresponding to a specific modal aspect, with respect to which analytical thought is applied to temporal experience (Dooyeweerd 1953, 39)? The question of the possibility of theoretical thought thus takes the following form: "*From what starting point is it possible to apprehend integrally in a synthetic view the diverse aspects of reality which are separated and opposed to one another in the antithetical relation?*" This is "the central... problem of our transcendental critique" (Dooyeweerd 1948, 36). The solution cannot be found in theoretical thought itself, because it is based on the very antithetical relation which constitutes the problem: "the point of departure of theoretical thought must transcend the opposed terms of the antithetical relation" (ibid., 51). This is the reason why theoretical thought cannot be autonomous, in Dooyeweerd's view. It needs a transcendent concentration point that alone can unite the logical and non-logical aspects. This concentration point is found in the human self: "It is the hidden player playing on the keyboard of theoretical thought" (ibid., 54). Transcending the temporal experience characteristic of modal aspects, the self is supra-temporal for Dooyeweerd, who identifies it with the *heart* in which the Creator has set eternity (Prov. 4:23; Eccles. 3:11; Dooyeweerd 1953, 31n1). Here lies the religious root of human existence, as "self-knowledge... is *always correlative to knowledge of God*" (Dooyeweerd 1948, 53). It provides the "central point where all the aspects of our conscious and empirical reality *converge in a radical unity*" (ibid., 49).

Dooyeweerd's Rejection of the *Logos* Speculation

For Dooyeweerd, the heart as the supra-temporal concentration point answers the problem of theoretical thought and thus guarantees the possibility of knowledge. As the antithesis characteristic of *Wissenschaft* is introduced by the theoretical activity of the *ego* itself, the latter is also able to guarantee its synthesis. Thus in Dooyeweerd, the supra-temporal self takes on the role that the *Logos* played in traditional Christian epistemology.[12] In older accounts, knowledge was grounded in the *Logos*. The correspondence between the knower and the world to be known was guaranteed by the double presence of the *Logos*: human knowers are created in the image of him who also provides the order for the world. This is the reason why creation order is both objective and—at least partially—accessible to humans.

Dooyeweerd is conscious of his departure from traditional Christian epistemology, and he condemns in very strong terms what he calls the "*logos* speculation" (Dooyeweerd 1953, 177–178, 181, 560n2; Dooyeweerd 1955, 506). He not only rejects not fully Trinitarian versions of the *logos* theory adopted during the first centuries of church history (he names the Apologists, Tertullian, Clement of Alexandria, and Origen), but he also critically scrutinizes its "definitive, 'orthodox' form in the thought of Aurelius Augustinus" (Dooyeweerd 1997, 80). In Dooyeweerd's analysis, it is inconsistent with a truly Christian worldview and was borrowed by Augustine from pagan philosophy:

> Augustine did adopt the Plotinian theory of degrees of reality, although he restricted it to the created cosmos. He also adopted the Stoic theory of germinal forms in the material world,

---

[12] My interest in the place of the *logos* in Dooyeweerd's thought arose from a comment on this replacement which Henri Blocher made in his course on contemporary thought (*Pensée contemporaine*, Faculté libre de théologie évangélique de Vaux-sur-Seine, 1991/1992, notes taken by Laurent Clemenceau). At and after the conference in Amsterdam, Rob A. Nijhoff and René van Woudenberg provided very relevant comments and references which helped me to get a better grasp of Dooyeweerd's critique of the *logos* theory.

albeit in a semi-Plotinian accommodation to the scriptural motive of creation. Most seriously, he adopted both the theory of the Logos as the seat of the divine creative ideas and the whole theory of the objective actualization of these ideas in the material world. All these speculative philosophical doctrines were inseparably tied to the ground motive of form and matter, whose religious nature was intrinsically pagan. Nevertheless, this was realized neither by Augustine and the scholastics who followed him, nor by Kuyper, Bavinck, and Woltjer, who followed Augustine in their logos theory.

We can grant that the accommodation of these pagan conceptions to the scriptural doctrines of the Trinity and of creation changed their original meaning to a certain extent. It is equally true, however, that because of this process of accommodation the Christian ground motive could no longer make itself felt in philosophical and theological thought in an unadulterated way. (Dooyeweerd 1997, 80)

An elaborate critique of the "speculative *logos* theory" can be found in his article "Kuyper's [sic] Wetenschapsleer" (Kuyper's philosophy of science). The *logos* theory results in a "logicistic-idealistic orientation" (Dooyeweerd 1939, 224[13]), which gives priority to the logical aspect instead of fully recognizing sphere sovereignty, which would allow all modal aspects to make their unique and irreducible contribution to human experience. He uncovers a deep harmony between the *logos* theory, realism about ideas (in the Platonic sense), and body–mind dualism (ibid., 197). It goes without saying that he wholeheartedly resists all of these. According to Dooyeweerd, the *logos* theory starts out with a conception of the divine essence as being rational, thus applying human thought categories to God and generating the antinomy between God's intellect and God's will, which lies at the origin of much debate between scholastic rationalists and voluntarists. Taking a realist position as concerns ideas, it includes these concepts of reason in the divine *Logos* so as not to allow them a status independent from God. They guide the creative act and are thereby expressed in the world (ibid., 214–215). As human beings are made in the image of God, this conception of the divine *Logos* leads to an understanding of the "soul as a unity [lit.: substance] centered in the intellect" (ibid., 200). Such a view of the soul replaces the religious centre of human existence, the heart, with "a theoretical abstraction from the full temporal existence of the human being," thereby generating mind–body dualism—whether Platonic, Aristotelian, Thomist, or Cartesian in kind (ibid., 204). Therefore, the *logos* theory no longer perceives the grounding of knowledge in the heart and looks instead to a correspondence between human reason and a rational order inherent in the world, independent of the logical aspect of human thought. By doing so, it reifies the logical aspect and neglects the mutual coherence and dependence of all modal aspects, abolishing sphere sovereignty (ibid., 223–225; cf. Dooyeweerd 1953, 560).

---

[13] For the English rendering of the Dutch text, I follow the translation in Bishop and Kok (2013, chap. 14). Page numbers refer to the Dutch original.

Back to Scripture

Those who have come to appreciate Dooyeweerd's insights—in particular his perspicuous analyses of different thought systems—will not easily dismiss his critique of the *logos*. Nevertheless, it is the concern, which we share with Dooyeweerd, for a truly Christian worldview, rooted in the divine Word-revelation, which leads us to pause. After all, it is John the Evangelist who first called the pre-existent Christ by this title. And it is the epistle to the Hebrews which affirms that the Son gives consistence to the world "by his word" (Heb. 1:3; cf. Col. 1:17). Certainly one should not read all later *logos* speculations into these texts. But the New Testament authors did not ignore the connotations which the term *logos* had for all those in contact with Hellenistic culture. Let us remember that, according to tradition, John wrote at Ephesus, and that the epistle to the Hebrews shows some affinities in vocabulary and ideas with Philo's synthesis of the Torah and Platonism. The concept of the *Logos* active in creation thus brings together the Old Testament tradition of creation by the word (Gen. 1:3, 6, 9, 11, 14, 20, 24, 26; Ps. 33:6) and of wisdom present in creation (Prov. 8:22–31), on the one hand, and the Greek concept of the *logos*, universal Reason which permeates the world and confers to it its harmony and order, on the other (Blocher 1974, 2011).

Dooyeweerd's critique of traditional *logos* theories can help us to discover traces of pagan thought in certain Christian theologians. This is most obvious for primitive versions which consider that the *Logos* is a divine being of inferior rank compared to the Father. But even Augustine, who as few others helped to formulate the orthodox doctrine of the Trinity, may not have succeeded in freeing himself completely from alien influences. Most notably, he resorts to Platonic ideas when explaining the rationality of creation. As they cannot exist independently from God, he includes them (as Philo had done before him) in the divine mind. God started from these ideas in his creation work; this guarantees that creation is neither arbitrary nor without reason (Augustine 1950, 287; cf. Wolfson 1956, 281–826). However, is this compatible with strict monotheism? If the divine mind already contains a prototype of the world (in the form of ideas), the world acquires a (quasi-)eternal status, if only in God's thought. It can thus claim some shared status with God. The inadequacy of Augustine's proposal becomes obvious in the difficulty it has in differentiating between creation and the Incarnation: such an understanding "must define the Creation as the material embodiment of the Word, and it can define the Incarnation in no other way" (Foster 1936, 9). As a Christian theologian, Augustine maintains the link between creation and God's will; but the dependence of creation on ideas in the divine mind makes it difficult to distinguish properly between intra-Trinitarian relations and God's *ad extra* works, leading John Milton to conclude that "this tension between a theology which emphasized the free-creative will of God and a philosophy in which explanations were ultimately grounded on the intrinsic necessity of the Ideas was never adequately resolved by Augustine" (Milton 1981, 188). In the end, this conception allows Platonic ideas to play, with regard to creation, the role which the New Testament gives to the Son of God. There is therefore, as Colin

Gunton observes, the risk "that the coeternal and personal mediator of God's creating work is effectively replaced by the almost eternal Platonic forms. The Logos is crowded out by the logoi" (Gunton 1997, 93; cf. Jaeger 2012c, 36–38).

But the existence of insufficient versions of the *logos* theory does not necessarily imply that we should give it up altogether. Its basis can be found in the New Testament; the Christian philosopher has the task to draw out its epistemological consequences. Errors of the past will certainly make us very prudent here. In particular, every effort must be made to safeguard the distinction between Creator and creation: *"The starting-point, taught by Scripture, is the Creator-creature pattern"* (Blocher 1990, 16). But this is the very reason why we have to listen to Scripture in order to get some grasp of the distinction between the Creator and his creation: our reason needs to be guided by the Creator's revelation if we want to speak truly about the distinction between him and the world. We will not yield to any preconceived conception, not even that of an opposition between the two.[14] For the biblical authors, the world-structuring activity of the *Logos* is obviously not incompatible with divine transcendence. And they did not fear that certain concepts shared with Greek philosophers would corrupt the purity of their teaching.[15] God being the Creator, all truth is his truth, even when it is found with a pagan. Thus the biblical worldview is not the opposite of Greek thinking.

Dooyeweerd's Rejection of Substance Dualism

As we have seen, Dooyeweerd's rejection of *logos* epistemology is coherent with the prominent role he attributes to the human self: the latter both brings forth the *Gegenstand*, apprehended according to different modal aspects, and allows for the unity of these perspectives. What can be considered to be an idealistic tendency in Dooyeweerd's thinking also shows up in his resistance to the traditional definition of truth as "adaequatio rei et intellectus," as correspondence between reality and thought (Dooyeweerd 1955, 566, 573; cf. Van Woudenberg 2005, 113–117). Instead, he defines truth as the *"correspondence of the subjective a priori meaning-synthesis as to its intentional meaning with the modal structure of the 'Gegenstand' of theoretical thought"* (Dooyeweerd 1955, 575). Without going into the details of

---

[14] It is one of the paradoxes of Dooyeweerd that he never enters into close dialogue with the biblical text, although he holds to the authority of the divine verbal revelation. (Historically, this silence may have been, at least partly, due to accusations raised against him by theologians at the Vrije Universiteit, which made him very cautious of moving into the explicitly theological realm, not being a professional theologian himself.) In addition, one may ask if, in spite of his better intentions, his idea of the human heart as supra-temporal unity does not jeopardize the Creator–creature distinction.

[15] Cf. Paul quoting a verse from the *Phenomena* of Aratos (third century BC) which expresses the continuity between God and humanity: "We are also His offspring" (Acts 17:28).

this (complex) definition, it is clear that it does not directly refer to a contact established between the knowing mind and the world to be known.[16]

In a similar vein, Dooyeweerd does not consider that the multiplicity of modal aspects, which also characterizes our encounter of other humans, points to a dualist human nature which could explain this non-reductionist experience. As we have seen, he quite strongly disapproves of traditional accounts of humans as being composed of a body and a soul. According to Dooyeweerd, they have no scriptural warrant and are inherited from Greek philosophy. Mind–body dualism is seen as a direct consequence of the necessarily antithetical, or dialectical, character of autonomous thought (or more precisely: thought which strives to be autonomous)[17]: as humans can think only with reference to the divine, they will always sacralise one aspect of the created realm for as long as they do not accept the transcendent foundation of all knowledge. But "the religious absolutization of particular aspects cannot fail to call forth their correlates, which in the religious consciousness begin to claim an absoluteness opposite to that of the deified ones" (Dooyeweerd 1975, 36). Thus autonomous thought is always caught up in irreducible dichotomies: "any idol that has been created by the absolutization of a modal aspect evokes its counter idol" (ibid.). Mind–body dualism is in particular indebted to the form–matter antinomy characteristic of Greek thought, which scholasticism also retains in a somewhat Christianised version.

In Dooyeweerd's analysis, mind–body dualism stems from a reification of modal aspects, confusing the *gegenstand*-relation with the structure of reality (Dooyeweerd 1942, prop. 5; cf. Dooyeweerd 1939, 204, quoted above):

> Whenever Scripture speaks in its *concise religious sense* about the *human soul* or *spirit*, it always shows it as the *heart* of all temporal existence, from out of which are all the issues of temporal life. Nowhere does Scripture teach a dichotomy between a "rational soul" and a "material body" within temporal existence. Rather, it views the body as *this temporal existence as a whole*. And this temporal body is to be laid down at death. In contrast, according to Scriptural revelation, the human soul or spirit, as the religious root of the body, is not subject to *temporal* death, because the soul in fact transcends all temporal things.

Thus Dooyeweerd replaces the traditional body-soul dichotomy with the duality between the heart and the "whole of man's temporal existence"; the latter is not be confused with an "abstract material body" (Dooyeweerd 1942, prop. 9; italics in original). The heart is "the central unity of our consciousness, which we call our I, our ego. I experience, and I exist, and this I surpasses the diversity of aspects, which human life displays within the temporal order." As it is "the central reference point" of all modal aspects, it cannot be determined by any of them (Dooyeweerd 1975, 180).

---

[16] Cf. his critique of Bernhard Bavink's critical realism (Dooyeweerd 1953, 560n2). I further develop the implications of the rejection of *logos* epistemology for the notion of truth in Jaeger (2012b). A critique of my article can be found in Coletto (2015).

[17] The antithetical character of autonomous thought is not to be confused with the antithetical structure of the *gegenstand*-relation: the latter is "*relative* and requires a theoretical synthesis developed by the thinking 'self'"; the former is "*absolute*" as it is the result of neglecting the transcendent grounding of all thought (Dooyeweerd 1948, 60–61).

Dooyeweerd's rejection of mind–body dualism is part of his general attack on the notion of substance. This concept has a long and complicated history in Western thought, and Dooyeweerd devotes a considerable amount of attention to the manifold understandings it has given rise to (e.g., Dooyeweerd 1953, 201–203; 1955, 11–14; 1957, 3–28). In his view, the subtle discussions it has generated point to a basic confusion: the metaphysical concept of substance is "a speculative exaggeration of a datum of naïve experience" (Dooyeweerd 1957, 3). We experience the persistence of things (or persons) in spite of sometimes considerable changes over time. Metaphysics reifies this pre-scientific experience into a "substance." In order to explain the persistence of entities over time, "metaphysics seeks a supra-temporal *substance*, possessing a permanence unaffected by the process of becoming and decay"; this "imperishable substance" is considered to be true "being" (ibid., 4). This view does not sufficiently distinguish between naïve, pre-scientific experience and theoretical thought, turning immediate experience into metaphysical speculation. In particular, it does not take into account that a *Gegenstand* is always constituted by the act of human thought. Instead, it postulates substances inherent in entities, without any link to the cognitive act: "'substances' are opposed as 'things in themselves' to human consciousness" (Dooyeweerd 1955, 11). Thereby, it does not see the constitutive role of modal aspects in the theoretical grasp of reality: "This view consequently breaks the integral coherence of all the modal aspects of our experience asunder" (ibid.).

Dooyeweerd's critique of the concept of substance has caused concern in theological circles. Not only does the concept feature prominently in traditional dualist descriptions of the human being, but (more importantly) it also plays a pivotal role in classic formulations of such central Christian dogmas as the Trinity and Christology. The standard formulation of the Trinity speaks of one divine essence in three persons. The Chalcedonian creed teaches that the Incarnated Son is "consubstantial" with the Father as regards his divinity and with us as regards his humanity, one person in two natures. Despite his severe critique of the concept of substance (synonymous in this context to *essence* or *nature*), Dooyeweerd certainly had no intention whatsoever of rejecting the historic Christian faith. Not considering himself a theologian, he did not provide an account of the Trinity or of the person of Christ which avoids the concept of substance and still succeeds in retaining the content of the traditional affirmations (Young 1969, 292, 300–301). The ancient Christological creeds emerged after painstaking struggles in which some of the early church's finest minds battled against heresies of many kinds. One may thus legitimately doubt whether orthodoxy can be preserved without the traditional language. The warning of another twentieth-century Dutch academic may be worth pondering. Concerning attempts to break from the formulations of the ancient creeds, G.C. Berkouwer made the following observation: "The result was nearly always that in contending with the words of the church the polemicist actually clashed with what the church intended, namely, to confess that Christ was truly God and truly man, and not to offer a scientific formulation of the mystery of the Incarnation" (Berkouwer 1954, 71).

The church fathers could only partly resort to conceptual tools provided by Greek philosophy in order to formulate the Trinitarian and Christological dogma in a way which would guard against heresy. Never before had human thought had to clearly distinguish between nature and person. Up to that point, it had been enough to define a person by a sufficiently detailed description of his/her substance. In antiquity, philosophers faced with the question "Who is he?" (the person) thought it enough to provide an answer to the question "What is he?" (his nature or substance). But monotheism implies that Father, Son, and Holy Ghost cannot be distinguished by refining the description of their nature. Being one God, they share one essence. As Augustine saw, the only possible distinction between them is relational, constituting them as different persons. Although early theologians brought over from Greek philosophy the concept of substance, they first had to forge the concept of person before they could provide a satisfactory formulation of the Trinity (and of the two natures of Christ, united in one person).

In this context, it is striking to observe that Dooyeweerd's notion of *heart* comes close to what in Trinitarian terminology can be designated as the *person*: the subject as the centre of personality, underlying all experience and acts.[18] Dooyeweerd thus replaces the dualism of mind (or soul) and body with the dualism between person and the "whole of man's temporal existence." He claims that in doing so, "the distinction between soul and body acquired a more sound formulation than in the traditional dichotomistic theory of substance" (Dooyeweerd 1939, 232). But can this claim be upheld? The distinction between substance and person is, historically, a unique contribution of Christian theology to human thought. Given that Trinitarian and Christological dogmas clearly seem to need both concepts, one may well wonder if anthropology can do with only one of them—the concept of person.

Emergence in the Context of Creation

It will certainly not be possible to reintroduce the concept of substance without engaging in a thorough revision of Dooyeweerd's whole thought framework and, in particular, the implied epistemology. Dooyeweerd's critique of the concept is too closely linked to major tenets of his system for a minor adaptation to be sufficient. Undertaking a thorough evaluation of Dooyeweerd's epistemology is beyond the scope of this paper. A particularly delicate aspect in the context of our discussion is his sharp distinction between naïve, pre-scientific experience and theoretical thought. Could it be that this distinction is itself an antinomy, ultimately linked to autonomous thought (turning one of Dooyeweerd's critical tools against his own system)? And why limit our investigation to understanding the unity of modal aspects? Given that they are generated by theoretical thought, it may be sufficient to look to a human anchor, the heart, in order to guarantee their coherence.

---

[18] There is a subtle debate in Trinitarian theology on the exact notion of the person. Traditionally, the person as "mode of being" played a central role. I use here a slightly more existentialist notion.

But despite all idealistic disclaimers, the deep question is the correspondence between human thought and created reality. Dooyeweerd himself points to an order inherent in the reality to be known when he speaks of the "resistance" of the *Gegenstand*: in the antithetical relation characteristic of theoretical thought, "any attempt to grasp ... [the non-logical aspects] in a logical concept is met with resistance on their part" (Dooyeweerd 1953, 39). But this resistance is a sign that human thought is in touch with a reality that goes beyond human cognition. Nobody less than the divine *Logos*, creatively active both in the world to be known and in the human knower, will enable us to understand how this contact is possible. As we have seen, *logos* epistemology has scriptural groundings. And there is no need to develop a *logos* theory of knowledge in the direction of rationalism, which Dooyeweerd was right to reject: human knowledge is the knowledge of created beings, and therefore always derived and limited. In addition, grounding knowledge in the biblical *Logos* makes ample room for the personal dimension of all human cognition, as he is the second person of the Trinity.

Since I have written elsewhere on the subject of developing a coherently Christian epistemology based on the *Logos* (cf. Jaeger 2004, 2012c, chap. 4), I will not attempt a thorough engagement with Dooyeweerd's overall system here. In conclusion, let us simply summarize the extent to which insights taken from Dooyeweerd in this article have shed light on contemporary debates about emergence, and where we cannot follow his lead (in my opinion at least) if we want to obtain a thoroughly non-reductionist view of mankind.

1. Typical analytical accounts of emergence are non-starters, because they try to build up non-physical levels of reality from a physical base level. Despite their best intentions, they do not radically break with physicalism, the latter being characteristic of so many strands of modern thought. Dooyeweerdian thinking can assist us here in two ways. Firstly, it shows us that no purely physical reality exists; any object can be analysed according to all modal aspects. Purely physical objects are a theoretical abstraction, useful in certain circumstances (when physics provides us with an increased understanding and control of reality, for example). But one should not try to build up other levels (life, psychology, history, etc.) from such an abstraction. Secondly, Dooyeweerd provides us with an explanation for the persistence of reductionist thought far beyond its practical usefulness: it is a nostalgic yearning for a united existence. Those who do not know of the unifying origin of the world in the one Creator have to look for a unifying point in creation: "the innate religious impulse of the human ego [is diverted] from its true origin and direct[ed] ... upon the temporal horizon of experience with its diversity of modal aspects. By seeking itself and its absolute origin in one of these aspects, the thinking *I* turns to the absolutization of the relative" (Dooyeweerd 1975, 27).
2. The anti-foundational account of emergence starting out from *autopoiesis* agrees with the foregoing Dooyeweerdian analysis in that there is no purely physical level from which an emergentist account can start. But the two radically disagree on how to link this fact to creation: whereas the former considers that foundational

views are correlated to "the creation-creator pairing" (Bitbol 2004, 28, quoted above), the latter finds in the transcendent grounding of the world the decisive argument against reductionism and its absolutization of one aspect of created reality. This radical difference concerning creation leads in turn to a significantly different outworking of the anti-foundational intuition. The *autopoiesis* account is understood in the sense of Buddhist emptiness: the reification of the physical base is avoided, as nothing can claim independent or inherent existence. In this way, the absolutization of the physical is forestalled. But this seems to come at a high price: since all levels (base and emergent) are relative to each other, without any transcendent foundation, they tend to fade into nothingness. Only creation can provide things with true reality, without absolutizing any of them.
3. Dooyeweerd has provided us with a very illuminating analysis of the necessarily antithetical character of (would-be) autonomous thought. He holds that standard philosophical dualisms about human nature are indebted to the form–matter antinomy of Greek thought, continued in a christened version in scholasticism (of both medieval and modern times), and I consider his line of argument to be decisive on this point. Platonic, Aristotelian, Thomist, or Cartesian dualism cannot be integrated into a Christian anthropology without a thoroughgoing revision, overcoming the form–matter antinomy. In particular, no form of dualism that depreciates the body can be accepted, as humans in their whole being are part of God's good creation.
4. Nevertheless, Dooyeweerd's own account of humanness does not make sufficient room for the duality of body and mind/soul which is taught in Scripture (for example, Matt. 10:28; 1 Cor. 5:3; 2 Cor. 4:16; James 2:26). In particular, it is not clear how Dooyeweerd can accommodate the traditional understanding of the intermediate state as a conscious disembodied existence of the individual which is *contemporaneous* with the body's lying dead (and rotting) in the grave.[19] If the body was the "whole of man's temporal existence," being "laid down at death" (Dooyeweerd 1942, props. 5 and 9, quoted above), how could the person enjoy conscious existence during the time leading up to the resurrection of the body? A kind of "minimal dualism" is needed to accommodate the duality of mind/soul and body, implied by the intermediate state (Cooper 2000, xxv–xxviii). Dooyeweerd's analysis warns us against expressing this minimal dualism in terms indebted to the form–matter antinomy. But his own account does not take into account all the data about human existence at our disposition.
5. The distinction between substance (or nature) and person is an original Christian contribution, which was historically accomplished by Trinitarian theology. Given the pivotal role that the concept of substance plays in Trinitarian and

---

[19] Despite the existence of alternatives (e.g., soul sleep and immediate resurrection), I consider that the traditional understanding of human existence between the individual's death and the eschatological resurrection has good biblical warrant. For a thorough defence of the continued disembodied existence in time, cf. Nicole (2009) and Buchhold (2009). One of the strongest arguments for the contemporaneity of the intermediate state with ordinary history is Jesus' existence between his death and his resurrection three days later, and his promise to the criminal crucified at his side: "Today you will be with me in Paradise" (Luke 23:43).

Christological formulae, it is to be expected—over against Dooyeweerd—that a Christian anthropology cannot do without it. Whereas most discussions in (analytical) philosophy of mind limit themselves to the choice between monism and dualism and are suspiciously quiet about the fact that humans are *persons*, Dooyeweerd's replacement of the substance of the soul with the person also leaves us with an overly restricted conceptual basis for a satisfactory account of humanness.
6. Granted, not all of the connotations of the concept of substance can be upheld in a Christian worldview. In particular, no created substance is capable of independent existence in an absolute sense: it is defined primarily by its relation to the Creator and secondarily by its relations to other creatures. But an epistemology based on *Logos* theology allows us to recover a workable concept of substance: Trinitarian creation provides the creature with "thickness," that is, substantial consistence as God's covenant partner, while at the same time affirming its continued dependence on the Creator. As the classic definition by Reformed theologian Louis Berkhof states: by his act of creation, God provided the world with an "*existence, distinct from His own and yet always dependent on Him*" (Berkhof 1953, 129; italics in original). Dooyeweerd clearly saw one side of the dual-faceted truth of creation, and he is right when he insists that the creature is "nothing *in itself*," when it is isolated from other creatures and above all from its transcendent Origin (Dooyeweerd 1975, 181). But his rejection of *logos* theology, and consequently of *substance*, leads one to doubt that he allowed enough space for the reverse of this truth; namely, that the act of creation confers the privilege of *being* to the creature. In the words of the apostle: "For in Him we live and move and have our being" (Acts 17:28).
7. An epistemology based on *Logos* theology that implies a recovery of created substances allows for a realist reading of the multidimensionality of human existence, highlighted by Dooyeweerd's modal-aspects analysis. Instead of concentrating on theoretical *thought* which brings to light the modal aspects, we can read them as a manifestation of a multidimensional *reality*—in line with Dooyeweerd's own contention that a modal sphere is *not* of a "purely epistemological character" but is "an aspect of cosmic reality" (Dooyeweerd 1955, 4–5). Thus the active powers which humans deploy in the different spheres reveal a multidimensional human nature. When multidimensionality is fully taken into account as characteristic of reality, and not merely of human theoretical experience, we have arrived at last at a truly non-reductionist view of humanity.

References

Augustine of Hippo. 1950. *Œuvres de Saint Augustin*. Vol. 12, *Les révisions*, Translated from Latin by Gustave Bardy. Paris: Desclée de Brouwer.
Beauregard, Mario, and Denyse O'Leary. 2007. *The Spiritual Brain: A Neuroscientist's Case for the Existence of the Soul*. New York: HarperCollins.

Berkhof, Louis. 1953. *Systematic Theology*. Grand Rapids: Eerdmans.
Berkouwer, G.C. 1954. *The Person of Christ*. Grand Rapids: Eerdmans.
Bishop, Steve, and John H. Kok, eds. 2013. *On Kuyper: A Collection of Readings on the Life, Work and Legacy of Abraham Kuyper*. Sioux Center: Dordt College Press.
Bitbol, Michel. 2004. Origine et création. In *Les origines de la création: Journée de la philosophie à l'UNESCO 2002*, ed. Guy Samama, 5–30. Paris: UNESCO. http://unesdoc.unesco.org/images/0013/001375/137529fo.pdf. Accessed 11 May 2017.
———. 2007. Ontology, Matter and Emergence. *Phenomenology and Cognitive Sciences* 6: 293–307. http://philsci-archive.pitt.edu/4006/. Accessed 11 May 2017.
———. 2010. *De l'intérieur du monde: Pour une philosophie et une science des relations*. Paris: Flammarion.
Blanke, Olaf, and Sebastian Dieguez. 2009. Leaving Body and Life Behind: Out-of-Body and Near-Death Experience. In *The Neurology of Consciousness*, ed. S. Laureys and G. Tononi, 303–325. Maryland Heights: Academic.
Blocher, Henri. 1974. La venue du Fils-Logos: Remarques sur le prologue de Jean. *Ichthus* 47–48: 2–7.
———. 1990. Divine Immutability. In *The Power and Weakness of God: Impassibility and Orthodoxy; Papers Presented at the Third Edinburgh Conference in Christian Dogmatics, 1989*, ed. Nigel M. de S. Cameron, 1–22. Edinburgh: Rutherford House.
———. 2011. John 1: Preexistent Logos and God the Son. In *Theological Commentary: Evangelical Perspectives*, ed. Michael Allen, 115–128. London: T & Clark.
Buchhold, Jacques. 2009. L'"âme" et la continuité de la personne dans la mort. Enquête sur le Nouveau Testament dans son contexte. In *L'âme et le cerveau: L'enjeu des neurosciences*, ed. Lydia Jaeger, 95–130. Vaux-sur-Seine/Charols: Édifac/Excelsis.
Clayton, Philip. 2004. *Mind and Emergence*. Oxford: Oxford University Press.
Clouser, Roy A. 1996. A sketch of Dooyeweerd's philosophy of science. In *Facets of faith and science*, ed. Jitse M. van der Meer, vol. 2. Lanham: University Press of America.
Coletto, Renato. 2015. Lydia Jaeger and Herman Dooyeweerd: Dialogues on the Foundations of Christian Scholarship. *Koers—Bulletin for Christian Scholarship* 80 (2): art. no. 2223.
Cooper, John W. 2000. *Body, Soul, and Life Everlasting: Biblical Anthropology and the Monism-Dualism Debate*. 2nd ed. Grand Rapids: Eerdmans.
Dooyeweerd, Herman. 1939. Kuyper's [sic] Wetenschapsleer. *Philosophia Reformata* 4: 193–232. English translation in Bishop and Kok 2013, chap. 14.
———. 1942. The Theory of Man in the Philosophy of the Law Idea (*Wijsbegeerte der Wetsidee*): 32 Propositions on Anthropology, Translated by J. Glenn Friesen. (Original text: De leer van den mensch in de wijsbegeerte der wetsidee, *Correspondentie-Bladen* 7, Dec 1942.) https://jgfriesen.files.wordpress.com/2016/12/32propositions.pdf. Accessed 12 May 2017.
———. 1948. *Transcendental Problems of Philosophical Thought*. Grand Rapids: Eerdmans.
———. 1953. *A New Critique of Theoretical Thought*. Vol. 1. Philadelphia: Presbyterian and Reformed. (All volumes of A New Critique can be downloaded at http://www.reformationalpublishingproject.com/rpp/paideia_books.asp. Accessed 25 Nov 2011).
———. 1955. *A New Critique of Theoretical Thought*. Vol. 2. Philadelphia: Presbyterian and Reformed.
———. 1957. *A New Critique of Theoretical Thought*. Vol. 3. Philadelphia: Presbyterian and Reformed.
———. 1975. *In the Twilight of Western Thought: Studies in the Pretended Autonomy of Philosophical Thought*. Nutley: Craig Press.
———. 1997. *Reformation and Scholasticism in Philosophy*. Vol. 2. Lewiston: Edwin Mellen Press. http://www2.redeemer.ca/dooyeweerd/seriesa-excerptsA6.php. Accessed 2 Nov 2011.
El-Hani, Charbel Nino, and Antonio Marcos Pereira. 2000. Higher-Level Descriptions: Why should We Preserve Them? In *Downward Causation: Minds, Bodies and Matter*, ed. Peter Bogh Andersen, Claus Emmeche, Niels Ole Finnemann, and Peder Voetmann Christiansen, 118–142. Aarhus: Aarhus University Press.

Foster, Michael B. 1936. Christian Theology and Modern Science of Nature (2). *Mind* 45: 1–27.
Garfield, Jay L. 1994. Dependent Arising and the Emptiness of Emptiness: Why Did Nagarjuana [sic] Start with Causation? *Philosophy East & West* 44: 219–250. http://www.thezensite.com/ZenEssays/Nagarjuna/Dependent_Arising.htm. Accessed 27 Apr 2013.

———. 1995. *The Fundamental Wisdom of the Middle Way: Nâgârjuna's Mûlamadhyamakakârikâ, translation and commentary*. Oxford: Oxford University Press.

Gregersen, Niels Henrik. 1998. The Idea of Creation and the Theory of Autopoietic Processes. *Zygon* 33: 333–367.

Gunton, Colin. 1997. The Trinity, Natural Theology, and a Theology of Nature. In *The Trinity in a Pluralistic Age: Theological Essays on Culture and Religion*, ed. Kevin J. Vanhoozer, 88–103. Grand Rapids: Eerdmans.

Jaeger, Lydia. 2004. *Cosmic Order and Divine Word*. Churchman 118: 47–51. Also published in Spiritual Information: 100 Perspectives, 2005, ed. Charles L. Harper Jr., 151–154. Philadelphia: Templeton Foundation Press.

———. 2007. *Lois de la nature et raisons du cœur: Les convictions religieuses dans le débat épistémologique contemporain*. Bern: Peter Lang.

———. 2012a. Against Physicalism-plus-God: How Creation Accounts for Divine Action in the World. *Faith and Philosophy* 29: 295–312.

———. 2012b. Herman Dooyeweerd, la "spéculation sur le logos" et la vérité. In *L'amour de la sagesse: Hommage à Henri Blocher*, ed. Alain Nisus, 299–310. Vaux-sur-Seine/Charols: Édifac/Excelsis.

———. 2012c. *What the Heavens Declare: Science in the Light of Creation*. Eugene: Wipf and Stock.

Kim, Jaegwon. 1993. *Supervenience and Mind: Selected Philosophical Essays*. Cambridge: Cambridge University Press.

———. 1999. Making Sense of Emergence. *Philosophical Studies* 95: 3–36.

———. 2006. Emergence: Core Ideas and Issues. *Synthese* 151: 547–559.

Milton, John R. 1981. The origin and development of the concept of the "Laws of Nature." *Archives européennes de sociologie* 22: 173–195.

Murphy, Nancey. 1998. Nonreductive Physicalism: Philosophical Issues. In *Whatever Happened to the Soul? Scientific and Theological Portraits of Human Nature*, ed. Warren S. Brown, Nancey Murphy, and H. Newton Malony, 127–148. Minneapolis: Fortress Press.

———. 2000. *Bodies and Souls, or Spirited Bodies?* Cambridge: Cambridge University Press.

Nicole, Emile. 2009. L'Ancien Testament enseigne-t-il la dualité du corps et de l'âme. In *L'âme et le cerveau: L'enjeu des neurosciences*, ed. Lydia Jaeger, 69–88. Vaux-sur-Seine/Charols: Édifac/Excelsis.

O'Connor, Timothy, and Hong Yu Wong. 2009. Emergent Properties. In *The Stanford Encyclopedia of Philosophy*, ed. Edward N. Zalta, Summer 2015 ed. https://plato.stanford.edu/archives/sum2015/entries/properties-emergent/. Accessed 11 May 2017.

Pascal, Blaise. (1670) 1976. *Pensées*, ed. L. Brunschvicg. Paris: Flammarion.

Thompson, Evan. 2004. Life and Mind: From Autopoiesis to Neurophenomenology. A Tribute to Francisco Varela. *Phenomenology and Cognitive Sciences* 3: 381–398.

———. 2007. *Mind in Life: Biology, Phenomenology, and the Sciences of Mind*. Cambridge, MA: Harvard University Press.

Van Fraassen, Bas. 1996. Science, Materialism, and False Consciousness. In *Warrant in Contemporary Epistemology: Essays in Honor of Plantinga's Theory of Knowledge*, ed. Jonathan L. Kvanvig, 149–181. Lanham: Rowman & Littlefield.

———. 2004. Transcendence of the Ego (The Non-Existent Knight). *Ratio* 17: 453–477.

Van Woudenberg, René. 2005. Two Very Different Analyses of Knowledge. In *Ways of Knowing in Concert*, ed. John H. Kok, 101–122. Sioux Center: Dordt College Press.

Varela, Francisco J., Evan Thompson, and Eleanor Rosch. 1991. *The Embodied Mind: Cognitive Science and Human Experience*. Cambridge, MA: MIT Press.

Weber, Andreas, and Francisco Varela. 2002. Life After Kant: Natural Purposes and the Autopoietic Foundations of Individuality. *Phenomenology and the Cognitive Sciences* 2: 97–125.

Wittgenstein, Ludwig. 1974. *On Certainty*. Oxford: Basil Blackwell.

Wolfson, Harry Austryn. 1956. *The Philosophy of the Church Fathers. Vol. 1, Faith, Trinity, Incarnation*. Cambridge, MA: Harvard University Press.

Young, William. 1969. Herman Dooyeweerd. In *Creative Minds in Contemporary Theology: A Guidebook to the Principal Teachings of Karl Barth, G.C. Berkouwer, Dietrich Bonhoeffer, Emil Brunner, Rudolf Bultmann, Oscar Cullmann, James Denney, C.H. Dodd, Herman Dooyeweerd, P.T. Forsyth, Charles Gore, Reinhold Niebuhr, Pierre Teilhard de Chardin, and Paul Tillich*, ed. Philip E. Hughes, 2nd ed., 270–301. Grand Rapids: Eerdmans.

Part III
Creation Order and the Philosophy of Religion

For the Love of Wisdom: Scripture, Philosophy, and the Relativisation of Order

Nicholas Ansell

Abstract A central tenet of creation order thinking is that right living requires that we align our lives to the God-given structure of existence. This conception of life as a going *with*, rather than a going *against*, what some have called "the grain of the cosmos" is widely believed to be grounded in, and supported by, the wisdom literature of the Old Testament, especially the book of Proverbs. Building on the work of Roland Murphy and others, however, this essay argues that what Scripture means by the life-giving *way* of wisdom is not best interpreted as conformity to a normative *order*. My central claim that turning to order to find orientation constitutes an imposition on the biblical writings of a hermeneutic that is indebted to the wisdom tradition we know as Western philosophy, is explored by means of an intratextual and intertextual study of the male-female relationship in Proverbs 30:18–20, as read within Proverbs 30:10–33 and the book as a whole. By paying special attention to facets of meaning and experience that a creation order reading tends to obscure, a wisdom that would celebrate the enigmatic ways of existence and a wisdom that would uncover a hidden order are distinguished, thereby allowing a fresh approach to creational revelation to take shape. In its attunement to the original blessing with which the biblical narrative begins, such a mystery-affirming appreciation of creation does not lead to an *anti*-nomian eradication of order but to what we might call its *ante*-nomian relativisation.

Keywords Ante-nomian · Book of Proverbs · Creation order · Creational revelation · Gender symbolism · Intertextuality · Intratextuality · Mystery · Normativity · Wisdom literature

---

N. Ansell (✉)
Institute for Christian Studies, Toronto, ON, Canada
e-mail: nikansell@hotmail.com

Introduction

Although Christians who subscribe to the notion of a *creation order* may disagree about how that order, and our access to it, might best be characterised, it is nevertheless fair to say that, whatever terminology is used, *creation order thinking* typically involves positing a close relationship between an order *of* creation that may be experienced and investigated and an order *for* creation in response to which true life is to be found. In such thinking, which goes beyond neo-Calvinism and beyond Protestant theology,[1] order, normativity, and religious direction—in the sense of finding our way wisely and well in God's world—are intimately related.

Despite the fact that there is considerable scholarly support for the claim that Scripture itself assumes a fundamental connection between a normative order and what the book of Proverbs calls the "way of wisdom,"[2] this essay will argue that creation order thinking is not as biblically rooted as it may first appear. That *discerning* and *pursuing a way or direction* should not be construed as *uncovering* and *conforming to an order*, I will suggest, not only resonates with the very different ways in which the call of Wisdom is heard in the biblical and philosophical traditions, but also promises to help us rethink the connections between creation, order, and Christian scholarship.

The Grain of the Cosmos?

The biblical explorations that form the central sections of this essay will hopefully address some of the main concerns that non-Calvinist Christians can have with creation order thinking. But before introducing the scope and focus of my biblical exegesis, I will first comment on why order is so close to the heart of the

---

[1] Arguably, the Christian natural law tradition can be considered an older (and more widespread) form of creation order thinking. For the transformation of the former into the latter and for the coexistence of the two traditions in Calvinism, see VanDrunen (2010). Both traditions posit a normative order but differ over how it may be accessed. Neo-Calvinists, especially in the Kuyperian tradition (inspired by the Christian cultural and scholarly vision of Abraham Kuyper), have often been especially critical of the natural law tradition's acceptance of natural reason, or rational autonomy, in this context. Furthermore, in contrast to the Thomist understanding of natural law, Kuyperians typically see normative order as rooted in the divine will rather than as located in the divine mind. For two succinct Kuyperian responses to natural law, both of which are sympathetic to the thought of Herman Dooyeweerd (on whom see n. 7 below), see Skillen and McCarthy (1991, 377–395) and Chaplin (2011, 318–320). For all the differences that Kuyperians can (rightly) point to, it is revealing that Chaplin, in the final sentence of this appendix, speaks of "affinities" between Dooyeweerd and natural law thinking while Skillen and McCarthy, at the beginning of their final paragraph, place "natural law" and "creation order" in parallel. For the claim that an almost identical *realist* understanding of creation order exists in the Roman Catholic and neo-Calvinist traditions, see Echeverria (2011).

[2] For the "way of wisdom," see Prov. 4:11. For the "ways" of wisdom, see Prov. 3:17 and 8:32. Biblical quotations will be from the NRSV unless otherwise stated.

neo-Calvinist, Kuyperian tradition in particular and will then indicate why I think that advocating the relativisation of order is consistent with that tradition's own deepest convictions.

One image that has been used to illustrate the correlation between the *God-given structure of* and *God-ordained direction for* the world[3] is that of the "grain of the cosmos."[4] Although those who think in terms of creation order are unlikely to see *going with the flow* as a metaphor to live by, they are usually more than willing to speak of life—and right living—as a matter of *going with the grain of existence*, as this image, in calling to mind the deep, perduring structure of the world, points to precisely the kind of normative discernment that creation order thinking entails.

Perhaps the most striking feature of social philosophy, or cultural analysis, in the Kuyperian tradition is its characteristically positive view of boundaries, contours, laws, limits, and structures.[5] Although this might be construed as a fixation on order for its own sake, a more charitable (and more insightful) reading becomes possible

---

[3] If this formulation is read in the light of the *order of/for* distinction expressed above, it becomes apparent that the *order* that is central to creation order thinking is typically seen as both *structural* and *directional*. Many neo-Calvinists understand what I have called the *order for* creation as God's law or law-order.

[4] Thus neo-Calvinist philosopher/theologian James K.A. Smith draws on an image that Stanley Hauerwas, in his 2001 Gifford lectures, appropriates from the Mennonite theology of John Howard Yoder, to observe the following in the context of Christian worship:

> The announcement of the law and the reading of God's will for our lives ... reminds us that we inhabit not "nature" but *creation*, fashioned by a Creator, and that there is a certain grain to the universe—grooves and tracks and norms that are part of the fabric of the world. And all of creation flourishes best when our communities and relationships run with the grain of those grooves. Indeed the biblical vision of human flourishing implicit in worship means that we are only properly free when our desires are rightly ordered, when they are bounded and directed to the end that constitutes our good. That is why the law, though it comes as a scandalous challenge to our modern desire for autonomy, is actually an invitation to be freed from a-teleological wandering. It is an invitation to find the good life by welcoming the boundaries of law that guide us into the grooves that constitute the grain of the universe and are conducive to flourishing. (Smith 2009, 176; his emphasis)

For earlier neo-Calvinist use of "against the grain" imagery, see Walsh and Middleton (1984, 67, 70) and Wolters (2005, 98). For John Howard Yoder's original, resurrection- and ascension-affirming (and not merely creation-oriented) "grain of the cosmos" image, see Yoder (1994, 246) and Yoder (1988, 58). The cruciform, Christocentric character of Yoder's image is not lost on Hauerwas. See Hauerwas (2001, 6, 17).

[5] Although this attention to structures and limits may be more or less *detailed* (here we might compare Dooyeweerd [1986] with any of the essays found in Griffioen and Verhoogt [1990b]), and although the correlative attention to the disclosure of normativity can be more or less politically *radical* (cf. the comparison between the social philosophies of Goudzwaard and Dooyeweerd in Wolterstorff [1983, chap. 3]), this positive view of structures and limits seems to be a defining feature of the Kuyperian tradition in its conservative and progressive forms. I suspect that what often worries critics of this tradition is its relative inattention to the fact that the deepest and most *normative* of structures may be experienced as oppressive—and not just as an affront to humanistic autonomy. A desire to address this apparent blind spot is evident in some of the essays in Walsh et al. (1995). Cf. Dooyeweerd's real yet limited application of the language of the fall to what he calls the "law-side" of creation in Dooyeweerd (1986, 107).

when we see this as reflecting the older, distinctively Calvinist conviction that finding our way in the world with God is made possible by accepting God's *law* as *grace*.⁶ The central significance of God's law within the neo-Kuyperian philosophy known as the *wijsbegeerte der wetsidee* (the philosophy of the law-idea)⁷ is also best understood along these lines. Right *order* is thus a privileged metaphor for translating the positive, biblical notions of blessing and shalom into academic discourse. In this religio-philosophical vision, to attempt to live life against the God-given grain of existence is to experience the curse of autonomy, while to live life in accordance with the order for creation is to find true freedom. To its advocates, therefore, a central argument for acknowledging the reality of a normative creation order in theory and in praxis is that there can be no human flourishing, indeed no human future, without it.⁸

Arguably, the most compelling biblical support for the existence and importance of a creation order—understood as what I am calling an *order of/for existence*—lies in Old Testament wisdom literature.⁹ Although the current consensus among scholars who study the relevant parts of the Hebrew Bible or Christian Old Testament (such as Proverbs, Job, Ecclesiastes, and the sapiential motifs in, e.g., Genesis and the Psalms[10]) would seem to support this claim, this essay will argue that OT/HB specialists, inasmuch as they accept this reading, have mistaken Scripture's portrayal of the mysterious, enigmatic *ways* of creation for a *hidden order* and have

---

⁶For a discussion of the distinctive "third use of the law" in Calvin and his followers that flows from this *law as grace* position, see Hesselink (2001, 134–136). For the often overlooked yet fundamental role of gratitude in Calvin, see Gerrish (1993). Calvinists will often refer to God as lawful where Christians in other traditions will speak of God's universal presence, grace, reliability, and care. Robert Sweetman provides a striking example, given his otherwise evocative language, when he writes, "I do not intend to give in to the allure of the generic. Rather, I am aiming to foster sensitivity to the language we assume because it seems to us, for whatever reason, to express those deep secrets of the universe whispering subliminally of the creation's encounter with its lawful Maker" (Sweetman 2016, 16).

⁷Here I refer generally to the neo-Calvinist philosophy developed by Herman Dooyeweerd, D.H.Th. Vollenhoven, and their followers and not only to Dooyeweerd (1935–1936), a seminal work best known in its expanded English edition (see Dooyeweerd 1953–1958).

⁸In addition to the works of Dooyeweerd cited in n. 7 above, the idea of an order that makes for freedom and flourishing is given systematic attention in Hart (1984) and Strauss (2009).

⁹Here I refer not to isolated (so-called proof-) texts but to what appears to be sustained support. Historically, Paul's reference to "against nature" in Rom. 1:26 (av/kjv)—understood apart from the only other NT occurrence of *para physin* in Rom. 11:24—has probably functioned as the clearest proof-text for the "against the grain" idea. For a reading that looks to connect Rom. 1:26 and 11:24, see Ansell (1997).

[10] The presence of sapiential material outside Job, Proverbs, and Ecclesiastes is a contested issue among OT wisdom specialists. James Crenshaw, who as much as anyone has argued for a restricted use of the "wisdom literature" designation (see, e.g., Crenshaw 2010, 33–34), nevertheless does accept the sapiential character of several psalms in Crenshaw (2010, 187–194). For a wider approach, see, e.g., Morgan (1981) and Sailhamer (2000, 15–35). Cf. Ansell (2011c, including 124n47).

thus imposed what is central to a rival wisdom tradition—that of Western philosophy—onto the biblical witness.[11]

This raises the possibility that creation order thinking, despite its importance in a reformational tradition that seeks its point of departure not in philosophy but in Scripture,[12] may be an instance—arguably the supreme instance—of the kind of syncretism that this tradition refers to, pejoratively, as "synthesis" thinking.[13] Presumably, for all Christian scholars who are committed to letting biblical wisdom shape their philosophising—and who are thereby opposed to letting the philosophical tradition dictate how Scripture is to be read—to recognise this possibility *as* a possibility is to call creation order thinking into question.

To that end, this essay will explore the difference between the enigmatic ways of creation and the hidden order of Western thought by examining Prov. 30:18–20 within the context of Prov. 29–31. This will pave the way for a new conception of biblical wisdom that, in helping us properly relativise order as the central metaphor for finding our way in life and in the academic enterprise, may be an important first

---

[11] In my view, a preoccupation with conceptually grasped order, already evident in pre-Christian natural law thinking, has been a dominant form of Western philosophy for most of its history and a persistent form for the rest of its history, despite the challenges of nominalism and antirealism. I do not think this preoccupation is limited to realism. The subjectivising of order associated with Kant and the modern period, for example, is still a subjectivising of, and thus preoccupation with, *order*. That said, the imposition on the biblical witness of the notion of a hidden order that wisdom may (attempt to) lay bare comes from a still influential way of thinking that is most indebted to Greek philosophy. The additional influence on early Christian thought of a Roman preoccupation with order, not least in later Stoicism, also merits our attention here.

[12] See, e.g., Zuidema (1972). The import of his title, "Philosophy as Point of Departure," is that Christian scholarship that rightly sees philosophical work such as the rethinking of ontology and epistemology as a *sine qua non* for any reformation of the various academic disciplines, can only hope to be Christianly reformational if philosophy itself is *not* the point of departure! Reducing the role of Scripture to generating a few philosophically oriented (or philosophically determined?) axioms to get such a project going (such as assertions about the place of God's law in relation to God and creation) would also be inadequate, on my reading. Here, Seerveld's "iceberg" image (see Seerveld [2003, 97] in the light of his title) is suggestive: Scripture speaks beneath, and not only above, the waterline of consciousness, thereby sensitising our whole selves to the live speech of creation. See nn. 14, 23, 26, 75, 77, and 87 below.

[13] Synthesis thinking, in Dooyeweerd's memorable phrase, is the result of "the monster-marriage of Christianity with the movements of the age" (Dooyeweerd 1968, 4). Attributing to Scripture notions that are the result of combining biblical and non-biblical ideas is one form such synthesis thinking takes. The immortality of the soul is often seen as a classic example of this phenomenon. One might wonder about the merits of a transformational synthesis. But the concern here, for many close followers of Dooyeweerd, is that it is Scripture that gets transformed rather than the thinking with which it is combined. On the phenomenon of "cisegesis-exegesis" (reading extra-biblical notions into Scripture to then read them out again as biblical), see van der Walt (1973). Sweetman's concern that "thetical" alternatives are not realistic possibilities if they are not already "available" (2016, 23n20) seems to assume a rather derivative view of the human person. He may also be (understandably) concerned about a "narrow purity ideal" here (reading the "purely" of 23n20 together with language used on p. 15 of the same work). For an alternative view of purity, however, see Ansell (2002).

step in helping us re-articulate the religious dynamics of existence that have played such a central role in neo-Kuyperian, reformational philosophy at its best.[14]

A Scholarly Consensus

My claim that what characterises creation order thinking typically involves positing a close relationship between an order *of* creation that may be experienced and investigated and an order *for* creation in response to which true life is to be found, may need to be further nuanced or qualified before everyone who sees themselves as subscribing to such thinking feels that the most important distinctions and connections are being named in the best possible way. Be that as it may, my claim that, biblically speaking, it is the wisdom literature of the Old Testament that would seem to provide the clearest support for this way of thinking is one that I think few would disagree with.[15]

If that is so, then it is surely significant that while different terminology may be used, the specific idea of an *order of/for existence* is easily detected in an important six-point summary of the central claims of biblical "wisdom theology" put forward by OT scholar Walter Brueggemann—this occurring in a section of his magisterial *Theology of the Old Testament* that is entitled "A Scholarly Consensus." Here, Brueggemann is not simply speaking for himself but capturing what he understands numerous wisdom scholars to be saying, whether they find themselves in agreement with the biblical witness or not. I take what follows, therefore, as strong evidence that my way of characterising creation order thinking is far from arbitrary.

Focussing on what I have referred to as the *order of* existence in his first three points before shifting to a focus on the *order for* right living, Brueggemann begins his summary by stating that biblical wisdom theology is (1) "a theology reflecting on creation, its requirements, orders, and gifts," (2) the "data for [which] is lived experience." While Brueggemann stresses that this experience of creation has what he calls an "enigmatic quality," he is also quick to point out that it is (3) "understood … to have a reliability, regularity, and coherence, so that one may make generalizing

---

[14] In speaking here of the "religious dynamics of existence," and by later referring to "attunement" to the "spirituality of existence," I am in part echoing James Olthuis' appropriation of Dooyeweerd, as found in the section entitled "Spirituality of Creation" in Olthuis (1985, 22–23). See, e.g., the covenantal answering—the Hebrew can also mean *singing*—of YHWH, the heavens, and the earth, together with the grain, wine, and oil in Hosea 2:21–22. Given the double-meaning of the Hebrew verb here, to speak of *attunement* to the spirituality of existence is to use an appropriate metaphor. In my understanding, the spirituality, directionality, and covenantal initiative and responsiveness of existence are so closely related that they are virtually synonymous. For the voice of creation/voice of Wisdom, see nn. 23, 26, 75, 77, and 87 below. For a fruitful development of *attunement* in Heideggerian phenomenology, see Levin (1985).

[15] Hence the attention to biblical and ancient Near Eastern wisdom literature at the beginning of Wolters (1995). As I have already indicated, I believe the affinity that many find between biblical wisdom literature and creation order thinking is more apparent than real. See also n. 9 above.

observations that can be sustained across a richness of concrete experience" (Brueggemann 1997, 680–681).

The relationship and the tension between these last two points will be central to the theme of this study. In particular, it remains unclear whether the "enigmatic" to which Brueggemann refers in point two is a manifestation of the order he emphasises in point three or whether its wisdom might call for a different kind of discourse. Whatever we think about this, Brueggemann himself seems to assume that the "enigmatic quality of experience" that the "wisdom teachers" of the OT "stay close to" makes way for, or is taken up into, a certain kind of understanding or "theology."[16] Whether the "enigmatic" gets lost in the process is something we should judge from what happens next.

Building on what he takes to be the wisdom tradition's portrayal of the *structure* of creation, Brueggemann goes on to address its inherent *religious direction* in points four to six by saying that (4) the "reliability, regularity, and coherence of lived experience has an unaccommodating ethical dimension, so that certain kinds of conduct produce beneficial outcomes, and other kinds of conduct have negative consequences," this "linkage of deed and consequence [being] intrinsic to the shape of created reality." Here Brueggemann, in his succinct characterisation of what OT scholars (and ethicists) have called the "act-consequence nexus" or "construct,"[17] comes close to the "grain of the cosmos" idea noted above.

In his penultimate point, the enigmatic character of experience and the reliability of order—the *foci* of points two and three—seem to come together, at least in his understanding, when he says that lived experience (5) "discloses to serious discernment something of the hidden character and underpinnings of all of reality." This hidden ground of being ultimately points to God, he concludes, as (6) "experience discerned theologically ... mediate[s] Yahweh, who is seen to be the generous, demanding guarantor of a viable life-order that can be trusted and counted on, but which cannot be lightly violated" (Brueggemann 1997, 681).

All in all, and especially in this final point, creation order proponents could hardly hope for a more eloquent scholarly endorsement of their concerns. For here Brueggemann, himself a senior OT scholar, is not only claiming to speak for wisdom specialists across the theological spectrum but, as is typical of his writing style, does so without shying away from what he sees as the kerygmatic implications of the biblical witness. Given Brueggemann's compelling six-point summary, the case for positing a close, normative connection between wisdom, creation, and order seems unassailable for all who accept Scripture as the starting point for any authentic Christian philosophy.

---

[16] Hence the language of his first point, which concerns the "wisdom theology" he sees behind (as well as in) the text, not the theology distilled or constructed from the text by the contemporary interpreter. In his second point, the lived experience of creation is described as "data" for this "theology."

[17] See nn. 29–30 below.

Misgivings

While Brueggemann's summary is now 20 years old, it still represents the dominant paradigm for interpreting biblical wisdom literature, especially the book of Proverbs.[18] Central to my thesis, however, is the claim that in its engagement with existence, Western philosophy, as itself a highly influential wisdom tradition,[19] has also tried to read the *enigmatic* "ways" that may or may not be found in Prov. 3:17 and 8:32 as a kind of *hidden order* analogous to the one that Brueggemann and the scholars he speaks for believe biblical wisdom would have us uncover. Part of what is being overlooked here is that there is a world of difference between what is inherently mysterious, wondrous, and in that sense enigmatic and what is merely hidden from, or hidden within, ordinary experience. This is the difference between a mystery that can never be exhausted or fathomed and a problem or puzzle that can be solved and thus transcended.[20]

This leads to the second part of the misconstrual: for unless we think that biblical wisdom celebrates and reveals what reason seeks to grasp and lay bare—and I for one do not believe this—to identify Wisdom's enigmatic *ways* with a hidden *order* is a serious conflation. For while the order or structure of the world does indeed have directional implications—the "grain of the cosmos" idea is not without insight—such creation order thinking, despite the scholarly consensus that Brueggemann represents, nevertheless consistently obscures the dynamics of meaning that biblical wisdom literature would want us to experience when it calls for attunement to the spirituality of existence.[21]

It would no doubt be revealing to address this conflation by tracing its presence and its resistance in the OT scholarship of the last 40 years or so. While this cannot be attempted in any detail here, we should certainly take note of von Rad's influential attempt, in his seminal 1970 work *Weisheit in Israel* (*Wisdom in Israel*), to see the female Wisdom figure of Prov. 1–9 as a mysterious order reminiscent of *Ma'at*, the Egyptian goddess of law, world order, and justice.[22] Also very significant is Roland

---

[18] Proverbs is widely viewed as exemplifying a conventional wisdom that is far more hierarchical and authoritarian in character than the edgier, questioning wisdom of Job and Ecclesiastes. For an important critique of this assumption, see Hatton (2008).

[19] This point is especially evident in the early history of philosophy, on which see Hadot (1995, 2002) and Cooper (2012). For the hope that contemporary philosophy might reconnect with its wisdom roots, see, e.g., Jaspers (1954), Rubenstein (2008), and Caputo (2013).

[20] Linguistically, the "enigma code" of World War II illustrates how an enigma as (potentially solvable) puzzle differs from enigma as (unfathomable) mystery. See my concluding discussion, including nn. 85 and 86, below. We should also distinguish between a creation-affirming appreciation for mystery and its otherwordly counterpart. Caputo (2013, 30) refers to the ancients' wise acceptance of the "unfathomable" and hopes that it might return to contemporary philosophy. I am not convinced that this has ever been developed (rather than hinted at) in a truly creation-affirming way in the Western philosophical tradition. But see, e.g., LaMothe (2009), which draws on a reading of Nietzsche (and Irigaray).

[21] For the spirituality of existence and for the language of attunement, see n. 14 above.

[22] See von Rad (1972), chap. 9, which is entitled "The Self-Revelation of Creation." My own references to Wisdom (with a capital *W*) in this essay refer to the personification and call of Wisdom as found in Prov. 1–9.

Murphy's 1984 presidential address to the Society of Biblical Literature entitled "Wisdom and Creation" in which he stressed both the creatureliness of biblical Wisdom and her capacity to speak of and for God—this joint emphasis representing an acceptance of von Rad's awareness that in the Hebrew Scriptures she is no longer seen as a divinity coupled with a rejection of his claim that her calling out to humanity represents nothing more than the "self-revelation of creation."[23] Even more noteworthy is the way Murphy sought to separate von Rad's insightful emphasis on the mystery of wisdom from his fixation on a world order. Thus while noting that "one need not deny that the presumption of regularity underlies the observations of the sage," Murphy argues that the metaphors used in Prov. 1, 8, and elsewhere, which often "indicate a wooing, indeed an eventual marriage," hardly suggest an understanding of Lady Wisdom as *Ordnung*. "Who has ever sued for, or been pursued by, order," he asks, "even in the surrogate form of a woman?" (Murphy 1985, 9).[24]

Although I would contend that Murphy's question has not been taken seriously enough in the last 30 years, the tension that I believe exists between the second and third points of Brueggemann's "scholarly consensus" has continued to be felt.[25] Thus, while wisdom specialists have yet to embrace what I would see as the central implication of Murphy's position—that the voice of Wisdom is the mysterious way in which *creation*, and not just creation's *order*, may speak of and for God[26]— enough light has been shed on the key texts to make a simple return to what Murphy

---

[23] Thus in the published version of his address, Murphy claimed:

> The call of Lady Wisdom is the voice of the Lord. She is, then, the revelation of God, not merely the self-revelation of creation. She is the divine summons issued in and through creation, sounding through the vast realm of the created world and heard on the level of human experience. (Murphy 1985, 9–10)

Murphy's "not merely" indicates that her voice is the voice of creation even though von Rad's category of "the self-revelation of creation" (on which see n. 22 above) does not say all that needs to be said; namely, that creation also reveals and speaks for God. Cf. Murphy's formulation:

> Creation speaks but its language is peculiar (Psalm 19). It is not verbal, but it is steady, and it is *heard* (Ps 19:2). It is parallel to the Torah, which gives wisdom to the simple (Ps 19:8). With fine perception both Karl Barth and Gerhard von Rad concur that the Lord allowed creation to do the speaking for him in Job 38–41 (will [the lightnings] say to you, "Here we are?" 38:35). Creation had a voice which spoke differently to Job than the chorus of the three friends. (Murphy 1985, 6; his emphasis)

[24] His next sentence is: "The very symbol of Lady Wisdom suggests that order is not the correct correlation."

[25] Clifford (1997) also connects biblical wisdom to order in several places, yet his frequent qualifying of this association is revealing.

[26] Because in some theologies, *creation* and *creation order* (or *the created order*) are almost synonymous, let me stress that by the voice of "creation" here, I also mean the revelatory capacity of creational wildness (which is distinct from chaos) and the revelatory capacity of history. On the former, see Ansell (2011b). If Prov. 8:22a is translated/interpreted as "YHWH acquired me as the beginning of YHWH's own way" (cf. Nouvelle Edition de Genève 1979 [NEG79]), then Wisdom not only speaks *of* and *for* God, but also *to* God. I have explored this further in Ansell (2011a).

calls the "ma'atizing" of biblical wisdom (Murphy 1998, 289) unlikely.[27] Von Rad's emphasis on a "primordial order" (von Rad 1972, 156) may be somewhat appropriate for Egyptian wisdom literature (especially if the focus is on primordial *harmony*) just as his claim that the world has a "wise orderliness" (ibid., 155) that calls for our "intellectual love" (166) may make sense as a perennial concern for those who wish to write footnotes to Plato. But for all our powers of *eisegesis*, it is extremely telling that such notions are not evoked by the language used in Prov. 8 and elsewhere *of* and—most importantly—*by* Wisdom herself.[28]

The staying power of this particular reading of Proverbs, therefore, would seem to lie in its apparent ability to make sense of other parts of the book. In this context, any up-to-date survey of OT wisdom scholarship would have to include the patient and, to my mind, persuasive critique of the "acts-consequence construct"—the subject of Brueggemann's pivotal fourth point—in an important monograph by British Proverbs scholar Peter Hatton.[29] Also indicative of a possible sea-change is the very firm rejection of this widespread notion in a recent textbook-style introduction to wisdom literature by Stuart Weeks.[30] I highlight these examples because it is precisely here that the assumption that biblical wisdom is the knowledge of creation's normative order must be faced head-on if this scholarly consensus is to be shaken.

What needs to be seen, in my view, is that what Scripture, in its wisdom traditions and beyond, refers to as the "way" of and to "life" and "blessing" (Prov. 6:23, 10:6, 11; Deut. 11:26–28) is indeed a *way* and not an *order*.[31] The conflation and misconstrual that has dominated OT scholarship here is, as I have suggested above,

---

[27] At this point, Murphy refers to the important, earlier discussion now found in Murphy (2002, 115–118 [the 1996 edition he cites is identical here]). This section, entitled "Wisdom as a Search for Order," begins with the following statement: "It is practically a commonplace in wisdom research to maintain that the sages were bent on discovering order, or orders, in the realm of experience and nature." Cf. Murphy (2002, 128n16).

[28] Although there are distinctions (as well as connections) to be made between (1) the Egyptian experience/conception of order-as-harmony as the opposite of, or counterpart to, chaos, and (2) the Greek philosophical conception/experience of order as the ground and goal of rational insight, both notions are absent in the language used of and by Wisdom in Proverbs. Whether Hellenistic ideas of order may be detected in Israelite wisdom literature outside the canon of the Hebrew Bible and Protestant OT may well be a different story. See, e.g., Collins (1997).

[29] See Hatton (2008, chap. 4). For the importance of Hatton's work, see also n. 18 above and nn. 34, 35, 42, and 62 below.

[30] See Weeks (2010, 112–113; a section on world order). While Lucas (2008) notes that many scholars have objected to the idea that the "act-consequence nexus" presupposed belief in an order that was mechanical or impersonal, Weeks (2010, 113) rejects the idea that Israel posited a world order in any theologically significant sense at all! As far as he is concerned, the evidence for "order" that OT scholars have made so much of in, e.g., Proverbs is nothing more that the "most commonplace expectations about causation."

[31] Once we see the distinction, we can appreciate the fact that while the Hebrew term translated as "way" or "path" (*derek*) occurs over 70 times in Proverbs, terms that might be translated as "order" are virtually non-existent. Several English translations (e.g., ASV, AV./KJV, JPS, and New Living Translation, 2nd ed. [NLT²]) have no instances of *order*. The one reference to an "order" in the NRSV, NET, and NIV (at Prov. 28:2) involves a translation issue (cf. Murphy [2002, 212] and Waltke [2005, 395, 408]).

the result of assuming that the connection between order and orientation that is so central to Western philosophy, and that is now so much a part of our common sense, is also key to Old Testament wisdom—even though this represents an act of "eisegesis-exegesis" (see van der Walt 1973) in which a rival wisdom tradition's sense of what it means to discern our way in life is being read into the very Scriptures that are thought to provide it with its biblical basis. Seen in this light, Murphy's observation is as pertinent as it is perceptive:

> The scholarly postulate of "order" is a reconstruction of Israel's mentality. It raises a question never asked by Israel: on what is your wisdom insight based? *Our* answer to this (for silent Israel) might possibly be, on the order of and in creation. It may be a logical and correct answer, but Israel never raised the question nor consciously assumed the answer we give to it. The orderly Greek *kosmos* (for which there is, in Hebrew, no exact verbal equivalent) was not the world in which Israel lived. (Murphy 2002, 116; his emphases)

Just because we find it difficult to think of matters of religious direction without turning to the order of the world, or to (what we think of as) *the nature of things*, when trying to find our bearings does not mean the biblical writers negotiated life in the same way. That being the case, we must be open to the possibility that the way in which creation speaks in the biblical wisdom literature (and in the wider canon) asks to be heard within a different paradigm.

To that end, much of the rest of the present study will explore a new way forward by taking Prov. 30:18–20 as its focus. Naturally, there are many other biblical passages inside and outside the book of Proverbs (such as Prov. 3:13–26, 8:22–31; Ps. 119; Job 28; and Isa. 28:23–29) that would need to be taken into account in a more comprehensive survey. But the merit of focussing on this passage (in addition to the fact that it has so often been overlooked) is its surprisingly cosmic scope coupled with its sustained attention to the *ways of wisdom* motif. After showing how central this passage is to the final section of the book, therefore, I will argue that its portrayal of the male-female relationship finds wisdom in the celebration of a *mystery* that may be *known* rather than *understood*.[32] In the light of what I see as this biblical alternative to rationalism and skepticism, I will offer some concluding thoughts on the nature and status of order within Christian philosophy.

Wisdom in Proverbs 30:18–20: Hidden Order or Enigmatic Ways?

The passage that I would like to look at—Prov. 30:18–20—reads as follows in the New English Translation (NET):

---

[32] While this reflects the meaning of Prov. 30:18, which is well conveyed in the contemporary translations that speak of not understanding or fathoming, rather than not knowing (cf. the improvement of the NKJV over the AV./KJV), I am also drawing on the distinction between a multidimensional view of *knowing* and a conceptually focused view of *understanding* (or comprehending) articulated by Hart (1984, 355–357).

> ¹⁸ There are three things that are too wonderful for me,
>   four that I do not understand:
> ¹⁹ the way of an eagle in the sky [*baššāmayim*],
>   the way of a snake [serpent] on a rock [*ᵃlê ṣûr*],
>   the way of a ship in [the heart of] the sea [*bᵉleb-yām*],
>   and the way of a man [*geber*] with [in; within] a woman [*bᵉ'almāh*].
>
> ²⁰ This is the way of an adulterous woman:
>   she eats and wipes her mouth
>   and says, "I have not done wrong."

I have set these verses out a little differently in Fig. 2 below to bring out some of the structural features that I will comment on presently. Clearly this passage has a 3 + 1 structure as v. 18 indicates. All commentators see v. 20 as in some way connected to vv. 18–19, usually by way of contrast. So I have set this out below as a (3 + 1) + 1 structure.

Proverbs 29–31

Before looking at this passage more closely, it will be helpful to see its place within the final three chapters of the book of Proverbs. Here a chiastic structure is evident, though it has yet to be widely recognised. Significantly, the passage in question falls in the middle of this literary structure, set out in Fig. 1 below. As we shall see, these verses have their own centre-point that is suggestive of a particular angle of interpretation that I hope to pursue.

The outer frame of the literary structure (A/A′) involves recognising that the alphabetic acrostic poem of Prov. 31:10–31 finds a counterpart in the incomplete alphabetic acrostic of Prov. 29. The full alphabetic acrostic at the end of the book is now universally recognised by scholars,[33] and a good case can be made for seeing its A to Z celebration of the human disclosure and embodiment of Wisdom as answering and rectifying the way in which Wisdom is so far from the heart of human life in the opening chapter of Proverbs that she calls out from the margins in language that deliberately echoes the prophetic tradition.[34] As for the incomplete alphabetic acrostic of Prov. 29, this too has been recognised to the point that there have been several attempts to try and get behind the extant Masoretic text to reconstruct a full *aleph-tav* (or A to Z) structure. Here, I find Hatton's proposal that the poetic structure is *deliberately incomplete* to be compelling. For while Prov. 29 recapitulates many of the proverbs and themes from earlier chapters, the reader

---

[33] For the history of interpretation, see Wolters (2001, chaps. 5 and 6).

[34] On Wisdom as marginalised and isolated in Prov. 1:20–33, in relation to 29:1, see Hatton (2008, 61–65). On her use of the language of prophetic denunciation, see Murphy (1998, 10–11). On the reintegration of Wisdom in Prov. 31:10–31, see Hatton (2008, 77–81).

A   **Prov. 29**, with its (incomplete) *aleph–tav* (A to Z) structure, recapitulates proverbs and themes from the earlier chapters (cf. Hatton 2008, 60–68).

B   **Prov. 30:1–9**, the words of Agur, an oracle.

Prov. 30:10–33, seven numerical sayings within ten-unit structure (cf. Waltke 2005, 463–501):

			numerical structure:
		(10)	non-numerical
	3	11–14, 15a, 15b–16	<4, 2, 3/4>
C	+	(17)[a]	non-numerical
	4	18–20, 21–3, 24–8, 29–31	<3/4 + 1, 3/4, 4, 3/4>
		(32–33)	non-numerical (despite threefold intensification)

B'   **Prov. 31:1–9**, the words of King Lemuel, an oracle.

A'   **Prov. 31:10–31**, which is a full alphabetic acrostic, concerns the reintegration of Wisdom (cf. Hatton 2008, 77–81); on Wisdom as marginalised and isolated in Prov. 1:20–33, in relation to 29:1, see Hatton (2008, 61–65).

Fig. 1 The chiastic structure of Proverbs 29–31 (Note a: For the significance of vv. 17 and 18–20 as both central to the ten-unit structure, see the final exegetical section on Prov. 30:17 and 30:18–19 as the double-centre to Prov. 30:10–33 below. This is enhanced if the 3 + 4 structure of C is seen as a 3 + 1 + 3 structure, as suggested in the section on Prov. 30:10–33 immediately below. The ten (numerical and non-numerical) units then display the following pattern: (1) + 3 + (1) + 1 + 3 + (1).)

needs to be aware that it does not represent the end of the book's message. Its incomplete structure, in other words, befits its penultimate status.[35]

Moving from the outer frame to the inner frame, B and B', we have two oracles in 30:1–9 and 31:1–9 associated with individuals otherwise unknown (by these names, at least) in the pages of the OT, even though one is a king. It may also be the case that the word translated as "oracle" in 30:1 and 31:1 could actually connect these two men to a North Arabian tribe descended from and named after Ishmael's son Massa.[36] This is an interesting possibility given the creation-wide presence of

---

[35] See Hatton (2008, 60–68). The contrast between the move from incomplete to complete acrostic at the end of Proverbs and the deterioration of the acrostic form at the end of Lamentations is instructive, not least because the fall of the Jerusalem Temple that lies behind the despair, lament, and letting go in the latter work forms such a sharp contrast with the establishment and expansion of the cosmic Temple in Prov. 3:19–20, 9:1, and 14:1. On the significance of acrostic deterioration in Lamentations, see Yett (2017). On the cosmic Temple motif in Proverbs, see the section on Prov. 30:18–20 (including n. 47) below.

[36] See Van Leeuwen (1997, 251, 258). Massa, who appears in the genealogies of Gen. 25:14 and 1 Chron. 1:30, is the seventh of Ishmael's 12 sons and thus a participant in the blessing of the covenant between God and Abraham, as promised to Hagar in Gen. 16:10 (cf. 12:2).

Wisdom alluded to in Prov. 8. Either understanding is consistent with the chiastic structure.[37]

Proverbs 30:10–33

This leaves us with the central section, Prov. 30:10–33, much of which is made up of a collection of explicitly numerical, multi-element sayings, as set out in Fig. 1 above. A good way in to appreciating the very careful way this material has been structured is provided by Bruce Waltke, who argues that there are seven such numerical sayings surrounded by the non-numerical sayings found in vv. 10 and 32–33 and punctuated by the non-numerical saying of v. 17.[38] This means that seven numerical sayings are embedded in a ten-unit structure, the multi-element sayings found in vv. 11–16 and 18–31 together forming a 3 + 4 structure.

As the last three numerical sayings (vv. 21–23, 24–28, 29–31) each contain an explicit reference to a king (vv. 22, 27–28, 31; cf. 31:1), we can also talk of a 4 + 3 structure, or a 3 + 1 + 3 structure as these last three sayings are balanced by the first three (vv. 11–14, 15a, 15b–16), each of which focusses on the violence of greed (vv. 14, 15a, and 15b). It is surely no coincidence that four of the numerical sayings (the third, fourth, fifth, and seventh) have a 3/4 structure while three do not, as this is itself another 4 + 3 structure.

Even more telling is that three of the four 3/4 sayings have been placed at the centre of the seven, the *central* saying of the three, and thus of all seven, being Prov. 30:18–20, this possessing a unique (3 + 1) + 1 structure (as noted above).[39] Proverbs 30:18–20, in other words, is deeply interwoven with the surrounding material, and is found right at the centre, and right at the heart, of Prov. 30:10–33 and Prov. 29–31.[40]

This interwovenness is not merely formal but also thematic. For example, it is widely recognised in the commentaries that the saying about the leech that has two daughters in v. 15a is a picture of greed that alludes to the horse leech which has two

---

[37] If Fox (2009, 852 and 884) is right to take the former as an oracle and the latter as a place name, the Hebrew term (or homonym) still suffices to indicate a chiasm, given the surrounding evidence.

[38] See Waltke (2005, 463–464 and 481–501). I see my own observations as building upon and supporting Waltke's analysis.

[39] Proverbs 30:18–20 is not only the (3 + 1) + 1 centre-point of the sevenfold (or 3 + 1 + 3) series of numerical sayings; it is also part of the twofold centre-point to the ten-unit structure mentioned above (cf. Fig. 1), together with v. 17. I comment on Prov. 30:17 and its connection to vv. 18–20 below. See also note *a* of Fig. 1 above.

[40] As Prov. 30:18–20 may also be construed as having a 4 + 1 structure (v. 19a–d; v. 20) and a 3 + 2 structure (v. 19a–c; vv. 19d–20, if vv. 19d and 20 are seen as related by way of contrast), this means that the various parts of its structure echo all of the non-3/4 structures that surround it—specifically vv. 11–14 and 24–28, which each have a fourfold structure, and v. 13a, which has a twofold structure.

mouths, one at each end of its body.⁴¹ When this saying is read together with adjacent sayings in this section, intratextual meaning is picked up subliminally. Once the reader arrives at the description of the adulteress who eats and wipes her mouth in 30:20b, therefore, this not only triggers an association with the devouring teeth of v. 14, but via the allusion to the horse leech in v. 15a, prompts the reader, at least subconsciously, to wonder which of her two mouths she is wiping. This is not just a temporary or accidental spark of connected meaning as the mouth and violence motifs return in vv. 32–33. Similarly, because vv. 18–19 are designed to be read after the references to insatiability and barrenness in vv. 15b–16, and not in isolation, the reader can take it for granted that the union of v. 19d occurs in the context of pleasure and fertility, without this being explicitly stated. Examples such as these add to the case, recently made by Knut Heim and Peter Hatton, against the atomistic approach that has dominated the popular and scholarly reading of Proverbs for centuries.⁴²

Proverbs 30:18–20

Once framed in this way, the treatment of Prov. 30:18–20 as relatively, if not completely, incidental in most commentaries on the book of Proverbs can be seen as most unfortunate. For this is the central saying of the seven. And, as we might anticipate, a little attention to its own structural complexity and subtlety will quickly indicate how profound it is.

One thing that the 3 + 1 structure asks us to discern is the relationship within v. 19 between the way of the eagle, serpent, and ship (19a–c) and the way of the man and the woman (19d). Here, many commentators look to v. 20 and the way of the adulterous woman for the key, and infer from her attempt to cover her tracks that the eagle, serpent, and ship do not leave a trace in the sky, on the rock, or in the sea. As an interpretation of what might be especially "wonderful" and beyond "understand[ing]" (v. 18) about the way of the lovers of v. 19d, I suggest that this reading is about as underwhelming and unconvincing as any reading can be.⁴³

---

[41] See, e.g., Waltke (2005, 486–487). Because the word translated as "horse leech" is a *hapax legomenon*, a note in the ASV suggests "vampire" as an alternative. Presumably the "two daughters" are to be viewed as fangs. Strictly speaking, this kind of leech (the referent is clear in, e.g., the Septuagint/LXX, Vulg.) has "two [blood] suckers" (NLT²). But as they cry out, the "daughters" are best seen as mouths, thus forming an apt introduction to the insatiability motif that follows.

[42] See Heim (2001, 2013) and Hatton (2008). There is a significant affinity here with the reading of Proverbs proposed by Calvin Seerveld in the 1970s in a series of articles in the Canadian publication *Vanguard*. For a clear, more recent articulation, see Seerveld (1998).

[43] Furthermore, the adulteress is not covering her tracks so much as denying wrongdoing. For examples of this *traceless* reading, see Fox (2009, 871–872); Garrett (1993, 241); and Perdue (2000, 263–264). Appeal is often made to Wisdom of Solomon 5:7–14. But the fact that the latter is a judgment passage (in which the way of the lovers is replaced by the way of an arrow) is not taken into account. (The possibility that Wisd. of Sol. 5:7–14 is inspired by reading Prov. 30:19a–c together with v. 17 merits further investigation.) For the way of the eagle/serpent/ship as unrecoverable rather than traceless, see the helpful comments of Murphy (1998, 235). Cf. my reference to the "singular, unrepeatable path" in the final section below.

A clue to better understanding what is "too wonderful" in v. 18a lies in the reference in v. 19b to "the way of the [serpent] on a rock," this being the midpoint of the initial threefold structure of v. 19a–c. The "sky" referred to above this part of the verse and the "sea" referred to below it reflect the heaven/earth/sea view of creation found throughout Scripture (see Prov. 3:19–20; Exod. 20:11; Amos 9:6; Pss. 96:11, 146:6; Acts 4:24, 14:15; and Rev. 10:6). For the description of the earth as a "rock" (v. 19b), we need look no further than Job 18:4. Once we see that the image of the serpent in this setting alludes to what Gen. 3:1 calls the *wisest* (not craftiest) *of the wild animals* before it is pulled into the vortex of human sin,[44] we are in a position to appreciate that v. 19b is a picture of the "way" of wisdom on earth. In the first part of the (3 + 1) + 1 structure, wisdom is seen as a "way" that is beyond understanding in the three major divisions of (what we would call) the cosmos.[45]

This means that Prov. 30:19d, which celebrates "the way of a man [*geber*] with [within] a woman [*bᵉ'almāh*]," is presented as an especially enigmatic, yet wonderful, example of the way of wisdom on earth. Here the focus is more on the woman than the man as the poetic structure and repetition of the Hebrew preposition for *in* or *within*—which is absent in the reference to the serpent "on" the rock[46]—associates her with the skies and with the seas, thus giving her a cosmic significance.

Read within the book of Proverbs as a whole, this cosmic significance befits this young woman's status as a daughter of Wisdom. One of the most easily overlooked features of the book of Proverbs, due in part to inclusive language translations, is the fact that Wisdom is portrayed as relating to her male and female children quite differently. For although in her calling out to creation (8:1, 4, 9:3), she gives guidance to kings (8:15–16) and gives "instruction" to her other "sons" (8:10, 32–33 [NIV]), she is *made present* by her daughters. This is why Prov. 9:1 tells us that "Wisdom has built her house [*ḥākᵉmôt bānᵉtā bêtāh*]," this being a reference to the macrocosmic Temple of creation of 3:19–20, while Prov. 14:1, in almost identical language, tells us that "Wisdom of women builds her house [*ḥakmôt nāšîm bānᵉtā bêtāh*]" (NRSV, margin), her daughters here participating in and extending both her work and her reality.[47] This incarnational, embodied understanding of the dynamic and expanding presence of Wisdom in our world is especially evident, I suggest, in the "song of the valiant woman" that closes the final chapter.[48] And this is why

---

[44] For a detailed defence and exploration of this reading, see Ansell (2001).

[45] As noted by, e.g., Waltke (2005, 490). On "cosmos" as having no equivalent in the HB/OT, see Murphy (2002, 116), as cited above.

[46] The verset uses *'al* (on or upon) rather than *b* or *ḇ* (in). This means that Van Leeuwen (1997, 254) is wrong to suggest that the serpent is a "phallic symbol" here.

[47] On the linguistic connection between 9:1 and 14:1, see Van Leeuwen (1997, 138). The divine/human parallel between 9:1 and 14:1 can be related to the divine/human parallel between 3:19–20 and 24:3–4, on which see the helpful discussion in Van Leeuwen (ibid., 101). On the cosmic Temple in 3:19–20, see Van Leeuwen (ibid., 53–54).

[48] For a form-critical analysis and survey of the history of interpretation, see Wolters (2001, 3–14 and 59–154).

	¹⁸ There are three things that are too wonderful for me,	
	four that I do not understand:	
	¹⁹ᵃ the way of an eagle in the **sky** [*baššāmayim*],	
	¹⁹ᵇ the way of a [*serpent*] **on** a *rock* [*"lê ṣûr*],	<u>Wisdom on earth</u>
3	midpoint: rock = **earth**ᵃ	cf. Gen. 3:1
	¹⁹ᶜ the way of a ship in [the heart of] the **sea** [*bᵉleḇ-yām*],	
+		
	¹⁹ᵈ and the way of a man [*geḇer*]	<u>Daughter of Wisdom</u>
1	with [in, within] a **woman** [*bᵉ 'almāh*].	cf. Prov. 5:19; Jer. 31:22
+	²⁰ This is the way of an adulterous woman:	<u>Daughter of Folly</u>
	she eats and wipes her mouth	cf. Prov. 5:20, 7:6–27,
1	and says, "I have not done wrong."	9:13–18

Fig. 2 Proverbs 30:18–20 (NET), the middle of the seven numerical sayings (Note a: For rock as earth, see Job 18:4; for creation as heaven/earth/sea, see Prov. 3:19–20; Exod. 20:11; Amos 9:6; Pss. 96:11, 146:6; Acts 4:24, 14:15; and Rev. 10:6 as discussed above.)

Wisdom, who is herself described as a lover and kindred spirit in Prov. 7:4, may be known by the man who is faithfully "intoxicated" by a woman's love (5:19).[49]

The portrayal of the young woman as a daughter of Wisdom in Prov. 30:18–20, therefore, places her at the centre of a three-part wisdom motif that is found in the (3 + 1) + 1 structure as set out in Fig. 2 above. The first instance of this is the *wisdom on earth* theme in v. 19b symbolised by the serpent on the rock as the midpoint between sky and sea, and between 19a and 19c. The second, central instance is the Wisdom that the man finds but the woman represents in 19d. If this recalls the *intoxicated love* found in Prov. 5:19 and 7:4, then the third and final instance, in v. 20, revisits the *toxic lover* motif of Prov. 7:5–27 and 9:13–18 to provide us with a striking image of the daughter of Folly.

Proverbs 30:17 and 30:18–19 as the Double-Centre to Proverbs 30:10–33

Literarily, in finding the Wisdom that the woman represents, the Wisdom through which she is associated with the skies and the seas that surround the earth (19a–c), the man of Prov. 30:19d is surrounded by the woman. As OT scholarship has been slow to point out the connections between the wisdom literature and salvation history,[50] we should pause here to note the subtle but powerful allusion to a motif in the book of Jeremiah, which is summed up in the proverb found in Jer. 31:22: "For [YHWH] has created a new thing on the earth: a woman encompasses [or surrounds]

---

[49] This paragraph parallels Ansell (2011c, 140–141). The "sister" imagery in Prov. 7:4 (we would say: kindred spirit) is reminiscent of Song of Sol. 4:9–12 and 5:1–2. It is also instructive to compare Prov. 5:15–18 with Song of Sol. 4:15, and Prov. 5:19 with Song of Sol. 2:7 and 3:5.

[50] As Grant (2008, 859) notes, "Wisdom communication is always subtle and skilful, and so a lack of conspicuous reference does not necessarily indicate the absence of a theme." For salvation history motifs within wisdom literature, see Ansell (2017, 95–101 and 105–110).

a man" (NRSV).⁵¹ While several different interpretations and translations have been suggested for this saying, partly because the Hebrew verb placed between "woman" and "man" is unusual in this context,⁵² a number of leading Jeremiah scholars have concluded that there is a clear reference to lovemaking here.⁵³ This coheres extremely well with the bride/bridegroom motif that itself frames this text in Jer. 25:10 (cf. 7:34 and 16:9) and 33:10–11. On this reading, the "new thing" on the earth would refer to a time in which patriarchal, androcentric ways of defining the covenant between the sexes are overcome. The new covenant between female and male would then symbolise—that is to say embody and point to—the new covenant between God and humanity well known from the following verses of Jeremiah (see 31:31ff.).⁵⁴ It is interesting that the Hebrew term for "man" that we find in Jer. 31:22 (*gāber*) and in Prov. 30:19d (*geber*) is a derivative of *gābar* (to be great or mighty) that has clear connotations of strength and virility. Yet it is the woman who represents the divine in both passages.⁵⁵

That the editor of the last chapters of Proverbs would know a proverb preserved in the prophetic tradition is, I suggest, not unlikely. But my proposal that Prov. 30:19d deliberately alludes to the "woman encompasses [or surrounds] a man"

---

⁵¹ According to the Proverbs *and* Jeremiah scholar, William McKane, "there is a convergence of opinion that [the saying] has the openness and indeterminacy of a proverb" (McKane 1996, 806). He also remarks that it has been "the happy hunting-ground for aspiring [!] exegetes." In addition to Jer. 31:22, we might also explore the thematic links between Prov. 30:19d and Gen. 49:25, on which see Ansell (2013, 357n137).

⁵² For a helpful listing of possible interpretations, see McKane (1996, 806–807); Keown et al. (1995, 122–123); Fretheim (2002, 437–438); and Thompson (1980, 576). The structure of the wider poem (vv. 15–22) actually has a male subject, Ephraim, surrounded by two sets of phrases spoken either of a woman or to a woman, as noted by Trible (1978, 47–50). So this passage has been very carefully composed suggesting that multiple dimensions of meaning are not unlikely. It is possible that the *surrounding* in this text means that the woman will give birth and that Rachel, who is bereaved in earlier verses, will be a mother again. If Israel's return to YHWH is the primary meaning—the "return" of 31:21 echoing and marking an end to the harlotry of Jer. 3 (cf. Prov. 30:20)—then God is being viewed as either a strong male presence to Israel as a young woman, or as a protective female presence to Ephraim. But in context, it is also very likely that 31:22 means that the oppression by Israel's enemies that earlier led to the description of Israel's male warriors as terrified women in labour (30:6) will be so reversed that the women of Israel can be seen in the protective, comforting role. The latter coheres well with the lovemaking interpretation, argued for below.

⁵³ Commentaries by Thompson and Fretheim, cited in the previous note, refer to this possibility. Thompson, a conservative scholar, seems convinced in this regard by leading Jeremiah scholar William Holladay, citing Holladay (1974, 117). Holladay's views have been more recently articulated in Holladay (1989, 192–195).

⁵⁴ As Holladay (1989, 195) notes with respect to the "new thing": "The only other occurrence of 'new' in Jer (other than 'New Gate' …) is that of the 'new covenant,' v. 31." For a discussion of the eschatologically new within history as irreducible to emergence, see Ansell (2013, chap. 5, including 234n77).

⁵⁵ The woman's association with the divine occurs via her association with the heavens and via her status as daughter of Wisdom, the spokesperson(ification) for God (see n. 23 above). As argued in the previous sub-section (see n. 46 above), the fact that the Hebrew preposition for *on* rather than *in* is used in Prov. 30:19b discourages us from associating the man (with)*in* the woman (19d) with the serpent *on* the rock (19b). This supports the gender symbolism I am arguing for. The male may find Wisdom, but it is the woman who represents Wisdom.

motif in Jeremiah, and not just to a saying that the book happens to cite, may require additional evidence and argumentation if it is not to be dismissed as the kind of intertextual relationship that exists only in the eye of the beholder. Although limitations of space prohibit a more extended discussion, the possibility that Prov. 30:17, which, together with Prov. 30:18–20, provides us with the double-centre to the ten-unit structure set out in Fig. 1, also alludes to a related motif in the book of Jeremiah, adds considerable weight to my proposal.

In the NET translation, Prov. 30:17 tells us that:

The eye that mocks at a father
    and despises [receiving instruction from] a mother—
the ravens of the valley will peck it out
    and the young vultures [literally: the sons of the vulture/eagle] will eat it.[56]

While we may wonder how this saying belongs alongside the verses that celebrate the way of a man within a woman, or the wisdom of a woman surrounding a man, when the *ravenous*, eye-plucking imagery of v. 17 (cf. the greed and violence of vv. 14, 15a, 15b, and 32–33) is taken together with the critical attention paid to kings in vv. 22, 27, and 31, then what we find ourselves being presented with, I suggest, at least in part, is an allusion to a *birds of prey* judgment motif found throughout Jeremiah (see Jer. 7:33, 12:9, 15:3, 16:4, 19:7, 34:20) combined with a pointed reference to the fate of the last king of Judah (Jer. 39:7, 52:11; cf. 2 Kings 25:7).

According to Murphy, the implication of Prov. 30:17 is that the person who is subjected to the eye-plucking judgment "dies out in the open, only to have the cadaver consumed by vultures" (Murphy 1998, 235). Although he refers to the historical judgments of 1 Kings 14:11 and 2 Sam. 21:10 in this context, the Israelite "fear of dying without burial" that he sees this verse deliberately evoking is nowhere more vividly portrayed than in Jer. 7:30–8:2.

Given that the birds of prey in Prov. 30:17 are "ravens of the valley," it is noteworthy that in Jer. 7:32–33, the birds that will feed on the human remains in the onslaught to come are explicitly located in the fiery valley of Slaughter (cf. Prov. 30:16), also known as the valley of the son(s) of Hinnom, or Gehenna.[57] Of particular significance to the interpreter of Prov. 30:17 is that Jeremiah not only connects this judgment to the fate of the kings of Judah in Jer. 8:1–2, but also associates it with the silencing of the joy of the bride and bridegroom in Jer. 7:34. This sad silencing of male and female itself becomes a motif that is repeated three times

---

[56] The last word of v. 17, *nešer*, can be translated as "eagle" or "vulture." It also appears in v. 19a. The phrase in the first set of brackets follows one of the notes to the NET. The noun in that phrase, often translated as "obedience," is in a form that occurs elsewhere only in Gen. 49:10. Given my discussion of Jer. 39:7 and 52:11 below, the fact that Gen. 49:10 concerns the future kingship of Judah merits further investigation. On Gen. 49:25, see n. 51 above.

[57] That the fire of the valley referred to in Jer. 7:31 would be naturally associated with the unquenchable fire of Isa. 66:24 and thus connected to the insatiable fire of Sheol in Prov. 30:16 (cf. the addition of *tartaros* to *hadēs* in the LXX) is also noteworthy in this context, as is the fact that whereas Isaiah sees this as a place of historical judgment for Israel's enemies, Jeremiah turns this judgment back on the people, and kings, of Judah (Jer. 7:30, 8:1–2). See the discussion in Ansell (2013, 316–40).

(7:34, 16:9, 25:10) before the poem that celebrates the woman surrounding the man (31:15–22) heralds its joyfully reversal in 33:10–11.

As for the fate of the last king of Judah, I refer to Jeremiah's account of the way the sons of Zedekiah are slaughtered in front of their father at the command of Nebuchadnezzar, who ensures that the vision of their death will stay with Zedekiah for the rest of his days as he then calls for his eyes to be poked out before he is led off to Babylon. Eye-gouging is associated with Israel's enemies in several passages, including Judg. 16:21 and 1 Sam. 11:2, both of which are quite close, in Hebrew, to Prov. 30:17. But in the case of Zedekiah, its enactment is especially cruel. This awful image of the brutal reality of Israel's exile is repeated twice in Jeremiah (at Jer. 39:7 and 52:11). We also find it in 2 Kings 25:7. Read intertextually, the fact that the vulture, or eagle, referred to in the last word of Prov. 30:17 was associated not only with Babylon, but with Nebuchadnezzar himself, as in Jer. 48:40 and 49:22 (cf. Ezek. 17:3, 12; Lam. 4:19), is not incidental.[58]

Once the numerical proverbs are read holistically and canonically, therefore, we can begin to appreciate how to an early, biblically literate, Hebrew audience attuned to chiastic structures and intertextual allusion—to those with ears to hear—the double-centre of Prov. 30:10–33, far from standing apart from the redemptive narrative(s) of Israel, would be heard to move, in the two-part structure of v. 17 and vv. 18–19, from exile to return-from-exile,[59] and from announcing the end of the old world order to celebrating the dawn of the new covenant.[60]

The Relativisation of Order

If the preceding intertextual-intratextual exploration has served to underline the central significance of Prov. 30:17–20, with its pointed juxtapositions between life and death, in vv. 19d and 17, and between Wisdom and Folly, in vv. 19d and 20, the

---

[58] To clarify: unlike the portrayal of Nebuchadnezzar and Zedekiah in Ezek. 17:1–10 (cf. vv. 11–21), Prov. 30:17 is not an allegory. My point is that this proverb participates in a web of intratextual and intertextual associations far beyond its initial reiteration of themes found in Prov. 30:11–13 by virtue of its final placement in Prov. 30:10–33. In my opinion, the editor/composer of this section of Proverbs would have been aware of the multiple allusions to the book of Jeremiah that I am highlighting (even though they have been lost by the time of the more ahistorically minded rabbinic traditions). My argument concerning the birds of prey, the valley, the fire, the kings of Judah, the bride/bridegroom motif, the eye-gouging, and the eagle-Babylon-Nebuchadnezzar connection, is a cumulative one. The parallel between the proverbs found in Prov. 30:18–19 and Jer. 31:22, read in context, is key.

[59] Concern with the threat of exile may be detected earlier in the book of Proverbs at 2:20–21 and 10:30.

[60] Consistent with this shift is the fact that the eagle/vulture of the valley in 30:17 makes way for the wondrous eagle of the skies in 30:19, while the king who is lampooned in 30:31 makes way for the king who embraces the wisdom he has learned from his mother (contrast 30:17a) in 31:1. For the parallel presence of both apocalyptic language and the return-from-exile theme in the book of Job, see Ansell (2017).

question that remains for the present discussion is whether the Wisdom that the man has found and that the woman represents right at the heart of all this—the Wisdom that embodies and discloses the life of the new covenant—is indicative or suggestive of a *creation order* or of *creation order thinking*. In this concluding section, I would like to highlight just how foreign this concern is to the passage we have been considering. For although one can no doubt read Prov. 30:18–19 within a creation order framework, there is nothing in the passage itself, I suggest, that would require us, or even encourage us, to think along those lines.

In fact, the phenomenon of thinking *along certain lines* is precisely the issue. Viewed within a "grain of the cosmos" perspective, the "ways" of existence that Proverbs refers to are invariably seen as pre-existing "grooves and tracks and norms" that can help us negotiate our lives wisely and well.[61] The "way" of wisdom, on this reading, may well be a *road less travelled*, but it does not entail any trailblazing. This view of discerning the way as finding a path that already exists (at least in principle) fits the conservative image that many scholars have of the book of Proverbs.[62] Yet the pre-existing protocols, principles, and precedents that a *going with the grain* approach would no doubt want to draw to our attention are hardly evident in the biblical account of the *aporia* that Solomon faces in the paradigmatic wisdom narrative of 1 Kings 3:16–28.[63] Furthermore, if the wisdom of discerning our way entails aligning our lives to pre-existing patterns, norms, or structures, then why, we may ask, is it that Prov. 30:19a would have us turn our gaze not to the "fixed order" of the stars (Jer. 31:35), but to the singular, unrepeatable path of an eagle, while v. 19b urges us to see the disclosure of wisdom on earth in the subtle and supple way of the serpent?[64]

This is not to say that considerations of order are irrelevant. For while the way of the ship in v. 19c cannot be derived from pre-existing patterns or limits, it will

---

[61] The language is taken from Smith (2009, 176) as cited more fully in n. 4 above. Cf. Van Leeuwen (1990). Given his conviction that "boundaries" are a "more fundamental" notion than the "ways," "houses," or "women" of Prov. 1–9 (1990, 116), Van Leeuwen inadvertently abandons the biblical image of the *two ways* or *directions* to stress the importance of staying within the limits of the one way, now reconceived as an *order*. Thus he writes, "Good behavior consists of staying on prescribed paths, evil actions are trespasses over forbidden limina. Folly is not staying where you belong, not walking on the path prescribed for you, not being in tune with the order of the cosmos" (ibid., 126–127).

[62] As noted in n. 18 above, this image of Proverbs is persuasively rejected by Hatton (2008). The proposal of Seerveld (1998) that there is a dialogue between Wisdom and Folly throughout the sayings collections in Prov. 10:1–29:27, which the reader is called to discern, would do much to counter this conservative image.

[63] For a brief account of this narrative, see Ansell (2008). An *aporia*, for me, does not represent a question that cannot be answered, but one that cannot be resolved via rational-conceptual or analytic-logical procedures.

[64] Although the path of the eagle displays both universality and individuality, the *singularity* of its *way* to which I am referring here (cf. n. 31 above) is not an intensification of individuality, in my view. In popular speech, we sometimes refer to singularity paradoxically when we say of someone that she is *one of a kind*, thereby intending a meaning prior to/beyond the individuality-universality correlation and prior to/beyond the confines of realism and nominalism.

wisely take the divinely established, fluid boundaries of the waters (see Ps. 104:6–9 and Job 38:8–11) into account. By the same token, we may even say that the way of the man within the woman, the way of the woman surrounding the man, *has* an order or structure. But that is not to say that it *is* an order or structure or that it discloses its primary meaning by *exemplifying* such an order. If that were the case, the sexologists of the world, those who have professional insight into the structure of sexuality, together with those who have the requisite know-how with respect to sexual technique, could explain or demonstrate things to us so well that the language of v. 18—"There are three things that are too wonderful for me, four that I do not understand"—would no longer be necessary.[65]

Can we say instead that this way that the man and the woman have found together is a *response* to an order? Again, I have to confess that this question is also foreign to the text, as I read it. For while it is true that the man and woman have found blessing, have found the way of life rather than the way of death (this "rather than" being the contrast between vv. 19b and 20, and between Prov. 5:19 and 5:20), it is, I suggest, not helpful to claim that they have done so because the way that they have found conforms to "the way it's supposed to be" (to borrow the language of Plantinga [1995]).

As blessing is grounded in precisely this kind of normed obedience according to creation order thinking, I will elaborate further why I think this, too, misses the point of the passage. Granted, in a fallen world, the difference between the way life *is* and the way life *ought* to be is often the distinction that is uppermost in our minds. And for good reason. Questions of normativity, of what ought to be, play a key role in redirecting a life or a culture that has gone awry. Norm-talk is thus indispensable. But it also has its limits. Given the way it is used to address what has gone wrong, and given the negative associations that it has by virtue of its capacity to negate a negative, the discourse of normativity is far more haunted by the fall than we realise. The tendency of creation order thinking to see the *creational* and the *normative* as virtually synonymous is thus highly problematic.[66]

It is most significant, therefore, that Scripture does not begin with commandment and obedience but with blessing and benediction (see Gen. 1:28).[67] And it is within

---

[65] Would this not also be the case with a sexologist who possessed Christian philosophical insight? Such a professional would surely seek to honour the mystery of sexuality in her explanations rather than explain it away. The wisdom of a wise farmer (Isa. 28:23–29) likewise includes but goes beyond know-how. For the (in my opinion, wrongheaded) view that biblical wisdom is a kind of "technical expertise" which provides us with "the skill of living," see DeWeese (2011, 53). On the mystery of order beyond explanation, see also n. 82 below.

[66] This is especially so in Kuyperian neo-Calvinism as redemption tends to be seen as the restoration of what is creational. See Ansell (2012).

[67] Thus I would reject Fretheim's claim that Gen. 1:26–28 is a "positive command" that exemplifies "creational law" (Fretheim 2005, 135). Those of us familiar with what neo-Calvinism calls the cultural *mandate* might want to read Gen. 1:28 again! Cf. Nicholas Wolterstorff's judgment, with which I concur: "I have come to think that there is no mandate there at all. God is *blessing* humankind. The sense of the words is, *May you* be fruitful and multiply" (Wolterstorff 2004, 296; his emphases). Cf. his focus on gift in his response to Wolters (1995) in Wolterstorff (1995, 64). For the relativisation of law even within Exodus and Leviticus, see the extremely helpful discussion of Sailhamer (1995). For Paul's understanding of Torah as supporting and opposing God's redemption, see the nuanced analysis of Wright (2013, 475–537). I take it that the "new commandment"

this "graced horizon"[68] and in this *ante*-normative and *ante*-nomian spirit that Prov. 30:18–19 has been written.[69] If we *start* with normativity, if the language of *ought* sets the direction from (or even, as) the very beginning, then creation and fall are subtly conflated as our experience of and response to fallenness, as our negation of a negative, come to redefine the benediction and blessing that precede and exceed the fall. Order is seen as the central meaning of shalom. Grace is reduced to law. Our sense of original blessing is pre-empted by our grasp of creational/moral/societal first principles. That it is meaningful to speak of love—the love that fulfils the very heart of God's "law" (Matt. 22:40; Rom. 13:10)—not only as a gift but also as a calling is not in dispute.[70] But when love of God and neighbour is construed as a response to a *norm*, as something we *ought* to do, love will elude us.[71]

Finally, a comment is in order about how we might think of the difference between our access to the "way" of "life" that Proverbs is concerned with (e.g., Prov. 3:2, 16, 18, 22, 4:10, 13, 22, 23, 6:23, 8:35, 9:11) and our access to a normative order. That we find life through Wisdom and not through Reason is a very deep assumption of all biblical wisdom literature. Reformational philosophy, aware of the privileged connection between reason and order in the Western philosophical tradition, and rightly wary of the rationalism and intellectualism of that tradition, has sought to counter the inflation of reason by saying that the normative order we need is more-than-rational and that rational-conceptual knowing does not have the privileged access we think it has.[72]

This is salutary. But does it go far enough? What if we were to say that reason does have privileged, though not exclusive, access to order—and then set out to

---

of John 13:34a aims to reorient a community used to commandments towards the gift/call and blessing/benediction of v. 34b.

[68] I am borrowing this phrase from Duffy (1993). In Catholic theology, grace is not limited to redemption as it tends to be in Protestant theology.

[69] This fits with the biblical wisdom tradition's focus on the giftedness of existence, on which see Murphy (1985, 5–6, and 8).

[70] For a discussion of gift/call (*Auf/Gabe*) and promise/call (*pro/missio*), see Ansell (2013, 263–272). In this light, creation order thinking can be seen as a reductionistic view of the call that has been severed from gift and promise. If it is helpful to connect gift and promise with God's Spirit and to connect the call with God's Word, then creation order thinking tends to sever Word and Spirit while reducing the Word of God to God's law. On the *realist* misreading of the political, covenantal metaphor of law in Scripture, see Hart (1995, 87–88n40). Cf. nn. 74 and 77 below.

[71] Thus, in my view, the discussion of "the norms for love" in Griffioen and Verhoogt (1990a, 11–12) remains problematic. While I appreciate their endorsement of Goudzwaard's directional rather than order-centred view of normativity here, I would rather use different terminology for spiritual direction as the language of normativity, in my opinion, cannot avoid evoking and invoking what *ought* to be. If, by contrast, we develop a different final vocabulary for those spiritual or directional realities (such as grace and benediction) that precede and exceed our norm-talk, this will also have the benefit of allowing our most cherished norms, principles, and values—all of which may participate in the spirituality of existence—to be stated more concretely. On the so-called love command of John 13:34, see n. 67 above.

[72] See, e.g., Wolters (1995, 35, 41). This is a very widespread view among contemporary reformational philosophers. But see the survey in Hart (2000) and see n. 73 below.

relativise both?⁷³ If religious direction is what we are after, and if wisdom is found by finding the *way* of wisdom rather than by finding an *order of* or *for* wisdom, it seems to me that while we may still see order as, in principle, one of God's good creatures, we are set free from an overinvestment in order as that which will guide us in the way of life and blessing.⁷⁴

Given our prior overinvestment, the relativisation, or decentring, of creation order might sound like the beginning of the end for Christian scholarship. But if Christian thinking finds its inspiration in attunement to the spirituality of existence, if it finds its way by listening to the voice of Wisdom in and through the enigmatic-yet-revealing ways of creation,⁷⁵ then maybe we can learn to begin again.⁷⁶

After all, (ante-nomian) relativisation is not (anti-nomian) eradication. Nothing that I have written is intended to deny that order—and our understanding of that order in the philosophical disciplines of ontology and epistemology—participates in the spirituality of existence along with the rest of creation, even though its inclusion does not entail that wisdom be fundamentally construed as our insight into, or experience of, the relation between the order of and order for existence. Put differently: if the covenant between God and creation is to be seen not as the relation between God's law and that which is subject to law but as the wise interplay between God's blessing and creation's participation in that blessing, then both law and order—both the order *of* exis-

---

⁷³ For reformational precedent for this suggestion, see Hart (1995, 2000). In the latter essay, Hart proposes an understanding of order in terms of a Dooyeweerdian analytic subject-object relationship.

⁷⁴ If we see order as a good creature of God, then, biblically speaking, we must also see it as a reality that is caught in the fall and is in need of redemption and eschatological fulfilment. To the extent that they have their ongoing origin in God's primordial blessing, order and law, along with the rest of creation, may be seen as participating in that blessing and may thus be experienced as a source of God's blessing. Cf. Gen. 8:22; Jer. 31:35–36, 33:20; and Ps. 19. But we should not overlook the ways in which order and law (that is, the order *of* existence and the order *for* existence) contribute to, as well as participate in, the groaning of creation referred to in Rom. 8:19–22.

⁷⁵ On the speech of creation, see nn. 14 (on Hosea 2:21–22) and 26 above, together with the insightful discussion of "creational glossolalia" in Middleton and Walsh (1995, 168–169). See also nn. 77 and 87 below.

⁷⁶ Although Sweetman also issues an invitation to "think again" about the nature of Christian scholarship in the context of his discussion of "spiritual exercise as imaginative starting point" (Sweetman 2016, 3), I see what I am calling for as significantly different. Given the close relationship that seems to exist between (1) the principle of "nature" seen by the ancient philosophers as "a principle of intelligibility in the cosmos that was deeper and truer than its articulation via social and cultural formation" (ibid., 4); (2) Augustine and Justin Martyr's later, would-be Christian understanding of philosophy as "a schooled practice of right living, of life lived in accordance with our deepest nature (creatures in relation with their maker)" (18n3); and (3) the neo-Calvinist, law-subject view of the covenant between God and creation (which I claim is the result of an absolutisation that I am seeking to relativise), my proposal that we think of Christian scholarship as a response to the gift/call of Wisdom rather than as an attempt to find our fundamental orientation via *order*—whether construed as conformity to the "deeper and truer. ... nature of things" (4, 7) or as alignment with the "lawful" nature of the divine presence (16, cited in n. 6 above) or will (see 94–96)—is, I suggest, to engage in a fundamentally different kind of "learning to think ... the [presently] unthinkable" (4). On the need to qualify Hadot and Sweetman's view of ancient philosophising as a "spiritual exercise," see n. 84 below.

tence and the order *for* existence that is associated with law—may be seen as avenues of blessing and avenues of grace and revelation, along with the rest of creation.[77]

This means that in the academic arena, if our understanding of what makes for a healthy organism, a viable, life-sustaining economy, or a faithful interpretation of a scriptural text is the fruit of genuine wisdom at work in the areas of biology, economics, or biblical theology, respectively, then that understanding—which includes our conceptual grasp of an order that holds, or that ought to hold, for the kind of phenomenon or area of life in question—may be seen as participating in and contributing to the wisdom that has given rise to it.

This is not an argument for reinstating the privileged connection between order and religious direction that characterises creation order thinking. In the biblical portrayal of wisdom, the discernment of what Scripture calls the "way" that is characterised by "life" rather than "death" (cf. Deut. 30:11–20 and John 14:6, in addition to the Proverbs references cited above) simply cannot be consistently interpreted as a matter of determining the deep, normative structures to which our lives ought to conform.[78] Neither is biblical wisdom fruitfully understood in terms of the related project of figuring out certain normative principles, which we then attempt to put into practice. Wisdom precedes and exceeds even the best possible understanding.[79] But the fact that wisdom, the *discernment* of the *way* to *life*, involves *more than* what we can understand or grasp—and *more than* what (creation) order can reveal—means that wisdom can give direction, and thus wisdom, to understanding.

Although this account of what we might call *understanding in wisdom* or *wisdom in understanding* emerges from a critique of a central tenet of Kuyperian neo-Calvinism vis-à-vis its overinvestment in order, it nevertheless closely parallels that tradition's construal of the relationship between faith and reason. The story of Western philosophy may be read as the attempt to relativise religion by placing it

---

[77] The same goes for law and order as avenues of wisdom and revelation (or as ways we may encounter wisdom and revelation). On this reading, there is no reason why my account of the voice of Wisdom cannot include the witness of what Jer. 31:35 calls the "fixed order" of the moon and the stars—provided that this is not taken as a paradigm for all forms of creational revelation. Cf. Ps. 19 in which the regular movements of the sun are associated with the speech of the heavens as well as with Torah. For law as an avenue and expression of God's blessing, see also n. 74 above. For a critique of the attempt to read the biblical texts listed there as an endorsement of a *realist* interpretation of law, see Hart (1995, 87–88n40). Such a critique (which I accept) does *not* claim that our experience of blessing in such contexts cannot be interpreted as *an experience of God's law*. Such language may be entirely appropriate. The problem (as I see it) lies (at least in part) with a philosophical realism that would subsume (or enclose) our faith experience of *law as an avenue of blessing* within an ontology that sees *blessing as an avenue of Law*. Admittedly, the former kind of discourse can, in a given context, make law a central metaphor, as is the case with Ps. 119. But this does not mean that the latter kind of discourse represents a biblical outlook.

[78] Here I will simply repeat my earlier references to the paradigmatic narrative of 1 Kings 3:16–28 and to the serpent as paradigmatic in Prov. 30:19b.

[79] This becomes even more evident in the NT where wisdom is related to the way of the cross, as in 1 Cor. 1:17–31. In this light, the wisdom background to the bread and wine of the Eucharist in Prov. 9:5 merits further attention.

"within the bounds of reason alone," to use Kant's memorable phrase.[80] But the preceding discussion of wisdom suggests that it is only when reason is seen within the dynamics of religion that mystery and understanding may be reunited.[81] No longer separated by the dubious, two-realm split between facts and values, our celebration of the enigmatic gift and promise of existence and our academic insight into the order of and for creation may enrich one another.[82]

We are sometimes reminded, not least by those who wish to see a radical renewal of the philosophical enterprise from within, that this is a "love of wisdom"—a *philo-sophia*—that has its beginnings in wonder.[83] But if we are attentive to the kind of wisdom that may be on offer—for an intellectual orientation to Greek *logos* will differ from discerning the way of the Word that, according to John 1:12–14, "became flesh and lived among [and within] us [*en hēmin*]"—we should also remember that wonder in the face of unfathomable mystery and the endlessly enigmatic cannot be reduced to wondering how we might solve or resolve a particular puzzle or enigma, whether this be as vast and deep as the Being of continental, or premodern, philosophy or as vexing and specific as the problems discussed in philosophy's more analytic traditions.[84] Should our thinking begin in "a-teleological w[o]ndering" only for that wonder to end in understanding?[85]

---

[80] For an English translation that contains helpful introductory essays, see Kant (1960). The first German edition was published in 1793.

[81] Wolterstorff's welcome reversal of Kant in Wolterstorff (1988), entitled *Reason within the Bounds of Religion*, may overlook the fact that it is Western philosophy that is focused on "bounds" in the sense of the limits of possibility. To transfer the bounds to religion may thus still concede too much to Reason. Hence my emphasis on dynamics.

[82] For the order of/for existence as a correlation that is the special focus of theoretical thought, see Hart (1995, 92–93n57). That we may understand and explain various phenomena in the light of conceptually grasped (networks of) order does not preclude seeing order itself (together with understanding) as also transcending understanding and participating in mystery.

[83] See the appeal to, and discussion of, Plato, *Theaetetus* 155d in this regard in Rubenstein (2008, 2–3).

[84] Similarly, if Hadot's thesis (see Hadot 1995, 2002) that the earliest schools of philosophy not only saw their love of wisdom as a "way of life" but engaged in "spiritual exercises" in the pursuit of disciplined self-knowledge is correct (for an acceptance of the former [from Socrates onwards] but not the latter [until the time of Seneca], see Cooper [2012, 17–23, 29–30, and 402–403nn4–5]), we still need to ask: what kind of spirituality was at work? Foucault's mediating position, that what he prefers to call pagan and Christian "technologies of the self" differ markedly (see Foucault 2007), is helpful in this regard. As a summary of what they actually did, Cooper might well accept Sweetman's account of "what the ancient philosophers called spiritual exercises" (Sweetman 2016, 3), while rejecting his (and Hadot's) claim that such pedagogical procedures were seen by the ancients (or should be seen by us) as "spiritual" in character. Others might recognise spiritual (self-)formation at work here while seeing it as misguided. But can structure and direction be so neatly separated? A third alternative is to see a way of life that is animated by a commitment to reason, and lived in accordance with the principle of "nature" that reason is thought to disclose (Sweetman 2016, 4), as a "quasi-religion" (see Smith 1994). Such "pursuits of wisdom" (Cooper 2012) might then be seen to exhibit a closed-down, quasi-spirituality.

[85] Here I allude to the "a-teleological wandering" that is seen as so problematic in Smith (2009, 176), as cited in n. 4 above. For Aristotle's influential view in, e.g., *Metaphysics* 1.983a, that the philosophical wondering about an effect comes to an end in the theological knowledge of its cause,

An appreciation for the endless nature of the enigmatic parallels what we may say about the ongoing dynamic of faith, which, as the neo-Kuyperian tradition has been right to insist, does not merely provide reason with its basic beliefs any more than it is confined to giving us access to supernatural truths we could never hope to know on our own. Rather faith, together with hope and love, may guide and accompany reason throughout its journey because it is in experiencing the world in faith, hope, and love that we may know more than we can comprehend.[86]

Such a perspective is well placed, I suggest, to allow creation order to find its future within a wider appreciation for, and experience of, creational revelation—here interpreted as the live speech of ongoing, creational "meaning" (cf. von Rad 1972, 148; Dooyeweerd 1953–1958, 1:4).[87] If we pursue the *love of wisdom* in the academic arena in a way that is attuned to the *wisdom of lovers*, as celebrated in Prov. 30:19d, Jer. 31:22, and elsewhere, therefore, we may continue, in the spirit of Anselm's famous "credo ...," to "*believe* [that is: exercise faith/trust/hope] in order that [we] may *understand*"[88] even though, and moreover, precisely *because*, the way of Wisdom is, in the words of Prov. 30:18, way "too wonderful" for that.

---

see Rubenstein (2008, 12–13). It is ironic that in an extract entitled "The Love of Wisdom," in McInerny (1998, 718–743), Aquinas appears to accept this in his commentary on the same passage. But is it only the miracle of divine, rather than natural, causation that is unfathomable, as Aquinas seems to believe? Is endless wonder (*admiratio*, in distinction from the settled gratitude of *complacentia*) only appropriate when we love God and not also when we love creation? See the discussion in Crowe (2000), Rubenstein (2008, 13), and Smith (2014).

[86] See n. 32 above. One way to put this is to say that there is a depth-meaning to persons and phenomena, and to structures and singularities, that even God does not understand, not because of a deficiency of knowledge but precisely because God knows perfectly. Knowing, not least in the context of being "in" the "truth" (Pss. 25:5, 86:11; John 4:24, 8:44; 2 Pet. 1:12; 3 John 3–4), is more/other than getting it right or being "almost wrong ... almost right" (Sweetman 2016, 152). Cf. Ansell (2013, 360, 385–387): "The rose flowers without a why" (Angelus Silesius); "I smell a rose, and smell the kingdom of God" (Arnold van Ruler).

[87] It is telling that in his discussion of Job 28, von Rad backs away from an emphasis on order (as he does in von Rad [1972, 106–110]) to connect biblical wisdom not only to "the divine mystery of creation," but also to creation's "meaning" (ibid., 148). Dooyeweerd, in his attempt to relativise a philosophical fixation on being, recommends the same central metaphor when he claims: "Meaning is the being of all that has been created and the nature even of our selfhood" (Dooyeweerd 1953–1958, 1:4). On the speech, or voice, of created existence, see nn. 12, 14, 23, 26, 75, and 77 above. While there are indications that Dooyeweerd tended to associate *meaning* (in this more than linguistic sense) with *expression* in what he called the foundational direction of time and *religion* (in a more than fiduciary sense) with *referring* in the transcendental direction (see Olthuis 1985, 23; cf. Ansell 2013, 238n85), I would look to connect the speech of creation to both temporal directions (on which see Ansell [2013, 228–256]). In my view, the conveyed meaning of *creation* goes beyond what many theologians understand by *general revelation*, as it includes, inter alia, "the voice of [Abel's] blood ... crying out to [YHWH] from the ground" (Gen. 4:10 [NET]), the voice of the thunder in John 9:28–30 (cf. Exod. 19:19), and the "groaning" of the "whole creation" as it gives birth in Rom. 8:22.

[88] Anselm's "credo ut intelligam" may be traced back to Augustine's reflections on Isa. 7:9, on which see the helpful discussion in Sweetman (2016, 23–37). In seeing faith as trust (believing in) rather than as (thinking with) assent, I am indebted to Smith (1979).

References

Ansell, Nicholas. 1997. Commentary: Romans 1:26f. *Third Way* 20 (7): 20.
———. 2001. The Call of Wisdom/The Voice of the Serpent: A Canonical Approach to the Tree of Knowledge. *Christian Scholar's Review* 31 (1): 31–57.
———. 2002. Commentary: Exodus 19:5–6. *Third Way* 25 (9): 22.
———. 2008. Jesus on the Offensive. *The Banner* 143 (10): 44–45.
———. 2011a. The Embrace of Wisdom: Proverbs 8, Genesis 1, and the Covenantal Dynamics of Existence. Paper presented to the Wisdom in Israelite and Cognate Traditions unit at the Annual Meeting of the Society of Biblical Literature, San Francisco, November 20, 2011.
———. 2011b. On (Not) Obeying the Sabbath: Reading Jesus Reading Scripture. *Horizons in Biblical Theology* 33 (2): 97–120.
———. 2011c. This Is Her Body … : Judges 19 as Call to Discernment. In *Tamar's Tears: Evangelical Engagements with Feminist Old Testament Hermeneutics*, ed. Andrew Sloane, 112–170. Eugene: Pickwick.
———. 2012. It's About Time: Opening Reformational Thought to the Eschaton. *Calvin Theological Journal* 47 (1): 98–121.
———. 2013. *The Annihilation of Hell: Universal Salvation and the Redemption of Time in the Eschatology of Jürgen Moltmann*. Milton Keynes/Eugene: Paternoster/Cascade.
———. 2017. Fantastic Beasts and Where to Find The(ir Wisdo)m: Behemoth and Leviathan in the Book of Job. In *Playing with Leviathan: Interpretation and Reception of Monsters from the Biblical World*, Themes in Biblical Narrative 21, ed. Koert van Bekkum et al., 90–114. Leiden: Brill.
Brueggemann, Walter. 1997. *Theology of the Old Testament: Testimony, Dispute, Advocacy*. Minneapolis: Fortress.
Caputo, John D. 2013. *Truth. Philosophy in Transit*. London: Penguin.
Chaplin, Jonathan. 2011. *Herman Dooyeweerd: Christian Philosopher of State and Civil Society*. Notre Dame: University of Notre Dame Press.
Clifford, Richard J. 1997. Introduction to Wisdom Literature. In *The New Interpreter's Bible*, ed. David L. Petersen et al., vol. 5, 1–16. Nashville: Abingdon Press.
Collins, John J. 1997. *Jewish Wisdom in the Hellenistic Age*. Louisville: Westminster John Knox.
Cooper, John M. 2012. *Pursuits of Wisdom: Six Ways of Life in Ancient Philosophy from Socrates to Plotinus*. Princeton: Princeton University Press.
Crenshaw, James L. 2010. *Old Testament Wisdom: An Introduction*. 3rd ed. Louisville: Westminster John Knox.
Crowe, Frederick E. 2000. Complacency and Concern in the Thought of St. Thomas. In *Three Thomist Studies*, ed. Michael Vertin and Frederic Lawrence, 71–203. Boston: Lonergan Institute at Boston College.
DeWeese, Garrett J. 2011. *Doing Philosophy as a Christian*. Downers Grove: IVP Academic.
Dooyeweerd, Herman. 1935–1936. *De wijsbegeerte der wetsidee*. 3 vols. Amsterdam: H.J. Paris.
———. 1953–1958. *A New Critique of Theoretical Thought*. 4 vols. Philadelphia: Presbyterian and Reformed.
———. 1968. *The Christian Idea of the State*. Trans. John Kraay. Nutley: Craig Press.
———. 1986. *A Christian Theory of Social Institutions*. Trans. Magnus Verbrugge, ed. John Witte, Jr. La Jolla: The Herman Dooyeweerd Foundation.
Duffy, Stephen J. 1993. *The Graced Horizon: Nature and Grace in Modern Catholic Thought*. Collegeville: The Liturgical Press.
Echeverria, Eduardo J. 2011. *"In the Beginning...": A Theology of the Body*. Eugene: Wipf and Stock.
Foucault, Michel. 2007. Subjectivity and Truth. In *The Politics of Truth*, ed. Sylvère Lotringer and Lysa Hochroth, 147–167. New York: Semiotext(e).
Fox, Michael V. 2009. *Proverbs 10–31: A New Translation with Introduction and Commentary*. The Anchor Yale Bible Commentary 18B. New Haven: Yale University Press.

Fretheim, Terence E. 2002. *Jeremiah*. Macon: Smyth and Helwys.
———. 2005. *God and World in the Old Testament: A Relational Theology of Creation*. Nashville: Abingdon Press.
Garrett, Duane A. 1993. *Proverbs, Ecclesiastes, Song of Songs*. The New American Commentary 14. Nashville: Broadman.
Gerrish, B.A. 1993. *Grace and Gratitude: The Eucharistic Theology of John Calvin*. Minneapolis: Augsburg Fortress.
Grant, Jamie A. 2008. Wisdom and Covenant. In *Dictionary of the Old Testament Wisdom, Poetry & Writings*, ed. Tremper Longman and Peter Enns, 858–863. Downers Grove: InterVarsity Press.
Griffioen, Sander, and Jan Verhoogt. 1990a. Introduction: Normativity and Contextuality in the Social Sciences. In *Griffioen and Verhoogt* 1990b, 9–22.
———, eds. 1990b. *Norm and Context in the Social Sciences*. Lanham: University Press of America.
Hadot, Pierre. 1995. *Philosophy as a Way of Life: Spiritual Exercises from Socrates to Foucault*. Trans. Michael Chase, ed. A.I. Davidson. Oxford: Blackwell.
———. 2002. *What Is Ancient Philosophy?* Trans. Michael Chase. Cambridge: Harvard University Press.
Hart, Hendrik. 1984. *Understanding Our World: An Integral Ontology*. Lanham: University Press of America.
———. 1995. Creation Order in Our Tradition: Critique and Refinement. In *An Ethos of Compassion and the Integrity of Creation*, ed. Brian J. Walsh, Hendrik Hart, and Robert Vandervennen, 67–96. Lanham: University Press of America.
———. 2000. Notes on Dooyeweerd, Reason, and Order. In *Contemporary Reflections on the Philosophy of Herman Dooyeweerd: A Supplement to the Collected Works of Herman Dooyeweerd*, ed. D.F.M. Strauss and Michelle Botting, 125–146. Lewiston: Edwin Mellen Press.
Hatton, Peter T.H. 2008. *Contradiction in the Book of Proverbs: The Deep Waters of Counsel*. Aldershot: Ashgate.
Hauerwas, Stanley. 2001. *With the Grain of the Universe: The Church's Witness and Natural Theology*. Grand Rapids: Brazos.
Heim, Knut Martin. 2001. *Like Grapes of Gold Set in Silver: An Interpretation of Proverbial Clusters in Proverbs 10:1–22:16*. Beihefte zur Zeitschrift für die alttestamentliche Wissenschaft. Berlin: Walter de Gruyter.
———. 2013. *Poetic Imagination in Proverbs: Variant Repetitions and the Nature of Poetry*. Bulletin for Biblical Research Supplements 4. Winona Lake: Eisenbrauns.
Hesselink, I. John. 2001. Law. In *The Westminster Handbook to Reformed Theology*, ed. Donald K. McKim, 134–136. Louisville: Westminster John Knox.
Holladay, William. 1974. *Jeremiah: Spokesman Out of Time*. Philadelphia: United Church Press.
———. 1989. *Jeremiah 2: A Commentary on the Book of the Prophet Jeremiah Chapters 26–52*. Hermeneia. Minneapolis: Fortress Press.
Jaspers, Karl. 1954. *Way to Wisdom: An Introduction to Philosophy*. Trans. Ralph Manheim. New Haven: Yale University Press.
Kant, Immanuel. 1960. *Religion Within the Limits of Reason Alone*, 2nd ed. Trans. Theodore M. Greene, Hoyt H. Hudson, and John R. Silber. New York: Harper & Row.
Keown, Gerald L., Pamela J. Scalise, and Thomas G. Smothers. 1995. *Jeremiah 26–52*. Word Biblical Commentary 27. Waco: Word.
LaMothe, Kimerer L. 2009. *What a Body Knows: Finding Wisdom in Desire*. Winchester: O Books.
Levin, David Michael. 1985. *The Body's Recollection of Being: Phenomenological Psychology and the Deconstruction of Nihilism*. London: Routledge & Kegan Paul.
Lucas, Ernest C. 2008. Wisdom Theology. In *Dictionary of the Old Testament Wisdom, Poetry & Writings*, ed. Tremper Longman and Peter Enns, 907–909. Downers Grove: InterVarsity Press.

McInerny, Ralph, ed. 1998. The Love of Wisdom. Exposition of Metaphysics, Preface and 1, 1–3 (1271). In *Thomas Aquinas: Selected Writings*. London: Penguin.
McKane, William. 1996. *Jeremiah*. Vol. 2, *Commentary on Jeremiah 26–52*. International Critical Commentary. Edinburgh: T. and T. Clark.
Middleton, J. Richard, and Brian J. Walsh. 1995. *Truth Is Stranger Than It Used to Be: Biblical Faith in a Postmodern Age*. London: SPCK.
Morgan, Donn F. 1981. *Wisdom in the Old Testament Traditions*. Atlanta: John Knox Press.
Murphy, Roland E. 1985. Wisdom and Creation. *Journal of Biblical Literature* 104 (1): 3–11.
———. 1998. *Proverbs*. Word Biblical Commentary 22. Nashville: Thomas Nelson.
———. 2002. *The Tree of Life: An Exploration of Biblical Wisdom Literature*. 3rd ed. Grand Rapids: Eerdmans.
Olthuis, James H. 1985. Dooyeweerd on Religion and Faith. In *The Legacy of Herman Dooyeweerd: Reflections on Critical Philosophy in the Christian Tradition*, ed. C.T. McIntire, 21–40. Lanham: University Press of America.
Perdue, Leo G. 2000. *Proverbs*. Interpretation. Louisville: John Knox Press.
Plantinga, Cornelius, Jr. 1995. *Not the Way It's Supposed to Be: A Breviary of Sin*. Grand Rapids: Eerdmans.
Rubenstein, Mary-Jane. 2008. *Strange Wonder: The Closure of Metaphysics and the Opening of Awe*. New York: Columbia University Press.
Sailhamer, John H. 1995. Appendix B: Compositional Strategies in the Pentateuch. In *Introduction to Old Testament Theology: A Canonical Approach*, ed. John H. Sailhamer, 272–289. Grand Rapids: Zondervan.
———. 2000. A Wisdom Composition of the Pentateuch? In *The Way of Wisdom: Essays in Honor of Bruce K. Waltke*, ed. J.I. Packer and Sven K. Soderlund, 15–35. Grand Rapids: Zondervan.
Seerveld, Calvin G. 1998. Proverbs 10:1–22: From Poetic Paragraphs to Preaching. In *Reading and Hearing the Word: From Text to Sermon; Essays in Honor of John H. Stek*, ed. Arie C. Leder, 181–200. Grand Rapids: Calvin Theological Seminary/CRC Publications.
———. 2003. *How to Read the Bible to Hear God Speak: A Study in Numbers 22–24*. Sioux Center/Toronto: Dordt College Press/Tuppence Press.
Skillen, James W., and Rockne M. McCarthy, eds. 1991. *Political Order and the Plural Structure of Society*. Atlanta: Scholars Press.
Smith, Wilfred Cantwell. 1979. *Faith and Belief*. Princeton: Princeton University Press.
Smith, John E. 1994. *Quasi-Religions: Humanism, Marxism, and Nationalism*. Themes in Comparative Religion. London: Palgrave Macmillan.
Smith, James K.A. 2009. *Desiring the Kingdom: Worship, Worldview, and Cultural Formation*. Grand Rapids: Baker Academic.
Smith, Randall B. 2014. "If Philosophy Begins in Wonder": Aquinas, Creation, and Wonder. *Communio: International Catholic Review* 41 (1): 92–111.
Strauss, D.F.M. 2009. *Philosophy: Discipline of the Disciplines*. Grand Rapids: Paideia Press.
Sweetman, Robert. 2016. *Tracing the Lines: Spiritual Exercise and the Gesture of Christian Scholarship*. Currents in Reformational Thought. Eugene: Wipf & Stock.
Thompson, J.A. 1980. *Jeremiah*. The New International Commentary on the Old Testament. Grand Rapids: Eerdmans.
Trible, Phyllis. 1978. *God and the Rhetoric of Sexuality*. Overtures to Biblical Theology. Philadelphia: Fortress Press.
van der Walt, B.J. 1973. Eisegesis-Exegesis, Paradox and Nature-Grace: Methods of Synthesis in Mediaeval Philosophy. *Philosophia Reformata* 38 (1): 191–211.
VanDrunen, David. 2010. *Natural Law and the Two Kingdoms: A Study in the Development of Reformed Social Thought*. Grand Rapids: Eerdmans.
Van Leeuwen, Raymond C. 1990. Liminality and Worldview in Proverbs 1–9. *Semeia* 50: 111–144.
———. 1997. The Book of Proverbs: Introduction, Commentary, and Reflections. In *New Interpreter's Bible*, ed. David L. Petersen et al., vol. 5. Nashville: Abingdon.
von Rad, Gerhard. 1972. *Wisdom in Israel*. Trans. James D. Martin. London: SCM Press.

Walsh, Brian J., and J. Richard Middleton. 1984. *The Transforming Vision: Shaping a Christian World View*. Downers Grove: InterVarsity Press.
Walsh, Brian J., Hendrik Hart, and Robert Vandervennen, eds. 1995. *An Ethos of Compassion and the Integrity of Creation*. Lanham: University Press of America.
Waltke, Bruce K. 2005. *The Book of Proverbs, Chapters 15–31*. The New International Commentary on the Old Testament. Grand Rapids: Eerdmans.
Weeks, Stuart. 2010. *An Introduction to the Study of Wisdom Literature*. London: T & T Clark.
Wolters, Albert M. 1995. *Creation Order: A Historical Look at Our Heritage*. In Walsh et al. 1995, 33–48.
———. 2001. *The Song of the Valiant Woman: Studies in the Interpretation of Proverbs 31:10–31*. Carlisle: Paternoster.
———. 2005. *Creation Regained: Biblical Basics for a Reformational Worldview*. 2nd ed. Grand Rapids: Eerdmans.
Wolterstorff, Nicholas. 1983. *Until Justice and Peace Embrace*. Grand Rapids: Eerdmans.
———. 1988. *Reason within the Bounds of Religion*. 2nd ed. Grand Rapids: Eerdmans.
———. 1995. *Response to Albert M. Wolters*. In Walsh et al. 1995, 62–66.
———. 2004. *Educating for Shalom*, ed. Clarence W. Joldersma. Grand Rapids: Eerdmans.
Wright, N.T. 2013. *Christian Origins and the Question of God*. Vol. 4, *Paul and the Faithfulness of God*. Minneapolis: Fortress Press.
Yett, Danielle. 2017. Passing Away: Despair, Comfort, and the Poetic Deterioration of the Book of Lamentations. http://ir.icscanada.edu/icsir.
Yoder, John Howard. 1988. Armaments and Eschatology. *Studies in Christian Ethics* 1: 43–61.
———. 1994. *The Politics of Jesus: Vicit Agnus Noster*. 2nd ed. Grand Rapids: Eerdmans.
Zuidema, S.U. 1972. Philosophy as Point of Departure. In *Communication and Confrontation: A Philosophical Appraisal and Critique of Modern Society and Contemporary Thought*, ed. S.U. Zuidema, 124–128. Toronto: Wedge.

A Contribution to the Concept of Creation Order from a Lutheran Perspective

Hans Schaeffer

Abstract In (recent) Lutheran systematic theology, the concept of creation orders is used as a heuristic tool in interpreting the world around us in order to contribute to one of the central tasks of theological ethics: the interpretation of the reality in which human beings are called to live responsibly. This article describes the opinions of several contemporary Lutheran authors on creation orders in order to compare them with the concept of creation orders in Reformed theology. It concludes with an investigation of the possibilities of a Lutheran contribution to the concept of creation orders within neo-Calvinist thinking.

Keywords Lutheran theology · Creation order · Oswald Bayer · Theological ethics · Social ethics · Neo-Calvinism · Marriage

Lutheran Theology on Creation Orders

Within recent Lutheran theology, several authors have made a contribution to the elaboration of the concept of creation orders.[1] The most pervasive work is done by Oswald Bayer (b. 1939), former professor of systematic theology at the Eberhard-Karls-Universität Tübingen. Bernd Wannenwetsch (b. 1959), former professor of theological ethics at the University of Aberdeen, has taken Bayer's use of Luther's concept into account in his own contribution. The concepts of Martin Honecker and Hans G. Ulrich play a minor role in this article. The discussion will be focused on the use of a concept of creation orders with its consequences for social ethics. The fact that I use the plural *creation orders* instead of the more general expression *creation order* is due to the specifically Lutheran interpretation of how to conceive of the created reality.[2]

---

[1] The plural *orders* is already a specific Lutheran refinement of the more abstract *creation order*.
[2] This is how Bayer (2003, vii–viii) explicates this specifically Lutheran way of thinking: "Systemdenker sind auf Einheit und Stimmigkeit versessen; alles soll sich umstandslos reimen.

H. Schaeffer (✉)
Theological University Kampen, Kampen, The Netherlands
e-mail: jhfschaeffer@tukampen.nl

Bayer's concept of creation orders derives its specific shape from Martin Luther's doctrine of the three estates.[3] It is Bayer's explicit intention to overcome the flaws that critics easily attribute to the concept of creation orders[4] without succumbing to a postmodern "anything goes." It is best to start the discussion by analyzing the following statement of Bayer: "Creation is instituted and by that addressing us" (Bayer 1995, 126; cf. Ulrich 2007, 234).[5] This dense and brief dictum summarizes Bayer's specific stance and provides us with the tools to interpret and analyze his reception of Luther.

Bayer's statement means that creation (*creatura*) is constitutively dependent on God's activity (*creatio*).[6] The relation between *creatura* and *creatio* is here framed as *institutio*, by which Bayer takes up the classic word from an important Lutheran confession, the *Confessio Augustana*, article V.[7] Institution is a term intended to conceptualize Bayer's specific (theological, Lutheran) contribution to current discussions in anthropology and sociology on how to conceive of social structures. Bayer wants to provide an alternative to both Hegel's focus on the institution and Gehlen's focus on nature by stressing the indissoluble bond between nature and institution. This indissolubility means that, from a theological point of view, creation (*creatura*) is conceived as God's speech act[8] (*creatio*) by which he addresses

---

Luthers Theologie provoziert dazu, dieses uns so geläufige Verfahren gründlich in Frage zu stellen. ... Luthers Theologie ist zu beweglich und zu komplex, als dass sie sich auf einen einzigen Begriff bringen liesse. Sie ist nicht apriorisch als System konzipiert, sondern erhält ihren inneren Zusammenhalt nur von der Bezogenheit auf die Dynamik des vom Gesetz unterschiedenen Evangeliums: der göttlichen Zusage (promissio)."

[3] The Lutheran triad of estates is the reformational form of doing social ethics. As such, this doctrine is far more important than the widely treated *Zwei-Reiche-Lehre* (Honecker 1999, 263–264; Bayer 1995, 121).

[4] The use of creation orders in the German Third Reich and its use in South African apartheid are the most compelling examples of such misuse (Link 1976; Van der Kooi 2011). Misunderstanding of the concept of creation orders as being static and medieval, not fit to be of help in the "rapid social change," have also to be countered (Maurer 1974; Schwarz 1978).

[5] "Schöpfung ist eingesetzte Welt und darin Zusage." Unless noted otherwise, all translations are my own.

[6] For a more systematic definition of the word *creation* and its many theological aspects, see Schaeffer (2006, 9–12).

[7] Cf. *Evangelisches Kirchenlexicon*, s.v. "Institution," 418: "Theologisch und religionsgeschichtlich relevant ist der Begriff 'institutio' in der Fassung, die sich in dem Grundtext lutherischer Theologie—der 'Confessio Augustana'—findet" (H. Dubiel). The text of *CA* V ("De ministerio ecclesiastico") goes as follows: "Solchen Glauben zu erlangen, hat Gott das Predigamt eingesetzt (*institutum*), Evangelium und Sakrament gegeben, dadurch er als durch Mittel, den heiligen Geist gibt, welcher den Glauben, wo und wenn er will (*ubi et quando visum est Deo*), in denen, so das Evangelium hören, wirket, welches da lehret, dass wir durch Christus Verdienst, nicht durch unser Verdienst, ein gnädigen Gott haben, so wir solches glauben. Und werden verdammt die Wiedertäufer und andere, so lehren, dass wir ohne das leibliche Wort des Evangelii (*sine verbo externo*) den heiligen Geist durch eigene Bereitung, Gedanken und Werke erlangen" (*Bekenntnisschriften der evangelisch-lutherischen Kirche* 58, 1–17).

[8] Bayer on this point draws on the interpretation of Luther by Johann Georg Hamann: "Schöpfung als 'Rede an die Kreatur durch die Kreatur'" (Bayer 1990, 9–32).

us, thus allowing us to respond. There is no creation apart from its being instituted by God. The concept of *creatio* as divine speech act also contains the possibilities for human beings to become God's created means by which he institutes (creates and upholds) the world. Bayer therefore stresses the importance of man's ability to order life (otherwise unstable and fluctuating) by speech, to which God calls us.[9]

This specific doctrine of creation provides us with the possibilities to overcome the usual pitfalls in the area of creation orders. For creation, in Bayer's opinion, is neither mere substance to be treated and ordered by man, nor a set of prefixed, static orders man has to be squeezed into. Creation calls us to live responsibly, answering God's life-giving act of creation. Creation is the God-given space to live, together with the Creator who comes to creation in his communicative word and grants us life together with him and all creatures. We have to use the freedom that God gives us to interpret and order the world by "giving names" (Gen. 2:19–20). It is in this context of the doctrine of creation that Luther uses the doctrine of the three estates.[10]

Luther's use of the three estates is paradigmatically formulated in his commentary on Genesis. Bayer in turn uses this to stress that the estates are a theological tool to outline the implications of the Genesis-account for everyday life (Bayer 1990, 35–56; 1995, 116–146; 2003, 110–139). According to Luther, God's work as creation is divided in three "hierarchies": *oeconomia, politia*, and *ecclesia*. Church (*ecclesia*) is the primal relationship between a human being as *creatura* and God the Creator. Economy (*oeconomia*) denotes everything that in current society is differentiated, such as marriage, family, economy, education, and science. The third is the state (*politia*), which is neither only—from a pessimistic point of view— the necessary boundary for sinful and evil human behavior (Chaos), nor only—from an optimistic point of view—the required regulation of human organized life (Cosmos). If human beings need to learn how to live in a gracefully restored relationship with God and one another, it is through God's power that he enables us to do so. Political power should embody this Godly power, enabling human beings to live according to God's promise (Bayer 1991). Bayer's position can be summarized as follows:

> Addressed by God through creation and challenged to respond, man lives either in belief or in unbelief. Even creation orders are not something that, as order of the world, could be

---

[9] Bayer (2003, 111) summarizes: "Für die Gestaltung der nicht festgelegten Natur des Menschen ist die Sprache—als umfassender Vorgang der Symbolbildung—entscheidend. Die Sprache gibt der Natur eine Verfassung, stiftet Bestimmtheit, ordnet die Handlungsabläufe und macht so menschliches leben als in Erinnerung und Hoffnung perspektivisch erst möglich. Den damit angesprochenen Sachverhalt hat Martin Luther mit seiner Dreiständelehre bedacht. Die Hauptpointe dabei liegt in der unauflöslichen Verschränkung von 'Element und Einsetzung.'" The importance of *speech*—and, more specifically, *promissio*—is prominent in Bayer's theology (Schaeffer 2006, 101–189).

[10] From this point onwards, I will use the vocabulary of *estates, hierarchies*, and *institutions* instead of the general term *creation orders* to denote the specific Lutheran interpretation of it. Cf. Brian Brock's remark on terminology (Brock 2007, 180).

understandable out of itself, but can only be rightly perceived through faith in the word by which these orders are instituted. (Bayer 2007, 241)[11]

Bernd Wannenwetsch (2000b) uses this Lutheran concept of creation orders to demonstrate its heuristic power in the field of theological ethics. Instead of searching for universally applicable rules of conduct (Kant), Wannenwetsch's concern is to find a way to do social ethics without reduction or abstraction of concrete life, and yet not leaving us bewildered by a multitude of unordered phenomena. His concern—which is also Bayer's concern—is to find a way to order creation that is not imposed on it. In social ethics, the field is usually differentiated and labeled along the lines of *Bereiche* or "spheres": medical ethics, bioethics, management ethics, sexual ethics, etc. In each of those fields, the logic of autonomous functionality prevails (utilitarianism) because transcending questions of how the fields relate to one another, or how they are able to deal with possible conflicts, or questions as to whether an overarching field of praxis exists, are not posed.[12] Wannenwetsch argues, however, that the discernment of different fields is not neutral. The process of discernment is ideologically biased from the outset as Foucault and Derrida have shown. In order to overcome and avoid these problematic features of theological ethics, Wannenwetsch is searching for "elementary life-forms" (*elementare Lebensformen*), as an alternative to, on the one hand, a concept of obviously present and therefore discernible *Lebensverhältnisse*, and, on the other hand, non-related *Bereiche* in need of a transcendental functionality to be applied into specific fields. It is all about God calling human beings to live accordingly in faithful response: he calls us[13] to live in concrete relationships that are called forth by God's creative power. We do not have to discern self-evident spheres—we have to listen to God's word to become faithful to our calling within specific relationships.[14]

In a similar way, Hans Ulrich uses the estates to establish social ethics.[15] He claims that, because "moral claims are expressions of beliefs about how the world

---

[11] "Als von Gott durch die Kreatur Angeredeter und zur Antwort Herausgeforderter lebt der Mensch entweder im Glauben oder im Unglauben. Auch die Schöpfungsordnungen sind nicht etwas, was als Ordnung der Welt aus sich selbst heraus verständlich wäre, sondern was nur im Glauben an das Einsetzungswort dieser Ordnungen recht wahrgenommen werden kann."

[12] "Je mehr das Teleologische, die Frage nach der Zuordnung der Praxisbereiche (Praktiken) zu grösseren Praxiskomplexen, aussen vor bleibt, drängt sich nämlich das utilitarische Paradigma hinein" (Wannenwetsch 2000b, 103).

[13] Wannenwetsch uses the concept of *Beruf* (vocation) here (Wannenwetsch 2000b, 123–125).

[14] "Weil die Erkenntnis der Lebensverhältnisse und ihrer respektiven Ethiken für die reformatorische Auffassung am Wort hängt und nicht selbstevident ist, lässt sich an einem Beispiel aufzeigen. Dass es Väter gibt und gewisse, kulturübergreifende moralische Standards des spezifischen Lebensverhältnisses, in dem Eltern zu ihren Kindern stehen, is das eine; der theologisch springende Punt für die evangelische Ethik ist freilich, dass die christliche Interpretation der Vatterrolle nicht aus diesem Lebenswissen ableitbar ist, sondern von der Funktion der Stellvertretung, die den Eltern in spezifischer Weise zugesprochen und zugemutet ist: dass sie bei ihren Kindern Christus vertreten, indem sie ihnen sein Wort ausrichten. 'Denn gewisslich ist vater und mutter der kinder Apostel, Bischoff, pfarrer, ynn dem sie das Euangelion yhn kundt machen' [Luther]" (Wannenwetsch 2000b, 126).

[15] "Es gehört zur Geschichte des geschöpflichen Lebens im Kontext der Ökonomie Gottes, dass Menschen *nicht* einem unabsehbaren Prozess oder einer Entwicklung ausgesetzt sind.... An dieser

is, and language about how the world is cannot be separated from the patterns of behaviour and thought (norms) which already in-form us" (Brock 2007, 181), we need some *in-form-ative* address of God that locates[16] us in the world as God's creation and by which we can—tentatively—discern its order. "Martin Luther also continually stressed that procedures for living are *demanded*, not as extant life necessities to which we must adjust or submit. They are given with the Word of God, which contains the promise that human life may be creaturely and will ever again be freed to be creaturely" (Ulrich 2007, 114; translated in Brock 2007, 195).

The role of the doctrine of the three estates, therefore, is a *heuristic* one. Wannenwetsch (2000b) lists four ways in which the life-forms contribute to such heuristic task. Firstly, this doctrine stresses that the estates are created simultaneously with human beings. Their being created by God implies that they mark not only the God-given space of human life but also God's means for sanctification of human life within the present, sinful context. The estates, then, do not represent any current state of affairs but signify God's concrete life-giving forms. Secondly, this doctrine prohibits the annexation of one area at the cost of another: church is not state, economy should not be governed by the state or the church, church should not be governed by the economy, family should be controlled neither by state nor by church, etc. Thirdly, the doctrine of the estates is about relating the individual fields to the three life-forms in such a way that these fields are questioned with respect to their political, economic, and religious meaning.[17] Finally, it provides opportunities for incorporating personal ethics as well. Ethics should not be about conceiving quasi-neutral technical skills for applying criteria in any area of human life but about developing a God-answering way of life with respect to reality's threefold structure.

Ulrich makes his use of the three estates concrete by relating it to Niklas Luhmann's theory of social systems. Apart from all kinds of fundamental critique on Luhmann's theory,[18] Ulrich frames his main question in terms of *seeing*. A

... Stelle ist in der christlichen Tradition von 'Ordnungen' die Rede gewesen: also nicht so, dass 'Ordnung' ein bestimmtes Verhältnis meint, ein so oder so Hingeordnet-Sein, eine Anordnung, eine Verhältnisbestimmung.... 'Ordnungen' [müssen] als in der Geschichte Gottes beschlossene Lebensvorgänge verstanden werden, in denen sich Leben als geschöpfliches vollzieht" (Ulrich 2007, 102–103).

[16] Framing the question in terms of the spatial metaphor of *location* allows us to incorporate the concept of creation as address from a different point of view—if location here is conceived as God's story locating us, instead of us locating ourselves into the grand narrative of God (cf. Bayer [1999, 2–3 and 21–40] on the difference between *Deus poeta* and *homo poeta*). The word *ethos* already hints at this metaphor, stemming from the Greek word for "living" in the double sense of the word: way of life, or habits, on the one hand, and place to live, on the other.

[17] "So empfielt sich, die heuristische Bedeutung der Trias elementarer Lebensformen auch dahingehend auszuloten, dass der politische, ökonomische und theologische Bezug wie ein quer laufendes Filter jeweils über die jeweils betrachteten Gegenstände gezogen wird" (Wannenwetsch 2000b, 132).

[18] For example, "Die systemtheoretische Perspektive ist von vornherein auf die Beobachtung und Beherrschung aller Lebensverhältnisse gerichtet. Diese Beobachter- und Verwalterperspektive

system like Luhmann's translates observable experiences (and the causes of such experiences) to functions within certain domains of life: for instance, trade in terms of money and economics, faith in terms of religion. But according to Ulrich, what we can see should not be translated and coded. Experiences cannot and should not be reduced to seemingly universal codes that exclude the possibility of seeing things in a different way than the codes predict (Ulrich 2007, 235–237). For Ulrich, ethics is neither about the reduction of experiences to general theories, nor about applying principles in the right way. Rather, ethics should be about the right interpretation of reality with the help of the three estates as the heuristic tool and the consequent concrete action of human beings.[19] In short, social ethics is about the concrete praxis of *politia Christi*, i.e. "the public expression, testing and communication of the human way of life as contained in Jesus Christ" (ibid., 44).[20]

Creation Orders in Reformed Theology

Within Reformed—more specifically, neo-Calvinist—theology, the concept of creation orders is often used. Theologians such as Abraham Kuyper and Herman Bavinck and philosophers such as Herman Dooyeweerd thoroughly treat with the orders of creation as grounded in the eternal decrees of God.[21] In Kuyper's theology, the leading principle of God's sovereignty is traced back to the comprehensive view of man's place in creation under God. As the sovereign Ruler, God orders all things according to certain constant and universally valid principles which concern physical, social, moral, and religious life, by *common grace*. Common grace can be thought of in terms of the eternal decrees of God. According to Dooyeweerd, common grace preserves the structural laws of creation and contains special gifts of God to ensure orderly human life.[22]

---

unterscheidet sich kategorial von der Perspektive derer, die *handeln*, das heisst der Perspektive einer Veränderung, die sich nicht absorbieren und nicht auf die zwingende Erhaltung der Systeme beziehen lässt" (Ulrich 2007, 455).

[19] Ulrich says, for instance: "Es ist dann nicht davon zu reden, wie gerecht oder ungerecht ökonomische Verteilungsvorgänge sind, sondern, in welcher Weise es *in* den ökonomischen Vorgängen und quer zu ihnen ein gerechtes Handeln geben kann, das anders denn als Neuanfang, anders denn als revolutionär oder überschiessend nicht zu verstehen ist" (Ulrich 2007, 254).

[20] Cf. Ulrich (2007, 461–470). "Eine Sozialethik jedoch, die die *politia Christi* zum Gegenstand hat, ist eine Ethik des bezeugenden und mitteilenden, nicht nur zeichenhaften Tuns, keine Ethik, die moralische Verhältnisse einfordert, keine Ethik moralischer Prozeduren und ihrer Verwaltung. Diese Sozialethik hat ihre Pointe darin, dass sie fragt, was dem Nächsten von dem mitzuteilen ist, was wir von Gott empfangen" (461).

[21] Jeremy S. Begbie (1991) has given an overview of neo-Calvinist creation theology in relation to his research on a "theology of art" in which he compares neo-Calvinist theology and the theology of Paul Tillich in this respect. For the treatment of neo-Calvinist theology, see Begbie (1991, 81–141); for the eternal, created orders, see especially Begbie (1991, 85–86); cf. Begbie (1989).

[22] "Central to Dooyeweerd's project is the recognition that all reality belongs to God as Sovereign, and that creation is characterised by a magnificent law-order, ordained by the will of God, and upheld through common grace" (Begbie 1991, 121).

The neo-Calvinist concept of creation order runs several risks. First, the lack of Christological (and, for that matter, hamartiological) content is demonstrated by several critics, such as Begbie (1991), Douma (1974, 264−269),[23] Van Woudenberg (1998a, b, 1996), and O'Donovan (1994, 50−52), with respect to both its epistemological features and the content of creation (*creatura*) itself (Schaeffer 2006, 193−194, 208−209). If we acknowledge that creation itself as well as our knowledge of it suffer from the consequences of the fall, the following question concerning social order is pressing: How and to what extent is the prefixed order of creation reflected and discernible in current dissolving and disintegrating pluralist society? The Lutheran concept of creation orders (plural) as outlined above is not about an order present in reality. It is about living faithfully in response to God's address to creation through creation, differentiated as living in the *ecclesia*, the *oeconomia*, and the *politia*, and interpreting all phenomena and spheres of action in relation to these estates.[24]

Second, the acclaimed self-evidence of the concept of creation orders in Reformed philosophy is (as Albert M. Wolters explains) aimed at the *structure* of reality. Theology, on the other hand, is concerned with the *direction* reality takes.[25] However, according to Begbie (1991, 122−123), this philosophical concept is immune to theological criticism. The danger of not taking theology into account is that any conflict arising from this particular philosophical point of view in dialogue with other interpretations of reality is not evaluated theologically, i.e., not evaluated with respect to the *future* direction of creation orders. They are, and can only be, bent back to original creation without taking future developments into account.[26] With respect to social theory, this gives rise to the question of how we should evaluate the rise of new sociological entities, such as same-sex marriages, without theological presuppositions.

---

[23] Douma comes to the conclusion that "In de theologie van Kuyper wordt ... een dualiteit (Scheppings- en Verlossingsmiddelaar) openbaar, die veel (onoplosbare) tegenstrijdigheden in zijn bepaling van de verhouding tussen de gemene gratie en de particuliere genade verklaart" (1974, 266; cf. Douma 1981).

[24] "So kann von jenen *Ordnungen* nicht affirmativ geredet werden, vielleicht als von etwas letztlich Gegebenem und Verpflichtendem, vielleicht als von grundlegenden (moralischen) Verhältnissen oder von etwas grundlegend Erforderlichem. Nicht eine derartige Vergewisserung ist die Strategie einer solchen Ethik, sofern *umgekehrt* alles darauf annkommt, ... sich aller Güter in Gott gewiss zu werden und dieses *Leben mit Gott* in die Verhältnisse hineinzutragen" (Ulrich 2007, 471; italics in original).

[25] "Christelijke filosofie beschouwt de schepping in het licht van de fundamentele categorieën van de Bijbel; christelijke theologie beschouwt de Bijbel in het licht van de fundamentele categorieën van de schepping" (Wolters 1988, 18).

[26] "In fact, the lines are all bent back to creation.... Is it not possible that with the teaching of sphere-sovereignty the presently existing relationships in society too easily become sacrosanct? ... Where can I ever find an answer to the question whether marriage structure can be replaced by communal relations or by homosexual partnerships, outside the Scripture ... ?" (Douma 1981, 58−60; cf. Douma 1976, 64−67).

Another question concerns the relation between social ethics and the concept of creation orders. Originally, the concept of creation order, central to Reformed philosophy according to Van Woudenberg (1992, 27–35), is intended to counter the apparent disorder in reality in that God provides norms for both nature and culture.[27] The creation order therefore bears consequences for human behavior. Within Reformed philosophy, God's order and the appropriate distinctions are crucial because they aim at establishing the proper sphere sovereignty of different areas in life in order to avoid both collectivism and individualism (Griffioen and Van Woudenberg 1996, 243–248).[28]

Furthermore, these social relationships have to be defined according to their own modal aspects. The way in which different social relationships are distinguished, however, is complex. The question I would like to pose is: What ethical power would be lost if we did not distinguish in such detail between social relationships as Reformed philosophy does? What does the differentiated concept of social relationships bring to social ethics compared to and above the Lutheran use of creation orders as outlined above? Let me illustrate this with the example of the relation between family and marriage.

According to Griffioen and Van Woudenberg (1996, 245–259), family and marriage should be identified as distinct areas in life. The reason for distinguishing marriage from family is that a family remains a family even if the concrete members change. If a child is born within the family, or a father dies, the family remains the family. A marriage, however, is confined to the two marital partners and ends whenever one of the partners is excluded from the marital relation—e.g., by divorce or death (ibid., 250, 258). By distinguishing between family and marriage certain structural differences are clearly marked. At the same time, when pointing to the qualification of the family, Griffioen and Van Woudenberg (1996, 254) quote Dooyeweerd using *marital* biblical terminology when he states that the *family* is an image (*afschaduwing*) of the loving relationship between God the Father and human race reborn in Christ, between Christ and his community. These phrases are clearly reminiscent of Ephesians 5 where Paul writes about marriage. It seems that marriage and family share the same spiritual dimension (direction) whereas they are structurally distinct in the sense that, in the theory of enkaptic interlacement, marriage is foundational for the family structure (cf. Van Woudenberg 1992, 151). But the social-ethical consequences of this enkaptic interlacement are not clear, although the concept of creation order was intended (among other things) to provide norms.

Recent Lutheran contributions to social ethics that focus on marriage and family, however, stress their unity in the creation order of the *oeconomia* (Ulrich 2007, 337–349; Bayer 1995, 224–246; Wannenwetsch 1993). Their description of marital

---

[27] In this respect it can be said that for both Martin Luther and Abraham Kuyper, the concept of creation orders is God's protection against the powers of chaos (Link 1991, 66n134; Van Egmond and Van der Kooi 1994, 18–19).

[28] "Dit beginsel verbiedt, of beter gezegd, maakt onmogelijk, de reductie van de verscheidenheid van structuurprincipes tot een enkel principe en daarmee tevens de verabsolutering van bepaalde gepriviligieerde gehelen" (Van Woudenberg 1992, 147).

life, family, labor, and their relation to the whole of human living together as *oeconomia* is explicitly intended to deal with (post)modern ethical questions.

To conclude, Reformed philosophy and Lutheran theology share an interest in God ordering creation. Some critics might say that the specific strand of thought about creation orders within neo-Calvinism conveys an essentialist character to both the philosophical and theological variant. However, Van Woudenberg (1992, 31) explicitly states that Kuyper did not sanction the status quo and that it is the structural norms for social relationships which are found in the creation order. Nevertheless, creation (*creatura*) is understood here as a given, preexistent reality. Griffioen and Van Woudenberg (1996, 257–258) describe the act of marriage as man and woman "entering" something God has given: "a community whose structure is independent of their subjective arbitrariness but a structure to which they are subjected as to an institutional law."

Lutheran theology on the other hand treats creation (*creatura*) not as something we can enter into, but as God's ongoing address to human beings. With Geertsema, I would contend that living is responding to God's address to us in creation.[29] Such an interpretation of creation includes the theological doctrine of divine providence and thus comprises not only "original" creation but also creation in its teleological history towards the *eschaton* (Schaeffer 2006, 243–260).

For the concepts of marriage and family, this notion means that man and woman do not "enter a structure to which they are subjected"; rather, they are called into a specific relationship which is in turn God's own creation. This notion is underlined by the biblical testimony of Jesus saying about marriage that what God has joined together should not be separated (Matt. 19:6).

Possible Benefits

The neo-Calvinist doctrine of creation orders could benefit from a thorough conversation with its Lutheran equivalent in three respects. This paper can only point at these benefits, and not elaborate on them.

Recent use of the Lutheran three estates functions as an interpretation of reality with respect to social ethics. As such it is an adaptation of a long tradition, dating back to Aristotle. The three estates are irreducible spheres. Is it too far-fetched to see a resemblance here between the concept of the three estates and the concept of sphere sovereignty in the neo-Calvinist body of thought? In both concepts, the proper distinction is a warrant that prevents totalitarianism.[30] The comments by

---

[29] "Schepping verstaan in termen van het belofte-bevel maakt duidelijk dat het schepsel bestaat in de constante scheppingsrelatie. Het heeft geen ankerpunt in zichzelf" (Geertsema 1992, 132; cf. Van Woudenberg 1996, 56–57).

[30] "Al deze kringen nu grijpen met de tanden hunner raderen in elkaâr, en juist door dat 'op elkaâr werken' en 'in elkaâr schuiven' van deze kringen ontstaat het rijke, veelzijdige, veelvormige menschenleven; maar ontstaat óók, in dat leven, het gevaar dat de ééne kring den naastliggenden

Wannenwetsch on the ongoing fragmentation of society—and corresponding fragmentation of fields of social ethics—make clear that, apart from the proper distinction, the interesting point is how the spheres or estates could be distinguished and how they are related to one another.[31] Such a discussion on the kind of relation and the normativity implied by the relation needs a conceptual framework. An elaborated concept of vocation and its underlying theological presuppositions on creation as God's address to us instead of prefixed spheres could be of help here. The discussion whether we need to discern *three* estates or *more* "social relationships" based on Dooyeweerd's concept should receive proper attention, too.

In addition, Griffioen and Van Woudenberg outline Dooyeweerd's distinction between natural and historical social relationships. Family and marriage are clear examples of a natural relationship, whereas state, school, and university are historical social relationships.[32] The developments in the twenty-first century and the deconstructionist analysis of natural relationships make clear that the alleged "naturalness" of family and marriage is contested. Lutheran theology, especially after the misuse of the concept of creation orders in the German Third Reich, has developed a framework in which nature and creation are brought in a fruitful discussion with one another (Schaeffer 2006, 214–260). To conceive of creation as God's ongoing address to us within the three estates makes it possible to avoid various essentialist and naturalistic problems. It is then all about how we are "in-formed" (Wannenwetsch 2000a) to discern God's address. A discussion between Dooyeweerd's transcendental criticism and Lutheran theology—especially Oswald Bayer's (1994) use of Johann Georg Hamann in conceptualizing theology as a "science of conflict" (*Konfliktwissenschaft*)—could clarify this "in-formation" of human beings (Bayer and Suggate 1996; Wannenwetsch 2004).

Finally, the future of creation order is at stake. We need to attempt to find an answer to the question "whether there is still room for affirmation of pregiven norms—or what one could call *ontic normativity*—while also acknowledging the particularity and 'situatedness' of our articulation of those norms" (Glas and De Ridder, chapter "Introduction to the Philosophy of Creation Order, with Special Emphasis on the Philosophy of Herman Dooyeweerd," this volume). A Lutheran contribution to this discussion may critique ontological, pregiven norms while at the same time affirming God's call which urges us to live in *ecclesia, oeconomia,* and

---

inbuige; aldus een rad horten doe; tand na tand stuk wringe; en dusdoende den gang store van het geheel" (Kuyper 1930, 11). According to Kuyper, the state is the protector of the sphere sovereignty. This warrant against totalitarian tendencies of one sphere or estate is not only important for the *state*, but just as much for the *church*. The concept of "Total Church" (Chester and Timmis 2007) is one example of a "totalitarian" church concept. An example of the use of Dooyeweerd's concept regarding the relation of church and family is provided by Brent Waters (2007).

[31] Especially Dooyeweerd's conception of the ongoing historical differentiation process needs clarification in relation to the alleged fragmentation in both current society and science.

[32] "De verbanden vallen in twee hoofdgroepen uiteen. 'Als hoofdonderscheiding in de verbandsstructuren', zegt Dooyeweerd, 'zullen wij vinden: zulke welke typisch in een *natuurzijde der werkelijkheid* en zulke welke typisch *historisch* zijn gefundeerd'" (Griffioen and Van Woudenberg 1996, 251).

politia and which gives rise to a clear articulation of concretely located norms and values. Furthermore, the future of creation order can only be thought of in relation to the theological (and teleological) concept of hope. In Bayer's theology, the context of the doctrine of creation is exactly the context of hope. Over against all the groaning of creation, God's promise is not to let go what his hands began (Bayer 1990, 60–61). Therefore, the only future that creation orders have is in the hope on the Creator, who will lead his creation to its fulfillment. And God's love for his creation makes sure that whatever the *eschaton* will provide us with, it will never be less than what he gave us in creation.

References

Bayer, Oswald. 1990. *Schöpfung als Anrede. Zu einer Hermeneutik der Schöpfung*. Tübingen: Mohr/Siebeck.
———. 1991. *Autorität und Kritik. Zu Hermeneutik und Wissenschaftstheorie*. Tübingen: Mohr/Siebeck.
———. 1994. *Theologie*. Gütersloh: Gütersloher Verlagshaus.
———. 1995. *Freiheit als Antwort. Zur theologischen Ethik*. Tübingen: Mohr/Siebeck.
———. 1999. *Gott als Autor. Zu einer poietologischen Theologie*. Tübingen: Mohr/Siebeck.
———. 2003. *Martin Luthers Theologie. Eine Vergegenwärtigung*. Tübingen: Mohr/Siebeck.
———. 2007. *Zugesagte Gegenwart*. Tübingen: Mohr/Siebeck.
Bayer, Oswald, and Alan Suggate, eds. 1996. *Worship and Ethics. Lutherans and Anglicans in Dialogue*. Berlin: De Gruyter.
Begbie, Jeremy S. 1989. Creation, Christ, and Culture in Dutch Neo-Calvinism. In *Christ in Our Place. The Humanity of God in Christ for the Reconciliation of the World*, ed. Trevor Hart and Daniel Thimell, 113–132. Exeter: Paternoster Press.
———. 1991. *Voicing Creation's Praise. Towards a Theology of the Arts*. Edinburgh: T&T Clark.
Bekenntnisschriften der evangelisch-lutherischen Kirche. 1992. Göttingen: Vandenhoeck & Ruprecht.
Brock, Brian. 2007. Why the Estates? Hans Ulrich's Recovery of an Unpopular Notion. *Studies in Christian Ethics* 20 (2): 179–202.
Chester, Tim, and Steve Timmis. 2007. *Total Church. A Radical Reshaping Around Gospel and Community*. Nottingham: Inter-Varsity Press.
Douma, Jochem. 1974. *Algemene genade. Uiteenzetting, vergelijking en beoordeling van de opvattingen van A. Kuyper, K. Schilder en Joh. Calvijn over "algemene genade."* Goes: Oosterbaan & Le Cointre.
———. 1976. *Kritische aantekeningen bij de wijsbegeerte der wetsidee*. Groningen: De Vuurbaak.
———. 1981. *Another Look at Dooyeweerd. Some Critical Notes Regarding the Philosophy of the Cosmonomic Idea*. Winnipeg: Premier Publishing.
Evangelisches Kirchenlexicon. 1989. s.v. "*Institution.*" Vol. 2. Göttingen: Vandenhoeck & Ruprecht.
Geertsema, Henk G. 1992. *Het menselijk karakter van ons kennen*. Amsterdam: Buijten & Schipperheijn.
Griffioen, Sander, and René van Woudenberg. 1996. Theorie van sociale wetenschappen. In *Kennis en werkelijkheid. Tweede inleiding tot een christelijke filosofie*, ed. René van Woudenberg et al., 236–266. Amsterdam/Kampen: Buijten & Schipperheijn/Kok.
Honecker, Martin. 1999. Von der Dreiständelehre zur Bereichsethik. Zu den Grundlagen der Sozialethik. *Zeitschrift für Evangelische Ethik* 43: 262–276.
Kuyper, Abraham. 1930. *Souvereiniteit in eigen kring*. Kampen: Kok.

Link, Christian. 1976. *Die Welt als Gleichnis. Studien zur Problem der natürlichen Theologie.* Munich: Kaiser Verlag.
———. 1991. *Schöpfung.* Handbuch systematischer Theologie 7/1. Gütersloh: Gütersloher Verlagshaus Gerd Mohn.
Maurer, Wilhelm. 1974. *Luthers Lehre von den drei Hierarchien und ihr mittelalterlicher Hintergrund.* Munich: Verlag der Bayerischen Akademie der Wissenschaften.
O'Donovan, Oliver. 1994. *Resurrection and Moral Order: An Outline for Evangelical Ethics.* Grand Rapids\Leicester: Eerdmans\Apollos.
Schaeffer, Hans. 2006. *Createdness and Ethics: The Doctrine of Creation and Theological Ethics in the Theology of Colin E. Gunton and Oswald Bayer.* Berlin: Walter de Gruyter.
Schwarz, Reinhard. 1978. Luthers Lehre von den drei Ständen und die drei Dimensionen der Ethik. *Luther Jahrbuch* 45: 15–34.
Ulrich, Hans G. 2007. *Wie Geschöpfe leben. Konturen evangelischer Ethik.* Berlin: Lit Verlag.
Van der Kooi, Kees. 2011. Gratia non tollit naturam, sed perficit. In *Creation and Salvation: Dialogue on Abraham Kuyper's Legacy for Contemporary Ecotheology*, ed. Ernst M. Conradie, 213–221. Leiden: Brill.
Van Egmond, A., and C. van der Kooi. 1994. The Appeal to Creation Ordinances: A Changing Tide. In *God's Order for Creation*, ed. P.G. Schrotenboer et al., 16–33. Potchefstroom: Institute for Reformational Studies, Potchefstroom University for Christian Higher Education.
Van Woudenberg, René. 1992. *Gelovend denken. Inleiding tot een christelijke filosofie.* Amsterdam: Buijten & Schipperheijn.
———. 1996. Theorie van het kennen. In *Kennis en werkelijkheid. Tweede inleiding tot een christelijke filosofie*, ed. René van Woudenberg et al., 21–85. Amsterdam\Kampen: Buijten & Schipperheijn\Kok.
———. 1998a. Greijdanus' kentheologie. In *Leven en werk van prof. dr. Seakle Greijdanus*, ed. George Harinck, 165–174. Barneveld: De Vuurbaak.
———. 1998b. Over de noëtische gevolgen van de zonde. Een filosofische beschouwing. *Nederlands Theologisch Tijdschrift* 53: 224–240.
Wannenwetsch, Bernd. 1993. *Die Freiheit der Ehe. Das Zusammenleben von Frau und Mann in der Wahrnehmung evangelischer Ethik.* Neukirchen-Vluyn: Neukirchener Verlag.
———. 2000a. Caritas fide formata. "Herz und Affekte" als Schlüssel zum Verhältnis von "Glaube und Liebe". *Kerygma und Dogma* 46: 205–224.
———. 2000b. Wovon handelt die "materielle Ethik"? Oder: warum die Ethik der elementaren Lebensformen ("Stände") einer "Bereichsethik" vorzuziehen ist. Oswald Bayer zum sechzigsten Geburtstag. In *Kirche(n) und Gesellschaft*, ed. A. Fritzsche, 95–136. Munich: Bernward bei Don Bosco.
———. 2004. *Political Worship. Ethics for Christian Citizens.* Oxford: Oxford University Press.
Waters, Brent. 2007. *The Family in Christian Social and Political Thought.* Oxford: Oxford University Press.
Wolters, Albert M. 1988. *Schepping zonder grens. Bouwstenen voor een bijbelse wereldbeschouwing.* Amsterdam: Buijten & Schipperheijn.

Out of the Ashes: A Case Study of Dietrich Bonhoeffer's Theology and the Orders of Creation

Annette Mosher

Abstract In this case study, I examine what the orders of creation meant for theology in Germany during the 1930s and '40s as the Nazis rose to power. I compare the theology of Dietrich Bonhoeffer to that of two well-known creation order theologians—Paul Althaus and Emanuel Hirsch. Finally, I consider Bonhoeffer's theology of the four mandates in *Ethics* as an alternative answer to *völkisch* creation order theology.

Keywords *Volk* theology · History as revelation · Mandates · Nationalism · Creation order · Dietrich Bonhoeffer · Paul Althaus · Emanuel Hirsch

Introduction

In the introduction that provides the framework for this volume, Gerrit Glas and Jeroen de Ridder discuss the role that divine order plays in modern theology. They refer to the state of affairs surrounding orders of creation as still undeveloped even after 75 years of discussion within reformational philosophy. In particular, they point out that "theologians have questioned the implicit assumption that it is possible to gain access to creation order independent of the cross and of mediation by the church" (Glas and De Ridder, chapter "Introduction to the Philosophy of Creation Order, with Special Emphasis on the Philosophy of Herman Dooyeweerd," this volume).

Glas and De Ridder's reflection on this theological question reminds us of the theology of Dietrich Bonhoeffer, who argued that only through the cross and the resulting church could humanity have any access to an understanding of creation. The societal context of Bonhoeffer's theology, i.e., the Third Reich, was during the 75 years to which Glas and De Ridder refer. At that time, creation order was an

---

A. Mosher (✉)
Vrije Universiteit Amsterdam, Amsterdam, The Netherlands
e-mail: annette_mosher@hotmail.com

important topic in theological discussion. However, then the understanding of creation order was perverted from a creation mandate into an understanding of divine providence that elevated *volk* and nation above others and placed people into particular groups based on blood, race, and "nationality."

In this case study I will examine what the orders of creation meant for theology during the time that Bonhoeffer was writing his manuscripts. I will then compare Bonhoeffer's theology in *Sanctorum Communio* and *Creation and Fall* to that of two well-known creation order theologians in Germany during the 1930s and '40s—Paul Althaus and Emanuel Hirsch.[1] Finally, I will consider Bonhoeffer's theology of the four mandates in *Ethics* as an alternative answer to *völkisch* creation order theology.

Post-War I Theology in Germany and Creation Order

Germany's defeat in World War I brought difficult financial and social problems to its citizens. Not only did its humiliation cause its people to question their identity, but many Germans were suffering financially as well. The economic situation of the country was severe. Hyperinflation made the German mark nearly worthless. Food was in short supply, and that led many young people to violence in order to obtain the items they needed for survival (Evans 2004). The general feeling held by German citizens was that the war reparations—demanded by Germany's victors—were the reason for so much financial hardship.

Additionally, the progression of modernity left many feeling uncertain. Some felt that the German culture and nation that they knew was dissolving around them. Traditional roles and sexual norms were changing and becoming more contemporary. For some conservatives this development went too far and was too liberal (Evans 2004). There was a growing longing for the restoration of a safe, ordered society and a positive renewal of the German civic identity.

As a result, we see that a reactionary movement occurred in Germany. The citizens were disillusioned with the Weimar government and felt that it was unable to securely rule the country. The idea that the country was leaderless and drifting caused many Germans to become susceptible to a politician with a strong message. Hitler had such a message. He was popular because he offered an alternative to reality—an alternative that supplied a narrative story of a superior, geographically

---

[1] Both Althaus and Hirsch were Nazi sympathizers in the years leading up to the Third Reich. Robert P. Ericksen argues in his book *Theologians Under Hitler: Gerhard Kittel, Paul Althaus and Emanuel Hirsch* that Hirsch escaped denazification by retiring from the University of Göttingen before the Allies were able to cleanse that university. Althaus appears to have been more moderate in his support of the Nazis, certainly in his publications; however, he was removed during the denazification process. Later, he was rehabilitated and allowed to teach at the University of Erlangen, from which he retired in 1966. See Ericksen (1985).

defined German nation that included a history of an imagined conservative, pure community.[2]

Pastors and theologians were also not immune to this unrest and malaise. Nor were they immune to the romantic notions offered by Hitler of a historical epic for each nation's citizens. Theologians desirous for a restored German people lent their own "Christian" narrative to this burgeoning identity. Paul Althaus and Emanuel Hirsch were two of the prominent theologians who were active in promoting a theology that supported German nationalism. To do so, they created their own version of creation order.

In the following section I will shortly describe the theology of Althaus and Hirsch and then move into the theological implications of their theology.

Paul Althaus

Paul Althaus was a prominent professor of systematic theology at the University of Erlangen, who specialized in Luther's theology (Ericksen 1985). He was a well-respected theologian in the German churches as well—serving not only as the president of the Luther Society for 30 years, but also as journal editor of several journals and a leader in the political processes of the churches (Ericksen 1985).

Robert P. Ericksen, in his book *Theologians Under Hitler: Gerhard Kittel, Paul Althaus and Emanuel Hirsh*, reports that in 1916 Althaus, then serving as a pastor in Poland, was already favorable towards *völkisch*[3] ideology. Althaus felt that the societal trend towards restoring the German pride in nationality was a positive occurrence. Although, it was not until after the First World War that Althaus tied the church to the national identity. After the war Althaus said that when the war began he thought it was the churches' responsibility to tamper any high-mindedness or enthusiasm that the people might have as the stronger power in the war effort. However, after the war he found himself "disgusted" because of the "shameless voluntary surrender" of the German army (Ericksen 1985, 83). He learned through that experience that it was "incumbent on the church to recall for our *Volk* the value of a good conscience and the defiance of confident faith" (Althaus 1934a, 84).

It is not surprising then that when Hitler came to power in 1933 Althaus felt a connection to the *völkisch* rhetoric coming from the Nazi party. It was not only that Althaus was acting from his own political convictions, even though those convictions

---

[2] This chapter does not deal intensively with the German history that led to the formation of the Nazi movement, or with the cultural particularities of German character. This description is simply to give a brief glimpse into the impulse that informed the theology of the scholars investigated within this case study.

[3] *Volk* is not an easy word to explain within English. The literal translation is "folk" or "the people." However, the understanding of the meaning as used here would be closer to "nationalistic, populist movement" in English.

were in sympathy with National Socialism, but his Lutheran theological convictions also played a part in his political sway.

Althaus developed a *volk* theology as an extension and outworking of Luther's theology of government as the will of God for the people. According to Luther there are two kingdoms that each person finds him or herself in—either the kingdom of God (Christians) or the kingdom of the world (non-Christians). However, all people—even Christians in the kingdom of God—must submit themselves to the temporal authority as God's plan for them (Luther [1523] 2005). Luther claimed that those who are obedient to God display their obedience by being obedient to the government (ibid.).

Althaus agreed with Luther's two sphere theology and this led him—through his interpretation of Luther—to develop a theory named *Ur-Offenbarung*. This theory recognizes Christ as God's superior revelation, but includes an additional understanding of recognizing revelation in natural events such as history, nature (creation), and the natural ordering that we find around us (Ericksen 1985). In this theory Althaus goes a step further than Luther's claim that God created the government as an order for the people. In addition to the government Althaus declares the *volk* as a God-given, natural ordering of a nation's citizens. This means that birth and placement within a community goes further than a simple chance of birth. Instead, it is a preordained choice by God for our existence. It follows that since *volk* and the placement in the *volk* is God-ordained and a creation of God's, the ordering—the *volk*—becomes a "holy" creation of God's as well (Althaus 1937). The national spirit that accompanies the *volk* is an extension of this holy creation—in essence making patriotism a holy activity. Naturally, the theology of *volk* includes the national boundaries that accompany and delineate the *volk* and its spirit, bringing an unspoken approval by God of nationalism. The difference is that Luther demanded obedience to the state one finds oneself in while Althaus added *volk*, patriotism, and national boundaries as part of that obedience. With Althaus' theology, the German state took on divine proportions: "The State—the authority in the form of law—is in Lutheran doctrine, although created and managed by people, an ordering by God whereby only through God are people preserved from the chaos in this world of sin and opposition and community is made possible."[4] In essence, under Althaus the *volk* becomes one collective person by the means of the state with one national spirit defining the collective person.

---

[4] "*Staat*, d.h. Heerschaft in der Form des Rechtes, ist nach lutherischer Lehre, obgleich überall durch Menschen entstanden und verwaltet, eine *Ordnung Gottes*, durch welche Gott in einer Welt der Sünde und des Widerstreites die Menschheit vor dem Chaos bewahrt und Leben in Gemeinschaft ermöglicht" (Althaus 1934b; translation mine).

Emanuel Hirsch

The second theologian is Emanuel Hirsch—a systematic theologian who taught at the University of Göttingen. In the 1920s Hirsch prepared a lecture for his students reflecting on the state of Germany after the First World War. Ericksen claims that Hirsch offered history and learning one's national heritage as the solution for the problems that the German people faced. Hirsch embraced the same theory of *volk* that Althaus argued for but Hirsch additionally emphasized the idea of an ordering through the historical record that revealed God's plan for the *volk*. Hirsch believed one could see revelation in history: "Human history can ... only be understood by those who see its metaphysical core and its religious connection. Human history and notions about God belong necessarily together" (Ericksen 1985, 14).

Hirsch also claimed the creation order in a Germanic way—the ordering brought with it a specific duty for the individual (Ericksen 1985). The individual should live as a dutiful citizen in the proper role assigned to them—which included internalizing the spirit of the *volk*. Hirsch rejected a universal collective person in favor of the individual *völker* and said that we cannot see history as a whole for humankind, but only in the local experience of the *volk*.

> If history has any sort of a heart and a sense, one must seek it not in the multiplicity of developments and destinies, but in what is common to each place and time in historical life. That is the direct relationship of the human spirit in each of its movements to that which is beyond history, to the eternal. (Ericksen 1985, 132)

Of course, neither Hirsch nor Althaus was the author of the romantic idea of a true nation with a collective human spirit. Pan-Germanism had existed since the late 1700s–early 1800s.[5] The difference is that Hirsch attached theology to the ideology and wholeheartedly ascribed to the idea of a salvation history for Germany (Ericksen 1985).

However, Althaus and Hirsch were not without contemporary critics. Other theologians, such as the dialectical theologians, were not convinced. Ericksen (1985) describes the relationship between Hirsch and the dialectical theologians as oppositional. He points out that Barth tussled with Hirsch with the result that Hirsch dismissed Barth as a Swiss and Barth dismissed Hirsch as a nationalistic ideologist. Unfortunately, Ericksen missed the contribution to the discussion from Dietrich Bonhoeffer. This is particularly problematic since the majority of Bonhoeffer's work is in opposition to *volk* theology.[6] In the rest of this chapter, I will discuss Bonhoeffer's theology in light of the theology that he found offered by Althaus and Hirsch in Germany at that time.

---

[5] Covering all the issues that accompanied Pan-Germanism and affected creation order theology in Germany during the 1930s and '40s is too large a topic for this paper. For further reading, please see Evans (2004, chap. 1), "The Legacy of the Past," and Burleigh and Wippermann (1991, chap. 2), "Barbarous Utopias: Racial Ideologies in Germany."

[6] In his work, Ericksen recognizes the theological differences between Althaus-Hirsch and Bonhoeffer. However, for the most part he misses Bonhoeffer's critique on Althaus and Hirsch and instead highlights similarities in the goals of their theology.

Volk as Theology

In his first academic work, *Sanctorum Communio*, we see Bonhoeffer entering into dialogue with Althaus and Hirsch regarding the collective person and the definition of *community*. Bonhoeffer accepted Luther's structure of two kingdom theology along with the creation order theologians, but Bonhoeffer gave them a different theological interpretation than what Althaus and Hirsh offered. In the following sections we will consider three areas where Bonhoeffer responded to the creation order theologians—the idea of the collective person (*volk* in Althaus' theology), two sphere theology, and the form of revelation.

In this early work (completed in 1927 and first published in 1930) we can see that Bonhoeffer is digesting what *history* means for a collective person. He apparently accepts the understanding of a *volk* but Bonhoeffer bases his understanding of the *volk* on a different platform than Althaus and Hirsch (Bonhoeffer 1998). Bonhoeffer begins from an open species-based argument (ibid.). He argues that each person belongs to the larger species called "humanity" simply by being born human. This is not a complicated argument, but an unusual one for a theologian. It would have fit into the scientific focus that theology found itself in at the university. However, arguing for the *volk* (collective person) from this biological standpoint delivered a blow to the creation order theologians' ideas. Bonhoeffer's argument removes the nationalistic parameters of identity or local ordering, and argues instead for a more comprehensive inclusion of humanity in total. Now the *volk* is not only the national community, but the whole of humanity.

Once Bonhoeffer widened the definition of *volk* and argued for the identification of *volk* as the collective person of humanity, he then begins to reclaim the Christian ethic. To do this, Bonhoeffer places this collective person in relation to the divine. He writes that the Christian understanding of humanity is that "*the human person originates only in relation to the divine; the divine person transcends the human person*, who both resists and is overwhelmed by the divine" (Bonhoeffer 1998, 49). In reading his claim, we see the weakness of Althaus and Hirsch. They argued that the human person becomes a person as a result of his/her relation to the *volk*, and rather than transcending the *volk*, the human needs to adapt to the spirit of the *volk*. This focus on the human, even though connected to "divine history," means that Althaus and Hirsch based their qualifiers on humanism and human history rather than on a relationship with God, in Bonhoeffer's opinion. Bonhoeffer criticizes this as "unchristian" and "attributing to the human spirit absolute value that can only be ascribed to divine spirit" (ibid.).

As part of the Christian ethic, Bonhoeffer returns further in history than either Althaus or Hirsch does. Bonhoeffer begins with the classic, Christian doctrine of original sin. Since he has already identified the species "human" it is easy for him to argue for application of Augustine's doctrine. Bonhoeffer uses the *volk* idea but then turns it on its ear. The now universally defined parameter "*volk*" means a collective person that is affected—or in the case of sin, infected—by that within it. Thus when Adam sinned, the whole of humanity sinned. With this next step in his dialogue with Althaus and Hirsch, Bonhoeffer clarifies that the unifying aspect of this German *volk* is not their nationalistic membership or salvation history, but their membership in sinful humanity—humanity-in-Adam—which must repent of its sin (Bonhoeffer 1998).

It is with this designation that Bonhoeffer also disagreed with the political National Socialism that Althaus and Hirsh befriended. National Socialism used a similar species argument, but that ideology used race theories to order the inclusion into the *volk* based on race, health, physical characteristics, etc. Natural physical traits not only brought inclusion into the German *volk*, but also determined one's value to the *volk*. Bonhoeffer levels this hierarchical valuation of the human by placing the whole of humanity at the base level of sinful and separated from God. Bonhoeffer calls this debased, universal state "humanity-in-Adam." He realizes that it is fractured through the many and their individualism but writes that humanity-in-Adam "really has one conscience" and "one heart"—which mirrors, in a Christian manner, Althaus' nationalistic argument for *volk* (Bonhoeffer 1998).

By building this humanity-in-Adam, Bonhoeffer reflects Luther's two kingdom theology—our second theme. This kingdom of sinful humanity that includes everyone is a juxtaposition of the kingdom of the world in Luther's theology. Just as Luther saw the rule of government as a sphere where each human had a place and role, Bonhoeffer changes it slightly and writes that each person has a place and role in the sinful humanity-in-Adam.

Bonhoeffer does not stop there. He also introduces us to the second sphere in *Sanctorum Communio*. It is this sphere where he attacks the creation order theologian's understanding of revelation. Just as Bonhoeffer's first sphere mirrored Luther's kingdom of the world, the second sphere he introduces mirrors Luther's sphere of the kingdom of God. Bonhoeffer claims the church as the second sphere. This appears to be similar to Luther at first glance, but closer consideration reveals that Bonhoeffer has again slightly redefined the definition of the sphere. The difference is that in Bonhoeffer's theology the second sphere "swallows" the first so that the two spheres no longer stand side by side. Instead, Bonhoeffer says that all that belonged to the first sphere has been assumed into the second sphere called "church."

It is Bonhoeffer's special understanding of the church that makes this possible. He argues the Christian understanding that "in Christ humanity really is drawn into community with God, just as in Adam humanity fell" (Bonhoeffer 1998, 146). He adds that through "vicarious representation" humanity is now brought into reconciliation with God and the sinful "humanity-in-Adam is transformed into humanity-in-Christ" (ibid., 147). Bonhoeffer does not claim that some are brought into Christ, but that all of humanity is brought into Christ. He believes that the "church is the presence of Christ" just as "Christ is the presence of God," and he calls this new situation "Christ-existing-as-church-community" (140–141).

This is Bonhoeffer's attack on Althaus' *volk* theology. We see with Bonhoeffer that he is saying that there is no longer an order of creation that exists outside of Christ. A nation's history may be their history, but it is not their identity. Nor is the history the plan for the *volk*. Christ is the identity and the plan for all people. His is the spirit that must guide the community and its will. If history or any other part of a nation's identity is not fully aligned with Christ, then it is outside of God's will—even though it may appear to have been ordered by God.

Bonhoeffer said it very sharply—"Community with God exists only through Christ, but Christ is present only in his church-community, and *therefore community with God exists only in the church*" (Bonhoeffer 1998, 158). Bonhoeffer leaves no room for a two sphere ideology or *volk* theology outside of Christ. Anything that does not comply with Christ's revelation is part of the old sinful sphere. Bonhoeffer does not do away with the old sphere—the old world—but he says that it is dead and has been overcome by Christ.

In the following section we will see the problems that a misunderstanding of the presence of the old world brings with it by considering Bonhoeffer's response to history as revelation.

History as Revelation

After Bonhoeffer postulated alternative theology to Althaus' *volk* theology, he turned his attention to Hirsch's ideas of history as revelation and social ethics. This discussion occurred in the winter of 1932–1933 during a lecture series that the young Bonhoeffer gave at the University of Berlin. The lectures were so popular with Bonhoeffer's students that they encouraged him to publish the series. Bonhoeffer's manuscript—*Creation and Fall*—appeared in 1937. Because of the publishing date it is necessary to remember that the body of *Creation and Fall* was written before Hitler's appointment to chancellor, and so the discussion that occurred was with Althaus and Hirsch rather than a reaction to Hitler.

In *Creation and Fall*, Bonhoeffer exegetes Genesis 1–4. In doing so he is not only providing a biblical exegesis of Scripture, but attacking the idea of history as revelation. From the first pages we notice that Bonhoeffer does not agree with Hirsch's understanding of history. In his manuscript, *Deutschlands Schicksal*, Hirsch argued that history points "to the eternal" through the spirit and history of the *volk* (Bonhoeffer 1997, 132). Bonhoeffer answers in the introduction of *Creation and Fall* that while the history of the *volk* looks at the old things of "the world" the "church of Christ" looks to the end and "witnesses to the end" (ibid., 21). However, Bonhoeffer is not talking about the end as in the eternal that Hirsch indicates, but instead Bonhoeffer focuses on the new world that came when Christ redeemed the world. He argues that the eternal has already come to us and therefore our focus should not be in the past, but on the here and now.

This idea of the old world under humanity-in-Adam and the new world of Christ-existing-as-church-community was a source of contention with the creation order theology of Althaus and Hirsch. Bonhoeffer recognizes the problem and argues that in the church we have "people who think as the old world does" (ibid.). Bonhoeffer is obviously arguing against Hirsch and to some extent Althaus because he refers to their dependence on the old world (history before Christ) while also claiming the church (world after Christ) for themselves.

Bonhoeffer says that this is not the place for those who know Christ. The church can know all things—both history and that which is to come—only through Christ (Bonhoeffer 1997). Any addition or subtraction from Christ is a denial of Christ for Bonhoeffer.

Continuing in the first chapter, Bonhoeffer immediately and bluntly attacks the desire to see history as revelation. First he describes the futility of looking to the beginning for answers. The futility is that we can never know what the beginning was—it is an infinite question that cannot be known. Each question we ask about the beginning begs the next question and leads us into another question that brings us back to the beginning. Thus we enter into a spiral that always returns to itself, and we never discover the answer that we crave. The only answer possible is the one that we imagine. This makes our history a product of our own thinking about the purpose that God had for us, whereas God made the purpose obvious in revelation in Christ.

This leads to the second criticism that Bonhoeffer had for history as revelation. This criticism is particularly critical on theologians who subscribe to a method of reasoned revelation rather than revelation through Christ. Bonhoeffer says that the only way to "make a beginning in philosophy" is by "the bold and violent action of enthroning reason in the place of God" (Bonhoeffer 1997, 27). He claims that history as revelation is not only a mistake but idolatry.

Bonhoeffer bolsters this by arguing that there are only two beings who speak about the beginning—the evil one who was not at the creation and is therefore a liar, and "the very God, Christ, the Holy Spirit" (ibid., 29). Since the Trinity was the only being present at the beginning only one of the members of the Trinity can speak truthfully about the beginning. Bonhoeffer's argument is that all three Trinity members testify to Christ as the key and chalice to knowing revelation. With this argument, Bonhoeffer claims that revelation can only be through Christ, and therefore, anything outside of this testimony comes from Satan, the false witness. This is obviously an attack on any idea of revelation outside of the testimony regarding Christ.

In the same argument Bonhoeffer also disputes the "spirit of the *volk*" argument that both Althaus and Hirsch embraced. Bonhoeffer argues that we can only know that God created—that is what Scripture offers us—and it is "impossible to ask why the world was created, *what God's plan for the world was*, or whether the creation was necessary" (Bonhoeffer 1997, 31; my italics). In fact, even asking about the plan or meaning of history is asking "godless questions" for Bonhoeffer because it means that the questioner wants to go beyond God's explanation for our purpose (ibid.).

Bonhoeffer is willing to offer yet another blow to the creation order theologians because their theology masks the most important aspect about creation. As we discussed in the previous section, Hirsch and Althaus wanted to know about God through humanity and what they found in the human experience and spirit. Bonhoeffer argues that the only thing to know about the relationship between the Creator and the creature is freedom. God created the creature in complete freedom to experience relationship with the Creator, and this freedom "rules out every application of causal categories for an understanding of the creation" (ibid., 32).[7] Bonhoeffer claims freedom for the individual to respond to life from a relationship with his/her Creator rather than a preordained structure of history or *volk* that includes an accompanying duty.

In claiming this freedom of action, choice, and determination, Bonhoeffer is not naïve. He realizes that this freedom makes the creature uncertain, and Bonhoeffer believes it was the uncertainty that caused the problem of dependence on predetermined history or duty in the first place. He writes that we find ourselves in the middle of the question of the beginning and the end, and that we can only know where we are. We do not and cannot know the beginning, as we have just learned, and we cannot and do not know the end. It is the uncertainty that causes the human to try and create a reality in place of Christ as our reality.

This reality of Christ was also in disagreement with salvation history notions. Hirsch argued that history leads to the eternal (Ericksen 1985). Bonhoeffer argues that our focus should be on the reality of Christ in the present, which is in the middle of the past and the future. He does not quite argue for the present over against the future—or the culmination of history—but says that we already have "the new" in the present (Bonhoeffer 1997).

The new—Jesus Christ—does away with the old, and he also does away with the striving towards the culmination of salvation history. The eternal has already occurred and Christ has brought the new world to us. This argument is the same argument that Bonhoeffer worked out in *Sanctorum Communio*. It is the second sphere that Bonhoeffer developed—Christ-existing-as-church-community. The old world has been judged, declared dead, and brought to new life in Jesus Christ. Now there is no goal that history develops towards, or that a country or people must have in its sight. Christ is the only goal in this new world, and the church "believes in Christ and nothing else" (Bonhoeffer 1997, 22). Again Bonhoeffer does not mince words with the creation order theologians who wish to return to revelation outside of Christ. After describing that even theologians can cling to the things before Christ, and can even be in the midst of the church, they only know the things of the old in place of the things of the new "and in that way they deny Christ, the Lord" (ibid.).

Now that Bonhoeffer has addressed what he found erroneous in German creation order theology, he begins his magnum opus and offers an alternative. In our final section we will see Bonhoeffer's solution to *volk* theology.

---

[7] By this statement Bonhoeffer disagreed that national identity, history, or—in Bonhoeffer's time—racial theory had a role in our relationship with God. For Bonhoeffer, history did not determine who God is or what God requires of his creature in the current moment. Only relationship with and a personal understanding of God could require obedience.

An Alternative: Bonhoeffer's Four Mandates

In Bonhoeffer's final and unfinished manuscript, *Ethics*, we find him again attacking creation order theology. The difference now is that the theology that was an erroneous theology while Bonhoeffer was writing *Sanctorum Communio* and *Creation and Fall* had, in 1940, become an accomplice to Hitler's murderous actions.

In his biography about Bonhoeffer, Eberhard Bethge—Bonhoeffer's student, best friend, and nephew-by-marriage—outlines Bonhoeffer's thought progression in relation to creation order. Bethge writes that in 1932 Bonhoeffer was grappling with the "orders of preservation"[8] over against the "dangerous" orders of creation (Bethge 2000, 717). By the time that Bonhoeffer was working out his *Ethics* (1940–1943) the idea of orders of preservation had become *mandates*. Bethge says it could only be "'mandate' so that the one claim of Christ could be preserved in each gift and duty of the world; but *precisely* 'mandate' so that those who bear the responsibility may have full independence while history is allowed to take its course" (ibid.).

In his formulation of the orders of creation, Althaus argued that there were binding natural laws on each person. He named three—family, *volk*, and race. Bonhoeffer answered him with four mandates—work, marriage, government, and church (Bonhoeffer 2005).

Even the title of Bonhoeffer's last work—*Ethics*—hints to the theological problems in his context and to whom he addressed his arguments. Althaus and Hirsch were using the idea of creation orders to bolster their focus on an ethical norm for the German people. Particularly for Hirsch, the historical record of the German people exhibited "moral values and discipline" to which Hirsch longed to return and to which he wanted to redevelop "the order and the style of the German *Volk*" (Hirsch 1933). Hirsch even argues that "Germans must become a pious *Volk*" and a "*Volk* in which the gospel has power over conscience" (Ericksen 1985). Hirsch sees the problem of society as one where the *volk* has turned away from their piety, albeit a piety based on Hirsch's interpretation of the gospel.

Just as Althaus and Hirsch believed society was experiencing problems, Bonhoeffer also begins by recognizing that there is a problem in his society. However, when he begins *Ethics* he points out that what he sees as a problem is that everyone is trying to be good.[9] Being good was the standard—especially within *volk* theology. Those Christians captured within that stream were there because they wanted the good—a good life, a good nation, a good society, and good behavior. In

---

[8] Bonhoeffer used the term "orders of preservation" in order to avoid using the term "orders of creation" since he realized this term had disastrous results in theological application.

[9] For Bonhoeffer "good" in this sense was a prescribed ethical formulation of action, or a standard of knowing whether one was acting according to God's will. Bonhoeffer considered God as living and relational, which meant that we would need to hear God's will in fellowship with God as each occurrence arose rather than depending on static ethics to determine whether we were in God's will.

his first sentences of *Ethics* Bonhoeffer (2005) condemns this desire to be good as unchristian.[10] He argues that asking what the good is and how to be good is asking the wrong question. It is not an ethical equation or action that will make one good. Good is not even the question. The real question is, "What is the will of God?" (Bonhoeffer 2005, 47).

It appears that Hirsch and Althaus were asking this same question. Certainly they wanted to know from the history and spirit of the *volk* what was God's will for Germany. Bonhoeffer saw, however, that there was a fatal flaw in their approach. He realized that the creation order theologians had still not realized that the world has been accepted by Christ. The entire struggle to be good exhibits a hatred of redemption through Christ. The focus is not on relationship with God who has accepted humanity through Christ as Bonhoeffer worked out in *Sanctorum Communio*. Instead, Germany is still dealing with the Lutheran two sphere theology that believes there is a worldly sphere that must be managed.

For Bonhoeffer this rejection of Christ's redemption of the world led to concrete problems in society. The most pressing problem was that by the time Bonhoeffer wrote *Ethics* the government had demonized its Jewish citizens, taken draconic actions against them, and had even started the death camps in order to handle "the Jewish problem." Because of two sphere theology of a church realm and a worldly realm and *volk* theology of duty, the church was confused and refused to use her prophetic voice in opposition to the government because "the Jewish problem" seemed to be a civil rather than a spiritual issue.

This is the history surrounding Bonhoeffer's introduction of the four mandates. Bonhoeffer needed to again stress that Christ had already redeemed the world because in his country, sphere theology had allowed the government to usurp the position that rightly belonged to the church. The government was exercising its power as a totalitarian state even over matters concerning the church. Bonhoeffer was appalled that there was only a meager resistance to this. He could not stress strongly enough that the world and Christ were reconciled (Bonhoeffer 2005). That is why he introduced the four mandates. The way that the orders of creation were formulated by Althaus and Hirsch divided people from one another and divided the church from the world by considering the world in one sphere and the church, or Christ, in a separate sphere—even though Althaus and Hirsch struggled desperately to try and connect the state to Christianity by *volk* theology.

The problem was that the creation orders brought exclusivism with them. In the *Deutsche Christen* formulation, Christians were divided through their nations and *volk*. In contrast to the exclusivism that the creation orders brought with them, Bonhoeffer claimed that the four mandates were binding on all people and at once. There was no possibility to have a sphere for the government and a sphere for the church. Government and church were responsible to and dependent on each other.

Not because the mandates were a form of being human, or that which makes us human—such as what Althaus and Hirsch claimed for their *volk* theology—but

---

[10] In *Creation and Fall* Bonhoeffer identified "wanting to know good and evil" as the sin of Adam which brought separation from God.

rather, Bonhoeffer saw the four mandates as tasks. Our being came from our relationship to God as reconciled in Christ (rather than as *volk*). Recognition of this reality that Christ has redeemed the world means that the mandates become "divine" tasks as a testimony to Christ's work.

We can see that here again Bonhoeffer flips the focus from the *volk* to Jesus Christ. He tells the creation order theologians not to get it confused. The four mandates do not exist because work, marriage, government, or church are historical evidences in our development. They exist because God commanded them, *and* he commanded them in order to point to Jesus Christ. The four mandates are for his sake and that is what makes them divine and from God. Recognizing this and "aligning" the mandates under Jesus means that "a true ordering"—over against Althaus and Hirsch's erroneous ordering—will occur (Bonhoeffer 2005, 70).

Bonhoeffer describes then the purpose for each of the four mandates. Whereas for the *Deutsche Christen* work recovered societal fabric and repaired the economy, Bonhoeffer argued that the purpose of work was to reflect "the heavenly world" which directs one's thoughts to Christ (ibid., 71). When one works one participates in creating a world that "expects Christ, is directed towards Christ, is open for Christ, and serves and glorifies Christ" (ibid.). The focus is no longer on the *volk* and the glory (or recovery of glory) for a nationalistic entity, but now lies on the reconciler of all of humanity.

Bonhoeffer moves on to evaluate marriage. Here again we see him elevate what *should be* and we think of what *was*. The state of marriage in Germany was a battleground. The Nuremberg Laws were passed which prohibited marriage between "Jews" and "Aryans" in the hope to purify the *volk*. Even within Bonhoeffer's family this law had impact. Bonhoeffer's twin sister, Sabine, had to flee to England with her husband, Gerhard Leibholz, whose father was classified as Jewish. Families at that time were being torn apart due to the pious ideology of the times.

Bonhoeffer counters with the idea of unity rather than division. The marriage mandate exemplified the union between Christ and the church. Just as Christ bought new life to humanity when he reconciled and redeemed humanity into the church, the marriage partnership also brings forth new life in the form of children. These children are not objects of a great nation-building campaign, but new humans created to glorify and serve Jesus Christ (Bonhoeffer 2005). In response, parents become the representative of God to the children, another way of reflecting God's reality in the world.

Once Bonhoeffer establishes these two alternatives in repudiation to Althaus' *volk* and race theology, he then explains the mandate of government. Here he can agree with Luther that the government is given to sustain that which exists. However, it is not order as a final goal but preservation of all that is reality in Christ. Therefore, government does not create work or marriage but only works to preserve the work and marriage that God commands.

Of course, this is a rebuke to the Nazi work camps and the Nuremberg Laws, but it was also a challenge to Althaus' declaration that the state was an "ordering of God" that subdued "chaos" and made "community possible" (Althaus 1934a). Bonhoeffer says no to Althaus. God has already created community and order through work and marriage. It is only the government's task to maintain that which has been established by God. Bonhoeffer calls this action of the government "justice."

As a final note Bonhoeffer again holds forth the community that God has already established. Just as he argued in *Sanctorum Communio*, true community is not a nationalistic division, but the church which embraces and envelopes all of humanity. The church mandate is that mandate which forms the unified whole and brings all the other mandates into it. Bonhoeffer says that this mandate creates unified humans who are no longer worn down "through endless conflicts" which come from the results of Althaus and Hirsch's creation orders (Bonhoeffer 2005). This mandate also refutes any notion of a *volk* and instead, as the all-encompassing mandate, protects the universal humanity that has become Christ-existing-as-church-community, and brings humanity into a fellowship that transcends all man-made or territorial identities.

Conclusion

In the beginning of this chapter we recognized the question of whether we could gain access to creation order outside of Christ and the church. Bonhoeffer's Christological approach was not able to be tested by application and seeing the lived results, while Althaus and Hirsch's creation order theology is recorded as a disaster in historical records. One could turn history around on Hirsch and use it to prove the dangers of this form of creation order theology.

However, I argue that it is not the historical record that has the final say in determining whether God can be found in creation order independent of the cross and church. The final determination is whether a *Christian* creation order can be found outside of the cross and church. If we use creation order alone as the measuring stick, are we not only left with a human-focused and human-devised theology such as Althaus and Hirsch have offered? Bonhoeffer would certainly not have argued against the fact that we can see the Creator in the order around us, but he would have argued that without the sole interpretation middle—Christ—we would have a faulty understanding of creation order and a faulty understanding of the Creator behind the order.

References

Althaus, Paul. 1934a. *Die Deutsche Stunde der Kirche*, 3rd ed. Göttingen. Quoted in Ericksen 1985.

———. 1934b. Zum gegenwärtigen Lutherischen Staatsverständnis. In Paul Althaus, Emil Brunner, and Vigo Auguste Demant, *Die Kirche und das Staatsproblem in der Gegenwart*. Berlin: Furche-Verlag.

———. 1937. *Völker vor und nach Christus*. Leipzig. Quoted in Ericksen 1985.

Bethge, Eberhard. 2000. *Dietrich Bonhoeffer: A Biography*, rev. ed. Minneapolis: Fortress Press.
Bonhoeffer, Dietrich. 1997. *Creation and Fall: A Theological Exposition of Genesis 7:3. Dietrich Bonhoeffer Works 3*. Minneapolis: Fortress Press.
———. 1998. *Sanctorum Communio: A Theological Study of the Sociology of the Church. Dietrich Bonhoeffer Works 1*. Minneapolis: Fortress Press.
———. 2005. *Ethics. Dietrich Bonhoeffer Works 6*. Minneapolis: Fortress Press.
Burleigh, Michael, and Wolfgang Wippermann. 1991. *The Racial State: Germany 1933–1945*. Cambridge: Cambridge University Press.
Ericksen, Robert P. 1985. *Theologians Under Hitler: Gerhard Kittel, Paul Althaus and Emanuel Hirsch*. New Haven/London: Yale University Press.
Evans, Richard J. 2004. *The Coming of the Third Reich*. London: Penguin.
Hirsch, Emanuel. 1933. *Das Kirchliche Wollen der Deutschen Christen*. Berlin. Translated and quoted in Ericksen 1985.
Luther, Martin. (1523) 2005. Temporal Authority: To What Extent It Should Be Obeyed. In *Martin Luther's Basic Theological Writings*, ed. Timothy F. Lull. Minneapolis: Augsburg Fortress.

Sergei Bulgakov's Sophiology as the Integration of Sociology, Philosophy, and Theology

Josephien van Kessel

Abstract Sergei Bulgakov (1871–1944) grew up in a time of rapid economic progress and increasingly despotic state authority in Russia. His adult life coincides with the Russian Silver Age (1890–1920), a period of tumultuous cultural and political development. Bulgakov's Sophiology, which is the study of the Wisdom of God, is a reaction to the time he lived in and to the exigencies of his contemporary world, culture, and science. As the integration of sociology, philosophy, and theology, Sophiology had to provide an answer and an alternative to the fragmentation, disintegration, and differentiation of life spheres in the increasingly modern societies of Russia and the Western countries. Although a topical theory, Sophiology is also concerned with the future. In fact, in this chapter I argue that Bulgakov developed his Sophiology to save the *future of creation order* by studying the relation of Sophia to the world as created order (what Palamas called the divine *energies*), which I call his *sociological* Sophiology, and the relation of Sophia to the Trinity (i.e., the order of creation itself—what Palamas called the divine *essence*), which I call his *theological* Sophiology. Both are complementary and essentially one, since Sophia is the object of both Sophiologies—but they use different perspectives.

Keywords Sophiology · Integration · Sophia · Divine Wisdom · Creation order · Christian sociology · Religious philosophy · Orthodox theology · Economy

---

This chapter is the result of my participation in the conference "The Future of Creation Order" at Vrije Universiteit Amsterdam in 2011. It is also incorporated as a chapter in my PhD dissertation on Sergei Bulgakov and his Sophiology (publication and defense expected at Radboud University in 2018).

J. van Kessel (✉)
Radboud University, Nijmegen, The Netherlands
e-mail: j.vankessel@ivoc.ru.nl

© Springer International Publishing AG 2017
G. Glas, J. de Ridder (eds.), *The Future of Creation Order*, New Approaches to the Scientific Study of Religion 3, https://doi.org/10.1007/978-3-319-70881-2_15

Introduction

Sophia is God's Wisdom, Love, and Providence[1] (*Providenie*) that guides human history after the fall. Sophia is the hidden order of creation that reveals itself in this world; as Beauty in art and in nature, as Truth in science and philosophy, and as the Good in society. In this paper, I interpret Sergei Bulgakov's Sophiology as an attempt to gain positive knowledge of this Sophia, which is the *gran'* (border) and *sviaz'* (connection) between transcendence and immanence, between the divine world and the human world.[2] Sophia is the ultimate between (*metaxu*)[3] (Bulgakov 1999, 193),[4] both absolutely transcendent to created order as its source and fully immanent in created order as its ontological root and principle.[5] The antinomic reality of this "and" has central importance for Bulgakov.[6] In his understanding of *antinomy* Bulgakov was clearly inspired by his good friend Pavel Florenskii, according to whom "truth is an antinomy" (Florenskii 1997, 109).[7] Sophia is exactly the between (*mezhdu*) of God and the world—i.e., of transcendence and immanence— which simultaneously connects and separates the two. This *metaxic* quality of Sophia is also the ultimate precondition of any form of human sociality.

In this chapter, I aim to understand the development of Sophiology from the perspective of this quality of Sophia as border, link, and between of divine and

---

[1] Such designations of Sophia as God's Wisdom, Love, Providence, etc., are capitalized here because Bulgakov capitalizes them in the original texts, although this is unusual in Russian.

[2] In its attempt to gain positive knowledge of this border and "between" of transcendence and immanence, Sophiology is *cataphatic*. It is thus opposed to traditional *apophatic* Orthodox philosophy and theology, which deny the possibility of positive knowledge of God, who is considered to be absolutely transcendent to human thought. According to apophatic theology, it is only possible to say what God is not. On apophasis as a characteristic of Orthodox spirituality, see, e.g., Van den Bercken (2011, 87–89, 125).

[3] The Greek term *metaxu* means "between." This makes it possible to interpret Sophiology as a form of metaxology, in William Desmond's sense of the term—see Desmond (1995, 2001, 2008).

[4] First published in 1917, *Svet nevechernii* (*Unfading Light*) has only recently been translated into English by Thomas Allan Smith (Bulgakov 2012). As I used the available translations and original Russian editions of Bulgakov's works from the start of my PhD research in 2005, I had no access to and did not use Smith's translation. All translations from *Svet nevechernii* are my own, unless indicated otherwise. I refer to one of the most recent Russian editions: Bulgakov (1999).

[5] Although Sophia is a common female name and Sophia is often endowed with a feminine nature in Russian Sophiology (and is sometimes viewed as the Eternal Feminine), I refer to Sophia as "it" to stress its nature as a principle. I capitalize the words *Sophia* and *Sophiology* but lowercase such adjective forms as *sophiological* and *sophianic*.

[6] Bulgakov (1999, 99) defines antinomy as "a contradiction for rational thought" (*protivorechie dlia rassudochnogo myshleniia*). On the central importance of "and" in Sophiology, see Zander (1948, 2:181–182) and Zwahlen (2010, 271).

[7] Florenskii takes this notion explicitly from Kant, but he aims to develop an alternative to the Kantian interpretation. See also Louth (2015, 33). Florenskii (1882–1937) became a priest in 1911 and published his sophiological theological treatise *The Pillar and Ground of the Truth* in 1914. In the sixth letter, or in chapter 7, on contradiction, Florenskii gives a definition of antinomy as "a proposition which, being true, jointly contains thesis and antithesis" (Florenskii 1997, 113).

human worlds. I think this link- and border-like quality of Sophia explains why Bulgakov put so much emphasis on the question as to how knowledge of this between is possible, in relation to both worldly and divine economy. Furthermore, it determines the character of Sophiology as *cosmological* during the Silver Age (1890–1920),[8] when Bulgakov focused on the relation of Sophia and the world, and as *theological* during his life in emigration, when he focused on the relation of Sophia to the Trinity.

As Sophiology is a *-logy*—namely, the study of Sophia as the between of God and world—it has to answer the question of how knowledge of its object is possible. Cosmological Sophiology therefore explains how knowledge of Sophia in the world is possible, whereas theological Sophiology explains how knowledge of Sophia in relation to the divine world is possible.[9] The first question is an important subject in Bulgakov's PhD dissertation in political economy, published in 1912 as *Filosofiia khoziaistva. Mir kak khoziaistvo*.[10] In this work, Bulgakov focuses on the world as human economy, as its subtitle reflects; and his treatment of this subject is continued in *Svet nevechernii* (*Unfading Light*), which Bulgakov planned as its second volume. The two books together contain therefore not only the philosophy of knowledge, but also the phenomenology/ontology, axiology, and eschatology of worldly economy.[11]

Bulgakov justified his work on Sophiology in these war years as the "preparation for a more important battle in the spiritual sphere" (1999, 22). The First World War in the real world only pre-reflected a battle that would be even more decisive: the fight for the future of humanity itself, its spiritual emancipation (*dukhovnoe osvobozhdenie*) from the dominant bourgeois mentality in modern society consisting of "phenomenalism, juridicism, economism, and their general foundation, the triumph of methodical thinking and rationality, the rationalism of thought and life," which is the "music of our times" (Bulgakov 2009, 392). This mentality, according to Bulgakov, results in the "general secularization and particularization of life" (ibid., 393). Sophiology was needed to solve this urgent problem of contemporary spiritual fragmentation and weakness.

---

[8] See Zen'kovskii (2011, 841, 845), who also refers to this period of Sophiology as *cosmological*. The Russian Silver Age is the revival of the Golden Age of Alexander Pushkin and his followers in the areas of culture, religion, and arts before the communist revolution of 1917. The Soviet Union expelled most of the representatives of the Silver Age in 1922/1923, who became active in Western Europe; for example, in Germany, Czechoslovakia, and France. Bulgakov became a teacher and dean of the St. Serge Theological Institute in Paris, participated in the ecumenical movement before World War II, and was an important theologian for the Russian Orthodox Church in exile.

[9] In considering this basic philosophical question, Bulgakov clearly places himself in the tradition of Kantian transcendental idealism.

[10] Catherine Evtuhov translated this work as *Philosophy of Economy. The World as Household*. I refer to her translation as Bulgakov (2000), and to the Russian edition as Bulgakov (2009).

[11] On the planned unity of the two volumes of his first Sophiology, see also Evtuhov's introduction in Bulgakov (2000, 11) and the preface to *Philosophy of Economy* (Bulgakov 2000, 38; 2009, 35; see also 1999, 306), in which Bulgakov announces his intentions of publishing a second volume on the philosophy of economy.

After his forced departure from the Soviet Union in December 1922, Bulgakov started on a second and *theological* elaboration of Sophiology in two theological trilogies.[12] I will argue that Bulgakov's elaboration of two Sophiologies was not a coincidence because both deal with the same subject—i.e., Sophia, the Wisdom of God—but study this in relation to different worlds; to wit, the immanent world of human culture, and the transcendent world of God. Whereas cosmological Sophiology studies how Sophia appears in this world as the condition of human history, knowledge, and economy, and how Sophia is oriented to worldly economy, theological Sophiology studies Sophia in relation to inter-Trinitarian *oikonomia*—i.e., divine economy.[13] The ultimate goal of Sophiology is to understand all relations of Sophia, both in themselves and in their connections.[14] It tries to see and understand Sophia *in actu* as well as *in principe*, and presupposes a relation between the phenomenal level of Sophia *in actu* and the metaphysical or *noumenal* level of Sophia *in principe*. From a post-Kantian perspective, of course, Sophiology is a deliberate critique of the Kantian theory of knowledge, while using the same Kantian transcendental-critical terminology.[15]

In the first section, I describe Sophiology as an integrative attempt at a *cataphatic* understanding of Sophia as God's Wisdom that is active in this world. This presupposes not only an understanding of the phenomenal aspects of Sophia from a scientific-empirical perspective, but also of the transcendental conditions of this understanding from a transcendental-critical perspective, and of the noumenal conditions of the being of Sophia from a religious-metaphysical perspective.

In the second section, I interpret cosmological Sophiology as philosophy of human economy and social action. The focus of my interpretation is the double meaning of *khoziaistvo* as "economy" and "household"; i.e., economic activity.[16] This active aspect is also crucial in theological Sophiology and is reflected in Bulgakov's use of the Palamite distinction between the essence (*ousia*) and energies (*energeia*) of God.[17] According to Vaganova (2011, 370) Bulgakov tried to "re-orient

---

[12] According to Arjakovsky (2006, 59ff.), Bulgakov wrote in fact three trilogies, as *The Bride of the Lamb* consists of three books/parts.

[13] Bulgakov's theological Sophiology is not the object of this research, which is restricted to his pre-emigration life and work and to cosmological Sophiology. For a good introduction to theological Sophiology, see, for example, Bulgakov (1993) and Arjakovsky (2006).

[14] Most researchers limit themselves to one of Bulgakov's sophiological perspectives—predominantly to his theology, which was translated more completely into English. In contrast, my study focuses on Bulgakov's social-theoretical and religious-philosophical perspectives that are part of cosmological Sophiology. Only a few researchers have used writings from both cosmological and theological perspectives; for instance, Regula Zwahlen (2010) and Natalia Vaganova (2011).

[15] See also Zwahlen (2012, 186–187).

[16] In Bulgakov (2000, 18), Evtuhov emphasizes in her introduction the dual meaning of *khoziaistvo* as "economy" and "household." This distinction, which comes close to German *Wirt* and *Wirtschaft*, is lost in the English translation of "economy."

[17] Georgii Palamas' distinction between the essence (or being) of God and his energies is important in Hesychast, an important Eastern Orthodox monastic movement, and in what Agamben calls "Trinitarian and economic theology" (Agamben 2011, 12).

Sophiology from a Platonic to a Palamite basis in his theological trilogy on Divine Humanity," but did not finish this project.

Scientific-Phenomenal, Transcendental-Critical, and Religious-Metaphysical Aspects of Cosmological Sophiology

Sophia—the Wisdom of God in this world, and the object of cosmological Sophiology—can and should be studied in three dimensions simultaneously: the empirical-scientific, the transcendental-critical, and the religious-metaphysical dimensions (Bulgakov 2000, 36; 2009, 33). Bulgakov considered his *Philosophy of Economy* as one part of a more encompassing project, and chiefly concentrated on the "general bases of the economic process, or its ontology" (Bulgakov 2000, 28; 2009, 33), but did this mainly from the transcendental-critical dimension.[18] The religious-metaphysical dimension is only partly explored as the ontology of economy, and the scientific-empirical dimension is merely referred to in chapters on German contemporary sociology, in which Bulgakov stresses the inherent limitations of both Marxist and neo-Kantian sociology, and of social sciences as such. In *Svet nevechernii*, Bulgakov developed the other aspects of economy in the religious-metaphysical dimension, the axiology (study of laws and ethics), and eschatology (philosophy of history) of worldly economy.

Bulgakov's concept of economy (*khoziaistvo*) is broader than that of classical economic theory:

> Thus economy is the struggle of humanity with the elemental forces of nature with the aim of protecting and widening life, conquering and humanizing nature, transforming it into a potential human organism. (Bulgakov 2000, 72; 2009, 79)

Economy for Bulgakov includes this sphere of classical economy, but also the spheres of agriculture, science, and art, and consists of the totality of social actions and relations of all *khoziainy* (masters of the household) together. The study of economy is concerned with this totality of social relations that constitute *narodnoe khoziaistvo* (national economy/*Volkswirtschaft*). In this respect Bulgakov clearly opposes his Marxist background in sociology and postulates a common economy of a nation against the Marxist differentiation into classes.

In *Philosophy of Economy*, he makes a distinction between the science and philosophy of economy: science of economy is the study of various economic phenomena in the world, while philosophy of economy is the study of the general foundations of economy as a whole. Science and philosophy are, however, not opposed in his hierarchy of knowledge: rather, philosophy incorporates science as its

---

[18] According to Uffelmann (2006, 490), *Philosophy of Economy* represents "eher eine Erkenntnislehre mit kollektivem Subjekt ... als eine Wirtschaftstheorie" (a theory of knowledge with a collective subject ... rather than an economic theory).

empirical-scientific dimension and provides knowledge on the presuppositions of science or its principles. Both study the same object, but approach it in different ways:

> Economy ... breaks down into phenomena and has its own phenomenology; we can understand its principles only from the perspective of a philosophy of economy, though any given moment could be the object of a particular scientific investigation, if only from a particular, specialized point of view. (Bulgakov 2000, 245; 2009, 311)[19]

Whereas philosophy of economy has the living whole of economy as its object and is interested in its general foundations and conditions of possibility, the science of economy merely analyzes specific parts and singular phenomena of economy and tries to discover causal relations. "There can thus be no single scientific picture of the world, nor can there be a synthetic scientific worldview" (Bulgakov 2000, 161; 2009, 203).

The philosophy of economy can approach its object on three levels or in three dimensions:

> For the same thing that, in the empirical sphere, constitutes the object of "experience" and poses problems for science, constitutes a "transcendental subject" when regarded from the standpoint of cognitive forms, and, finally, descends deep into the metaphysical soil with its ontological roots. (Bulgakov 2000, 36; 2009, 33)

In the scientific-empirical dimension, Sophiology describes and analyzes the historical conditions of the phenomenal aspects of economy as it appears in reality and answers the question, what is economy? In the transcendental-critical dimension, it reconstructs the transcendental conditions of the cognition of economy and answers the question, how is knowledge about economy possible? And finally, in the religious-metaphysical dimension, it answers the question, how is economy possible? In this dimension, Bulgakov tries to discover the essential and necessary conditions of economy.

The main task of cosmological Sophiology is to describe the phenomenology of economy, its transcendental conditions of knowledge, and its ontology. Furthermore, the ontology of economy corresponds in some way to its present state or appearance, i.e., its phenomenology. Bulgakov took care to incorporate the phenomenology of economy into Sophiology, but at the same time he distanced himself from its one-sidedness and merely outer understanding. The title of his doctoral dissertation already indicates his inner alienation from the science of political economy. Although Bulgakov continued to teach as a professor in political economy, *Philosophy of Economy* is his goodbye to science. In fact, already in 1912, he was not interested in the empirical-scientific dimension of economy and only pointed to the relative value of science as a tool. Political economy as a specialized social science is a tool that can be used to look at economic collectivities (2000, 174, 252;

---

[19] This comes remarkably close to Michael Oakeshott's distinction between the conditional (science) and unconditional (philosophy) search for knowledge: "Thus, a theorist is not provoked to this enterprise by his recognition of identities as compositions of characteristics ... but by what in such identities he does not yet understand; namely, their conditionality.... A platform of conditional understanding is constituted by its conditions which, from different points of view, may be recognized as assumptions or as postulates" (Oakeshott 1990, 9).

2009, 219, 320), but "Sophia ... cannot be understood by science" (Bulgakov 2000, 155; 2009, 195). For Bulgakov, his *Philosophy of Economy* was the logical place to introduce Sophia as the necessary precondition of worldly economy and human economic activity.[20] Philosophy has to confront "the problem of economy in its essence" (Bulgakov 2000, 37; 2009, 34).

Interesting in this evaluation of science is Bulgakov's reference to and acceptance of the differentiation of science from other life realms, which is characteristic of modern society. He acknowledged the differentiation and relative autonomy of life spheres but not their absolute autonomy. Science is always a means, and never an end: the autonomy of science is always a relative autonomy, while its end is out of reach of science itself.[21] Bulgakov repeats his normative claim: theoretical constructions need to have "orienting value—the criterion of pragmatism—or are to be considered misjudgments" (Bulgakov 2000, 174; 2009, 219). Although the sciences have more to do with utility than with truth, they bear the mark of truthfulness, because they are children of Sophia and sophianic (*sofiinii*) in nature. Since knowledge is not only the activity of classification but also of projecting and implementing, "a pure theory of cognition is not enough, indeed it is impossible, for we need a theory of action based on knowledge, a *praxeology rather than an epistemology*" (Bulgakov 2000, 178; 2009, 223; italics in original).

In the course of *Philosophy of Economy*, Bulgakov deconstructs the question of how economy is possible into the questions of (1) how production is possible and (2) how consumption is possible. He is, however, less interested in the Kantian transcendental-critical conditions of the cognition of economic phenomena than in the religious-metaphysical conditions of their existence. Sophia is the answer to these questions. Sophia is not only the transcendental-critical but also the religious-metaphysical condition of human economy and society. Moreover, Sophia is the collective and transsubjective subject of economy and history: economy, history, and knowing are functions of the transcendental subject that is Sophia (Bulgakov 2000, 131; 2009, 158).[22] Bulgakov speaks of Sophia as the single subject of economy, the world soul and/or primordial unity, and identifies Sophia with the higher unity and harmony that actually exist in the metaphysical world (Bulgakov 2000, 136, 141; 2009, 164, 172).

---

[20] This is the first appearance of Sophia in Bulgakov's publications, but not its first appearance in Russian religious philosophy: Vladimir Solov'ëv (1853–1900) introduced Sophia to the Russian public, although most explicitly in his poetry, and not his philosophical publications. Bulgakov was one of his self-proclaimed heirs. Other Russian Sophiologists are Nikolai Berdiaev (1874–1948) and Pavel Florenskii (1882–1937), who refused emigration from the Soviet Union. Symbolist poets Viacheslav Ivanov (1866–1949) and Alexander Blok (1880–1921) were also heirs of Solov'ëv and "knights" of Sophia in the Russian Silver Age. See Cioran (1977).

[21] For Weber, too, science does not provide the tools to make existential decisions: every individual has to decide autonomously between God and devil, or which value to adhere to in other life spheres. See Buijs (1998, 20).

[22] *Transcendental* is a category of the Kantian philosophy of knowledge. *Transcendent* is a theological category of things that are trans-mundane (not of this world). Bulgakov in a sense conflates these terms in the functions of Sophia. See also Van Kessel (2014, 82).

In the third, religious-metaphysical dimension, Bulgakov singles out Schelling as the true founder of the philosophy of economy because of his natural philosophy that presents "a clear account of 'objective action' that is the object of the philosophy of economy" (Bulgakov 2000, 79; 2009, 89). West-European philosophy hardly found appeal in the natural philosophy of Schelling, but it was developed further in Russia by Vladimir Solov'ëv in his *Lectures on Divine Humanity* (1995). Bulgakov's conclusion is that "the philosophy of economy, as a philosophy of objective action, must necessarily be a conscious continuation of Schelling's enterprise" (Bulgakov 2000, 93; 2009, 109).

In conclusion, *Philosophy of Economy* does not give a systematic account of the ontology of economy from the religious-metaphysical dimension, but it does contain important indications as to where to start from; namely, Schelling's natural philosophy of objective action and Solov'ëv's spiritual materialism.[23] Furthermore, it firmly establishes Sophia as the single core, root, and basis of economy, and as the subject of knowledge, economy, and history. Human creativity in knowledge, economy, culture, and art is sophianic—that is, "partakes of the divine Sophia" (Bulgakov 2000, 145; 2009, 178).

Sophiology as Philosophy of Social Action

According to Bulgakov, any knowledge of economy is only possible because knowledge is an economic activity. Economy and knowledge are intrinsically connected, and their presumed connection is Sophia. The philosophies of economy, knowledge, and history are therefore intrinsic parts of Sophiology, which studies Sophia in its activity and through its functions. Sophiology as a theory of action combines, according to Bulgakov, a phenomenology that describes the *what* of action, an ontology that focuses on the *how* of action, an axiology (ethics, or the study of values) of action, and an eschatology (history of philosophy) of action.

Bulgakov started his cosmology with the problem of Christian creation versus Neoplatonic emanation. Emanation can only lead to monism "that sacrifices the many or the relative for the absolute," and does not allow for any real plurality (Bulgakov 1999, 166). He insisted on the relative autonomy of this "originally dependent world," an existence "outside of the absolute," which is only possible with the concept of creation. The creation of the world is only understandable as self-sacrifice of the absolute for the relative that becomes its other. God is love, and the life of love and its biggest joy is self-sacrifice (ibid., 168). God created the world not from *ouk-on* (nothing) but from undetermined *mè-on* (something; i.e., potentiality[24]) that is the matter, or all-mother, of the world (170–171). In this act of

---

[23] According to Khoruzhii (in Florenskii 2014, 215), Florenskii, Solov'ëv, and Bulgakov are religious materialists.

[24] See Zwahlen (2012, 189) on Bulgakov's meontology and its implications for his theory of the person.

creation, the Godhead also put a border (*gran'*) between itself and the world that both unites them and separates them from each other (193).

In Sophiology as praxeology and theory of action, various perspectives on human kind are presented, not only in phenomenology but also in ontology, axiology, and eschatology. According to Bulgakov (2000, 178; 2009, 223; italics in original), a "pure theory of cognition is not enough, indeed it is impossible, for we need a theory of action based on knowledge, *a praxeology rather than an epistemology*," which implies that Sophiology as praxeology is fundamentally normative. Objective human action is free, but it nevertheless has to answer both to necessary and to freely chosen, self-imposed conditions and rules. Here Bulgakov clearly protests to the Kantian attempt to formulate a theory of pure reason and a theory of pure action, as if these were separate spheres of human being in the world. Praxeology is not only epistemology (i.e., theory of knowledge) but also axiology (i.e., theory of values). Sophiology is thus not only phenomenology of social action from the empirical-scientific perspective, but also philosophy of cognition of social action from the transcendental-critical perspective, and theology of social action (the world) from the religious-metaphysical perspective.

All social—i.e., human and objective—action in history, knowledge, and economy is a function of the "single subject of economy, the world soul" (Bulgakov 2000, 136; 2009, 164). In empirical history, Bulgakov admits, a general transcendental subject or ideal humanity is hardly visible or imaginable. In fact, it is only intelligible since it belongs to the realm of ideas that preexist in the divine Sophia, or God's Wisdom. Here, he extends his Platonic theory of two worlds, already expressed in his collection of articles entitled *Dva grada* (Two cities—Bulgakov 1997), and presents social ideals as "efforts to reunite [humanity] with this metaphysical realm of all-unity or Sophia," saying that "social ideals are the hypothetical formulation of the higher unity and harmony that actually exist in the metaphysical world" (Bulgakov 2000, 141; 2009, 172). With this distinction between historical and metaphysical worlds, Sophia is distinguished into heavenly and empirical Sophia, similar to Plato's heavenly and popular Aphrodite (Bulgakov 2000, 151; 2009, 188).[25] Together with Sophia, humanity is distinguished into metaphysical and historical humanity.

Metaphysical Sophia is guiding historical Sophia; this guidance is what Hegel called "the cunning of Reason." The awareness of Sophia guiding historical development is the only guarantee for humankind that history has meaning, although this is not a meaning that can be grasped by rational analysis or science, since truth is only given in revelation:

> This revelation can take on different forms: religious, as myths and symbols; philosophical, as the brilliant intuitions of philosophical geniuses; artistic, as works of art. Sophia reveals itself, finally, in the mysteries of personal religious life. (Bulgakov 2000, 155; 2009, 196)

---

[25] Both Solov'ëv and Bulgakov distinguish two Sophias and compare this distinction with Plato's (or Socrates') distinction between heavenly (*ouranos*) and popular (*pandemos*) Aphrodite.

Humanity partakes in both realms. Humanity is created (and therefore conditioned) by God in his image and likeness, is co-creator of God's creation, and has a free will. Only the task and effort to become divine humanity (*bogochelovechestvo*) can overcome this antinomy of necessity and freedom for humanity. This demands a continuous effort—i.e., the religious deed (*podvig*) (Bulgakov 2000, 156; 2009, 196). This is a clear continuation of the theme of asceticism and spiritual fight (*podvizhnichestvo*) in his 1909 Vekhi article "Geroizm i podvizhnichestvo" (Heroism and asceticism—in Bulgakov 1997), and a complete reversal of Hegel's hierarchy of expressions of Absolute Spirit (*absoluter Geist*), with religion as its lowest, and philosophy as its highest expression. In this article, Bulgakov identified mangodhood (*chelovekobozhie*),[26] the opposite of divine humanity, as a mentality; but in *Philosophy of Economy* he explicitly called it a religion. Bulgakov characterized the religion of divine humanity—Christianity—as asceticism, and distinguished it from the religion of man-godhood—paganism—which he characterized as heroism. This distinction between the religion of man-godhood, in which "man is not created but creator," and the religion of God-manhood, "in which man receives his task of re-creation, of economic activity, from God," is central in Sophiology (Bulgakov 2000, 149; 2009, 183).

Therefore, anthropology is an important part of the philosophy of social (objective) action. Bulgakov's anthropology centers on the meaning of creation for humanity: the creation of Adam and Eve as the image and likeness of the triune God. In his view, the fall was not an event in time, but a metaphysical catastrophe, caused by human free will that is part of the "image and likeness" of God in humankind. God created humanity as free; that is, as having the possibility *not* to choose God.[27] The first Adam, or first humanity, did not choose God, which resulted in the fall. Human life in history after the fall is human labor "in the sweat of one's face," seeking redemption after the fall.

Bulgakov unmasked Marxist economic materialism as a philosophy of history. It inherited Hegel's monism in its asking for the "*single* regularity that connects the tangled multiplicity of immediate reasons and constitutes their foundation" (Bulgakov 2000, 267; italics in original), as well as Hegel's dialectical method, which Marxist economic materialism turned into a naïve evolutionary scheme: "Thus economic materialism is a metaphysic of history that, not realizing its true nature, considers itself a science but never actually becomes one or the other" (Bulgakov 2000, 272; 2009, 346). This materialistic monism tends to include the transcendent in the immanent, like pantheism or pan-materialism, and does not recognize the absolute border between immanence (this world) and transcendence (God) that is constituted by creation. Both monism and immanentism are the denial

---

[26] Bulgakov uses both *chelovekobozhie* and *chelovekobozhestvo* for "divine humanity," even within one text. Coates (2013, 305) notices a shift in Bulgakov's use from *chelovekobozhie* to *chelovekobozhestvo*, but does not explain this shift.

[27] Bulgakov's concept of freedom is very similar to the Augustinian concept of the good will as oriented towards God. Bulgakov rejected, however, the Augustinian notion of the total destruction of human freedom after the fall, and its complete dependence on divine grace. See also Tataryn (2000, 66–97).

of transcendence and therefore forms of *khlystovstvo*,[28] which Bulgakov identifies with man-godhood. They are the prime adversaries of his religious philosophy.[29]

In *Svet nevechernii*, Bulgakov is primarily interested in the nature of the connection and border between immanence and transcendence, which is Sophia as *metaxu*, or between, and medium. He does not choose between immanence and transcendence, but proposes the sophianic solution of living the antinomy[30] of the equal necessity of absolute transcendence and absolute immanence of the Godhead (Bulgakov 1999, 99, 180). Thus, rather than choosing either one of the sides of this rational—but not real and existential—antinomy, he stresses that the only possible choice is to live this antinomy of the between that establishes the relation of two mutually exclusive but also constitutive sides of the Godhead; namely, its unity and identity (its transcendence), and its plurality and difference (its immanence).

Bulgakov defines the goal of economy as the transformation of nature into culture, or as the "humanization of nature" (2000, 147; 2009, 180). Economy has a function in the process of salvation—the eschatology of economy. Humanity is predestined in Sophia as ideal humanity to play its role in this process—the axiology of economy. He promised in the preface to *Philosophy of Economy* that the eschatology and axiology of economy were to be the subjects of *Svet nevechernii*. However, in the latter he formulates its task as a search for "religious unity of life, and to become conscious of myself in my historical flesh in Orthodoxy and via Orthodoxy, and to reach its eternal truth through the prism of contemporary life and to see the latter in its light" (Bulgakov 1999, 21).

According to Bulgakov, the German spirit of immanentism and monism was expressed not only in Protestantism but also in socialist man-godhood. In connection with immanentism and monism, he refers to his above-mentioned 1909 Vekhi article "Geroizm i podvizhnichestvo" in which he developed the idea of oppositional inner-worldly ethics of man-godhood and God-manhood, and to other essays in *Dva grada* (Bulgakov 1997). In "Na piru bogov" (Bulgakov 2010), Bulgakov identified Rasputin and Bolshevik socialism as Russian forms of *khlystovstvo* and man-godhood. He furthermore considered the theories of Luther, Eckhart, Boehme, Kant, Fichte, Hegel, Feuerbach, and Marx as examples of "German immanentism" and man-godhood. The general problem with these various forms of man-godhood is that the distance between Creator and creation is only faintly felt and the border between the transcendent and the immanent world—i.e., Sophia—is not recognized (Bulgakov 1999, 23).

---

[28] The *khlysty* were members of a religious sect that came up in the late seventeenth–early eighteenth century in Russia and claimed the possibility of direct communion with the Holy Spirit and of its incarnation in the most righteous of people.

[29] Bulgakov (1999, 5) explicitly called the *chelovekobog* (man-god) *khlyst-chelovekobog*, which indicates his negative attitude towards both *khlyst* and *chelovekobog*.

[30] See Kuvakin (1994, 2:630) on Bulgakov's antinomical rather than dialectical thought. In his use of antinomy Bulgakov followed his friend Florenskii's *The Pillar and Ground of the Truth* (see Florenskii 1997, 113).

Bulgakov proposes instead a relation of *syzygy* between Creator and creation as well as between transcendence and immanence: "The leading idea of the philosophizing in *Svet nevechernii* is not united in a system, but is a kind of *syzygiia* [syzygy]; an organic unity, or symphonic connection" (ibid., 22). Bulgakov explains *syzygy* as an expression from classical Greek, meaning "union" or "togetherness." In fact, however, this term clearly points to Solov'ëv as its source, who used it in his "Smysl liubvi" (The meaning of love—Solov'ëv 1991, 178). Its opposites are the divinization of the world and man, man-godhood,[31] and *mirootritsanie* (world denial), which completely denies the value of the world. Both are mistaken attitudes to the world and to God. According to Bulgakov, the right interpretation of the relation between God and the world is neither man-godhood nor world denial (which can easily become Manichean dualism): rather, it is a *syzygy* of God and world.

Bulgakov elevates religious consciousness above abstract rationality and human reasonability:[32] not rationality, but belief (*vera*) provides "universality and all-humanity—catholicity [*kafolichnost'*[33]], which is another expression for objectivity" (Bulgakov 1999, 67). It is this belief that is at the basis of every human union or togetherness (ibid.). This identification of objectivity with *kafolichnost'/sobornost'* (catholicity) is characteristic of Bulgakov: it is not scientific rationality that provides certainty and objectivity, but belief which is shared with others. Sociology does not understand this universal nature of religion when it presupposes that humankind socializes because of political, juridical, and economic communications and actions. These particular associations are only possible because "humankind is already connected and cemented through religion, i.e., this shared experience of belief. If national spirit [*narodnost'*][34] is the natural basis of state and economy, this same *narodnost'* is in the first place belief: only religion can be really social, and in this quality lies the basis of any sociality" (ibid.).

Bulgakov explored the presence of monism[35] in the history of apophatic thought: from Plato's classical and religious thought, which he considered to be a prototype of Sophiology, to the Neoplatonic patristic writers of early Christian thought; and from Western mystical religious visions of the late Middle Ages to German idealist thought of the eighteenth and nineteenth centuries. In his account, Bulgakov con-

---

[31] Bulgakov referred to this monism and immanentism as *khlystovstvo,* typifying it according to its religious appearance in the sect of *khlysty*.

[32] Boris Jakim, who translated Florenskii (1997), comments on the correspondence of Russian *rassudok* with German *Verstand*, which he translates as "rationality," and of *razum* with German *Vernunft*, which he translates as "reasonability" (see Florenskii 1997, 7ne).

[33] In his note, Bulgakov explains *kafolichnost'* (from the Greek word *katholikos*) as Greek for "universality," which is translated in the Nicean Creed into Russian as *sobornost'*. Evtuhov renders it as "the conciliar principle" in Bulgakov (2000, 24).

[34] *Narodnost'* is, like *sobornost'*, untranslatable. Its closest translation is the German *Volkstümlichkeit*, which is closer in meaning to "ethnicity" or "national spirit" than to "nationality," which is the most common English translation.

[35] *Monism* is Bulgakov's term for every thought system that searches for one explanatory cause that is necessarily "of this world." He rejects monism, but he also rejects its opposite, i.e., dualism. He names his alternative *mono-dualism* (see Zander 1948, 1:192).

centrated on *antinomies* because an antinomy is a clear sign of a certain transcendence of the object of thought for thought. At the same time it is a "crash [*krushenie*] for rational gnoseological immanentism" (Bulgakov 1999, 100). An antinomy is a rational impossibility, but not a real impossibility, according to Bulgakov. In order to conquer an antinomy, we need belief which is love: "Belief is love, because the truth is only apprehensible for one who loves her" (ibid., 66). Belief, not science, can give certainty to knowledge.[36] He identifies this certainty with objectivity and catholicity (*kafolichnost'/sobornost'*):

> In as far as the content of faith has the quality of objectivity it also obtains the quality of universality and all-humanity or catholicity ... that is only a different expression for objectivity. (Ibid., 67)

In this quote, Bulgakov points to the limitations of science and human knowledge, as he did in *Philosophy of Economy*. He consciously replaces Hegel's scheme of climbing stages of rational consciousness in religion, art, and philosophy with his own hierarchy of possibilities to experience and express the absolute or divine revelation through levels of myth making (*mifotvorchestvo*). In myths, "the meeting of the immanent world with the transcendent world is registered" (72). In his hierarchy, philosophical myth making is on a lower level than the myth making of art, because art is based on *umnoe videnie* (reasonable vision). The content of a myth does not consist of concepts but of realities, and therefore religious myth making is the highest stage of myth making as the expression of the transcendent in the immanent sphere. Bulgakov connects his hierarchy of myth making with different levels of transcendence. These different levels of transcendence have access to different kinds of revelation that find expression in myths: natural, divine, and even demonic revelations. Only religious myth making is concerned with the transcendent in the absolute sense (74).

Bulgakov agrees with apophatic theology and metaphysics that "God is inaccessible to human rationality [*ratsional'nost'*]," but he also firmly believes that God is accessible to human reason (*razum*) in a kind of reasonable vision, because of divine revelation (74, 85).[37] The religious philosophy in *Svet nevechernii* is a "free artistry on religious motives" (21). In this work Bulgakov presents the history of apophatic philosophy that is concerned with the problem of transcendence and immanence and is therefore part of the history of religious philosophy that started with Plato, who put the "world of ideas accessible to reason," which Bulgakov identifies with Sophia, between God and world (197).

This border between transcendence and immanence—the *metaxu* that Bulgakov identifies with Sophia as the object of God's love—has to be more than an abstract idea or a dead mirror. Sophia can only be a living being that has a *lik* (countenance),[38] and therefore is a *litso* (person) and a kind of hypostasis (194). Sophia is not part of

---

[36] Bulgakov (1999, 48) quotes Nicolas of Cusa: "credere est cum ascensione cogitare."

[37] Bulgakov refers here to Plato's *Phaedrus*.

[38] *Lik* (*Anlitz*/countenance) is related to *litso* (face/person), but also to *lichina* (mask; also caterpillar). See also Florenskii (2014, 26ff.).

God as fourth hypostasis, next to Father, Son, and Holy Spirit, but as God's nature it is the principle of "hypostasis-ity" (*ipostasnost'*) that is conditional for the creation of the hypostases (persons), such as angels and human beings. Sophia does not have the quality of existence: it is beyond time and space, transcendent in the full sense. And yet, Sophia is the root of the world and of human existence and is therefore fully immanent. This means Sophia is both created and uncreated.[39]

To explain this immanence of divine Sophia, Bulgakov refers to Aristotle's concept of *entelekheia*: Sophia is the *entelekheia* of the world, and the creation of the world in this principle (i.e., in Sophia) is a "particularization in its potentiality from its eternal actuality" (202); in other words, the actualizations of the potential of sofianicity are the content of history.[40] This concept was incorporated in Eastern Christian patristic theology by church father Georgii Palamas (†1359) with his distinction between the essence and energies of God. In the cosmic countenance of Sophia, it is the world soul. The created world, however, partakes in different levels of Sophia, and its sophianic nature is varied. In its highest aspect, it is the church, the Godmother, heavenly Jerusalem, the new heaven and the new earth; but in its outer peripheral aspect, it is the outer universal unity of the world. Human success in science, technology, and economy as well as humanity's possibilities to rule the world as a household[41] rest on this sophianic nature that is manifest in the outer universal unity of the world. The outer manifestations in the world of human activity are sophianic in their foundations, but not in their actual historical being (204).

Because Sophia reveals itself in the world as Beauty, art knows Sophia in a more direct and immediate way than philosophy does (199). Religion, philosophy, and art are forms of myth creation because they are activities of creation (*tvorchestvo*); but myth is the natural habitat of religion, while ideas and concepts belong to philosophy, and images to art. In all forms of creation, an encounter with Sophia is possible (74). Religious myth, however, is the most direct expression of the encounter of the transcendent and immanent worlds, where Sophia reveals itself as between—i.e., as border and connection.

---

[39] As Gerrit Glas observed in one of his comments on a previous draft of this paper, this notion of Sophia, in its connection with the idea of border that separates as well as connects, differs from Dooyeweerd's idea of law (or creation order) as the boundary between God (Origin) and the cosmos. Superficially, however, they share quite a few commonalities. Consider, for instance, how God is viewed as one, and the world as existing in manifold ways, and how the independence of the world is rejected—the world does not exist as such, but is a reference to and an expression of God's power. According to Glas, one possible explanation is that, for Bulgakov, every form of differentiation has a connection with evil (differentiation happened only after the fall), whereas for Dooyeweerd, it is the telos of the cosmos to differentiate further and further (insofar as this process is guided by faith). In my understanding, however, there is no connection in Sophiology between evil and differentiation or inner-worldly activity. Bulgakov stresses the various gifts from God to his creation, the importance of human activity in the world, and the task of humanity to be co-creator.

[40] The transformation of being (essence) to energy (possibility/activity) is central in Palamite and Hesychast theology.

[41] Economy is thus clearly connected to *oikos*, as Agamben (2011, 17) confirms.

In the dimension of art, Bulgakov stresses the difference between pure matter and body-form. Sophia can only appear because of a particular kind of corporeality (*telesnost'*) that has nothing to do with materiality but is in fact the body of ideality. This corporeality is the distinguishing characteristic between Bulgakov's Sophia and Jacob Boehme's *Jungfrau* (virgin) Sophia, who, according to Bulgakov, is not only innocent (i.e., without sexual relations, and a virgin) but also without a face, impersonal, and a mere mirror of God. Boehme's Sophia is only the abstract principle of divine Wisdom that has no relation to gender (*pol*), except in her name (240). This denial of the femininity of Sophia made it impossible for Boehme to see the church as bride and body of Christ (243). Bulgakov sees Boehme's abhorrence of sex and gender (*brezglivost' k polu*) artistically expressed in Tolstoi's *Kreutzer Sonata*, which similarly leads to the thought that the world does not deserve transformation (*preobrazhenie*) but only destruction (242). In essence, Tolstoi's representation of love denies the facts of the incarnation and of the Godman Christ.

In contrast to both Boehme's and Tolstoi's kind of abstract hatred of the flesh of the world, Bulgakov positively viewed the anthropology of Anna Schmidt (1851–1905).[42] The union of the sexes, the sexual act, and childbirth are the *norm* of gender, according to the initial idea of God (261). For this reason, Bulgakov strongly disagreed with the promoters of a third, androgynous sex—such as Zinaida Gippius (1869–1945), poet and wife of Merezhkovskii; Berdiaev, his long-term colleague and friend; and Solov'ëv[43]—and referred to them as *skoptsy* after the orgiastic Russian sect of self-castrates, who recognize infatuation and spiritual marriage but feel aversion to physical marriage, and especially childbirth. Bulgakov considered Solov'ëv in "Smysl liubvi" to be the main ideologue of the third sex and the androgynous, and his vision of love not "a spiritual enlightenment of gender, which is based on a victory over sexuality, but the sublimation of gender" (266).[44] On the other hand, Bulgakov fully agreed with Solov'ëv that Plato's theory of love in *Symposium* is a denial of the world in its corporeal aspect, and that *salvation* is not only salvation of humanity *from* the world but also salvation *of* the world.

The salvation of humankind (*chelovechestvo*) for Bulgakov is only possible through God's self-sacrifice, through his incarnation in the Godman Christ. Christ's task on earth was expressed in his foundation of the church that is not just a "community [*obshchina*] of followers" but a "Being, a living organism" and at the same time "a hierarchically organized society [*obshchestvo*]" (300). The church is the new Adam—that is, true humanity as a positive power—whereas fallen humankind has lost original and true humanity (*chelovechnost'*).[45] In historical humankind,

---

[42] Bulgakov admired Anna Schmidt as a mystic. After her death in 1905, he obtained her manuscripts and published *Third Testament* and her correspondence with Solov'ëv together with Florenskii in 1914. See also Gollerbakh (2000, 206ff.).

[43] As Lossky (1952, 338) confirms, "the ideal of personality is for Merezhkovsky, as for Soloviev and Berdyaev, an androgyn, a man-woman."

[44] Bulgakov treats this subject in a polemical way, probably to stress his own more "orthodox" interpretation of marriage and sexual relations.

[45] Bulgakov mainly used *chelovechestvo* for "humanity" in the genetic sense. *Chelovechnost'* is used to indicate "humanity" in the moral sense.

every individual is enclosed in itself. Humankind consists in the replacement of individuals and generations: historical humanity exists as a collective, but not as *sobornost'* (302). For humankind to become catholic humankind (*sobornoe chelovechestvo*), every individual has to affirm his personality and be ready to sacrifice it. This is the goal of human history, and of economy, art, and knowledge—i.e., human culture, which is the eschatology of human economy.

Before the fall,[46] the life of Adam and Eve was life in a synthesis of economy and art: an active, creative life in beauty. After the fall, according to Bulgakov, a process of secularization and differentiation took place—that is, a separation of economy and art, and of political and religious powers. In the contemporary world, this separation of life spheres—their secularization and differentiation—has developed completely and is succeeded by a total secularization[47] of life and of power (342), because every life sphere is one-dimensional and has been alienated from its sophianic core. The philosophy of history of modern society is in this sense a philosophy of tragedy (306).

Bulgakov envisioned a process of gathering (*soborovanie*) to counter secularization and to overcome this alienation of humanity from its sophianic core. This is a process in which the church broadens its sphere of influence in society and culture and is not only a place of prayer (*tserkov'-khram*), but also of humanity (*tserkov'-chelovechestvo*), of culture (*tserkov'-kul'tura*), and of society (*tserkov'-obshchestvennost'*) (Bulgakov 1997, 348). This is not a vision of a super-church to replace a super-state, but a vision related to a new revelation of power (*vlast'*). According to Bulgakov, power is the skeleton of historical human society (*obshchestvo*) (Bulgakov 1999, 345) and its destiny after the fall. It is the task of Christianity to transform power and to overcome politics in a religious way (344). Bulgakov called for a new ethic and a new understanding of the church as *sobornost'* and soul of the world.

Conclusion

To have knowledge of Sophia as between implies having knowledge of both God and world—that is, of divine *oikonomia* and human economy. Sophiology is the study of these two meanings of *economy* and *oikonomia*, both as separate phenomena and in their relation to each other.[48]

---

[46] As Gerrit Glas has rightly commented, Bulgakov spoke earlier of the fall as a "trans-historic" event. The nature of this "before the fall" and "after the fall" is not temporal, but existential for Bulgakov.

[47] Bulgakov sees secularization as a weakening of the religious bond that binds power and subjects together. This is just one understanding of secularization—Taylor (2007, 20) gives three clearly distinguished basic meanings of secularization: (1) secularized public spaces, (2) decline of belief and practice, and (3) new conditions of belief. See also Casanova (1994, 211).

[48] For a thorough analysis of these two meanings of *economy*, see Uffelmann (2006). See also Van Kessel (2012) on Sophiology as economic theology.

Cosmological Sophiology offers a social-theoretical and social-philosophical perspective on God's Wisdom as present and revealed in creation, whereas theological Sophiology offers a theological perspective on Sophia in relation to the Godhead (Arjakovsky 2006, 60). Cosmological and theological Sophiology are complementary, and study Sophia in its different relations as border and connection between God and world. Cosmological Sophiology is interested primarily in the world as it is and in Sophia as God's presence in the world. It consists of Bulgakov's Christian sociology of the first decade and his religious philosophy of the second decade of the twentieth century. Theological Sophiology on the other hand is primarily interested in the inter-Trinitarian divinity, which is accessible to human reason only in its relation to Sophia as the manifestation of Trinitarian Wisdom in the world. Bulgakov developed theological Sophiology after he became a priest in 1918 and after his exile, when he had settled in Paris and worked as dean of the St. Serge Theological Institute.

In this chapter, I have reconstructed the development of Bulgakov's Sophiology and the tasks it had to fulfill to remedy the consequences of modern society as presented by modern sociology. As described above, Bulgakov's Sophiology is an integration of modern sociology, religious philosophy, and Orthodox theology, and is meant to be such. It opposes the fragmentation and differentiation not only of society but also of sociology, philosophy, and culture. Furthermore, I discerned two main periods in Bulgakov's Sophiology, to wit, a sociological period (1904–1918) and a theological period (1919–1944). As argued above, these periods are not exclusive or consecutive—rather, they are one and complementary.

In the section on Sophiology as philosophy of social action, I have explored the reasons why Bulgakov felt the urgency to develop Sophiology as a theory of action. This was chiefly prompted by the exigencies and problems of his contemporary world—e.g., the disintegration and fragmentation of the Russian Empire, church, and society, and of European culture, science, and society. Sophiology was intended not only to explain and diagnose Bulgakov's contemporary world, but also to provide an alternative for it. This alternative is formulated from the perspective of Sophia—i.e., God's Wisdom—as both border and connection between this world and God. It consists in the reaching out towards this Sophia by human asceticism (*podvizhnichestvo*) in the sense of economic activity in this world, and through the *sobornost'* of all humankind in the church of Christ.

References

Agamben, Giorgio. 2011. *The Kingdom and the Glory. For a Theological Genealogy of Economy and Government*. Stanford: Stanford University Press.
Arjakovsky, Antoine. 2006. *Essai sur le père Serge Boulgakov (1871–1944): Philosophe et théologien chrétien*. Plans-sur-Bex: Parole et Silence.
Buijs, Govert. 1998. *Tussen God en duivel. Totalitarisme, politiek en transcendentie bij Eric Voegelin*. Amsterdam: Boom.

Bulgakov, Sergei. 1993. *Sophia, the Wisdom of God. An Outline of Sophiology*. Hudson: Lindisfarne Press.
———. 1997. *Dva grada. Issledovanie o prirode obshchestvennykh idealov* [Two cities. A research into the nature of social ideals], ed. V.V. Sapov. Saint Petersburg: RKhGI. First published in 1911.
———. 1999. Pervoobraz i obraz [Proto-image and image]. Tom 1, *Svet nevechernij sozertsaniia i umozreniia* [Vol. 1, Unfading Light. Observations and Reflections]. Moscow: Iskusstvo; Saint Petersburg: Inapress. Originally published in 1917 as *Svet nevechernij sozertsaniia i umozreniia* (Moscow: Put').
———. 2000. *Philosophy of Economy. The World as Household*. Trans. Catherine Evtuhov. New Haven/London: Yale University Press. Originally published in 1912 as *Filosofiia khoziaistva. Chast' pervaia Mir kak khoziaistvo* (Moscow: Put').
———. 2009. *Filosofiia khoziaistva*, ed. O. Platonov. Moscow: Institut russkoi tsivilizatsii.
———. 2010. Na piru bogov: Pro i contra; Sovremennye dialogi [On the feast of the gods: Pro and contra; Contemporary dialogues]. In *Intelligentsiia i Religiia*, by Sergei Bulgakov. Saint Petersburg: Izdatel'stvo Oleg Abyshko, Satis'. First published in 1918.
———. 2012. *Unfading Light. Contemplations and Speculations*, translated and with a foreword by Thomas A. Smith. Grand Rapids/Cambridge: William B. Eerdmans Publishing Company.
Casanova, José. 1994. *Public Religions in the Modern World*. Chicago: University of Chicago Press.
Cioran, Samuel. 1977. *Vladimir Solov'ev and the Knighthood of the Divine Sophia*. Waterloo: Wilfrid Laurier University Press.
Coates, Ruth. 2013. Feuerbach, Kant, Dostoevskii: The Evolution of "Heroism" and "Asceticism" in Bulgakov's Work to 1909. In *Landmarks Revisited: The Vekhi Symposium 100 Years On*, ed. Robin Aizlewood and Ruth Coates, 287–307. Boston: Academic Studies Press.
Desmond, William. 1995. *Being and the Between*. Albany: State University of New York Press.
———. 2001. *Ethics and the Between*. Albany: State University of New York Press.
———. 2008. *God and the Between*. Oxford: Blackwell.
Florenskii, Pavel. 1997. *The Pillar and Ground of the Truth: An Essay in Orthodox Theodicy in Twelve Letters*. Trans. Boris Jakim. Princeton: Princeton University Press. Originally published in 1914 as *Stolp i utverzhdenie Istiny* (Moscow: Put').
———. 2014. *Ikonostas*. Saint Petersburg: Azbuka-Attikus. First published in 1972.
Gollerbakh, Evgenii. 2000. *K nezrimomu gradu: Religiozno-filosofskaia gruppa "Put'" (1910–1919) v poiskakh novoi russkoi identichnosti* [Toward the invisible city: The religious-philosophical group "Put'" (1910–1919) in search of a new Russian identity]. Seriia Issledovaniia po istorii russkoi mysli [Series of researches into the history of Russian thought], ed. M.A. Kolerov. Saint Petersburg: Izdatel'stvo 'Aleteiia'.
Kuvakin, Valery A., ed. 1994. *A History of Russian Philosophy: From the Tenth Through the Twentieth Centuries*. 2 vols. Buffalo: Prometheus Books.
Lossky, Nicolai O. 1952. *History of Russian Philosophy*. London: George Allen and Unwin Ltd.
Louth, Andrew. 2015. *Modern Orthodox Thinkers: From the Philokalia to the Present Day*. London: SPCK.
Oakeshott, William. 1990. *On Human Conduct*. Oxford: Clarendon Press.
Solov'ëv, Vladimir. 1991. Smysl liubvi [The Meaning of Love]. In *Smysl liubvi: Izbrannye proizvedeniia* [The Meaning of Love: Selected Works], ed. N.I. Tsimbaeva, 125–182. Moscow: Sovremennik. First published in 1892–1894.
———. 1995. *Lectures on Divine Humanity (1878–1881)*, ed. Boris Jakim. Hudson: Lindisfarne Press.
Tataryn, Myroslaw I. 2000. *Augustine and Russian Orthodoxy: Russian Orthodox Theologians and Augustine of Hippo: A Twentieth Century Dialogue*. Lanham/New York/Oxford: International Scholars Publications.
Taylor, Charles. 2007. *A Secular Age*. Cambridge, MA: The Belknap Press of Harvard University Press.

Uffelmann, Dirk. 2006. Oikonomia – ikonomija/ėkonomija/ėkonomika. Die doppelte Geschichte des Ökonomiebegriffs in Rußland zwischen Wirtschaftstheorie und orthodoxem Kirchenrecht und einige literarisch-kulturelle Weiterungen. In *Russische Begriffsgeschichte der Neuzeit. Beiträge zu einem Forschungsdesiderat, Bausteine zur slavischen Philologie und Kulturgeschichte* 50, ed. Peter Thiergen, 477–515. Cologne/Weimar/Vienna: Böhlau.

Vaganova, Natalia A. 2011. *Sofiologiia protoiereia Sergiia Bulgakova* [The Sophiology of Father Sergeii Bulgakov]. Moscow: Izd-vo PSTGU.

Van den Bercken, Wil. 2011. *Christian Fiction and Religious Realism in the Novels of Dostoevsky*. London/New York/Delhi: Anthem Press.

Van Kessel, Josephien H.J. 2012. Bulgakov's Sophiology: Towards an Orthodox Economic-Theological Engagement with the Modern World. *Studies in East European Thought* 64: 251–267.

———. 2014. Das Verhältnis von Wissenschaft und Politik bei Sergij Bulgakov. Zum siebenten Kapitel von Sergij Bulgakovs Philosophie der Wertschaft. In *Sergij Bulgakovs Philosophie der Wirtschaft im interdisziplinären Gespräch*, ed. Barbara Hallensleben and Regula M. Zwahlen, 78–88. Münster: Aschendorffsche Verlagsbuchhandlung.

Zander, Lev. 1948. *Bog i Mir. Mirosozertsanie ottsa Sergiia Bulgakova* [God and World. The Worldview of Father Sergius Bulgakov]. 2 vols. Paris: YMCA Press.

Zen'kovskii, Vasilii. 2011. *Istoriia russkoi filosofii* [History of Russian philosophy]. Moscow: Akademicheskii Proekt.

Zwahlen, Regula. 2010. *Das revolutionäre Ebenbild Gottes. Anthropologien der Menschenwürde bei Nikolaj A. Berdjaev und Sergej N. Bulgakov.* Syneidos. Deutsch-russische Studien zur Philosophie und Ideengeschichte 5. Vienna/Berlin: Lit Verlag.

———. 2012. Different Concepts of Personality: Nikolaj Berdjaev and Sergej Bulgakov. *Studies in East European Thought* 64: 183–204.

Errata to: The Future of Creation Order

Gerrit Glas and Jeroen de Ridder

Errata to:
G. Glas, J. de Ridder (eds.), *The Future of Creation Order*,
New Approaches to the Scientific Study of Religion 3,
https://doi.org/10.1007/978-3-319-70881-2

This book was inadvertently published without updating the following corrections:

Chapter 10:
The original version of the chapter "Creation Order and the Sciences of the Person" was published with the following incorrect footnote and reference:

> Footnote 4, page 207: see der Hoeven and Johan (1981)
> Reference page 228: der Hoeven, Van, and Johan. 1981. Wetten en feiten. [...]

The footnote has been corrected as:

> see Van der Hoeven (1981)
> The reference has been corrected as:
> Van der Hoeven, Johan. 1981. Wetten en feiten. [...]

Chapter 11:
The original version of the chapter "Beyond Emergence: Learning from Dooyeweerdian Anthropology?" was published with the following incorrect text on page 242: [lit.: Substance]. This has now been changed to [lit.: substance]

The updated online version of this book can be found at
https://doi.org/10.1007/978-3-319-70881-2_10
https://doi.org/10.1007/978-3-319-70881-2_11
https://doi.org/10.1007/978-3-319-70881-2

Index

A

Absolutization, 8, 16, 18, 19, 58, 60, 177, 189, 245, 248
Abstraction, 27, 107, 140, 143, 174–176, 178, 212, 236, 239, 240, 242, 292
Ackermann, 108
Adorno, T., 18
Aebersold, P., 143
Agamben, G., 320, 330
Alexander, D., 22, 28, 43, 160, 167, 223, 319
Alon, U., 162
Althaus, P., 26, 302–309, 311–314
Analysis, 9, 10, 18, 44, 52, 61–63, 73, 77, 78, 85–89, 91–93, 99, 101, 102, 105, 106, 108, 110, 112, 115, 121, 122, 136, 137, 141, 145–147, 175, 182, 205, 233, 236, 238, 241, 245, 248–250, 259, 298, 325
Ansell, N., 25, 27–29, 258
Anselm, 283
Ante-nomian relativisation, 25, 280
Anticipation, 21, 22, 110, 120, 125–127, 145–147, 227
Antinomy, 27, 59, 60, 109, 179, 185, 198, 208, 236, 240, 242, 245, 247, 249, 318, 326, 327, 329
Anti-realism, 174, 176, 178
Appleby, J., 55
Approximation, 14, 120, 139, 142, 162
Aquinas, T., 19, 36–41, 43, 45, 53, 84, 282
Aratos, 244
Archimedes, 99
Aristotelianism, 3, 198
Aristotle, 16, 41, 53, 65, 84, 102, 152, 153, 220, 282, 297, 330
Arjakovsky, A., 320, 333

Armstrong, D.M., 3, 204
Artifacts, 13, 120, 124, 131–133
Aspect, 8, 57, 65, 73, 104, 138, 166, 173, 208, 232, 236, 296, 307, 320
At once infinite, 101–103, 105, 111, 113–115
Atkins, P., 137
Augustine, 152, 173, 209, 227, 241–243, 247, 280, 283, 306
Autopoiesis, 216, 237, 248
Axiology, 319, 321, 324, 325, 327

B

Bacon, F., 152
Badiou, A., 83
Barbour, I., 6
Bardeen, J., 140
Batterman, R., 136, 141
Bavinck, 242, 294
Bayer, O., 289–293, 296, 298, 299
Beauregard, M., 232
Beauty, 2, 13, 14, 28, 129, 225, 318, 330, 332
Bechtel, W., 204
Becker, O., 102
Bednorz, G., 140
Begbie, J., 294, 295
Berdiaev, N., 323, 331
Berger, P., 55
Bergson, H., 156
Berkhof, L., 250
Berkouwer, G.C., 246
Bernays, P., 58, 107, 108, 111, 115
Berry, A., 153
Beth, E., 99
Bethge, E., 311

Bhaskar, 137
Biblical wisdom, 261, 262, 264–267, 278, 279, 281, 283
Biological evolution, 21, 127, 130, 151
Biological order, 183, 193, 195
Biological species, 6, 182–185, 192
Biologism, 18
Biology, 3, 8, 12, 18, 19, 22, 23, 28, 34, 41–43, 78, 123, 128, 136, 138, 145, 153, 158, 160, 164, 166, 167, 204–206, 221, 223, 232, 234, 281
Bird, 43, 44, 120, 161, 165, 275, 276
Bishop, S., 145, 242
Bitbol, M., 24, 237–239, 249
Blackburn, S., 33, 34, 48
Blanke, O., 232
Blessing, 25, 27, 29, 260, 266, 269, 278, 280, 281
Blocher, H., 241, 243, 244
Blok, A., 323
Boehme, J., 327, 331
Bohatec, J., 56
Boltzmann, 127
Bonhoeffer, D., 26, 301–314
Boulter, S., 186
Boundary (boundaries), 3, 5, 6, 14, 21, 24, 27, 28, 56, 71, 144, 146, 172, 174, 179–185, 187–191, 193–196, 206, 207, 209, 212, 216, 221, 223, 225–227, 259, 277, 278, 291, 304, 330
Bourbaki, 107
Boyd, R., 186
Boyer, C., 100, 101
Brock, B., 291, 293
Brouwer, L., 99, 102
Brownian motion, 123, 142, 144, 147
Brueggemann, W., 262–266
Brüggemann-Kruyff, A., 190
Buchhold, J., 249
Buijs, G., 323
Bulgakov, S., 27, 318–333
Bunge, M., 184, 191, 192
Burleigh, M., 305
Butterfield, H., 131

C
Calvin, J., 56, 260
Cambrian explosion, 127, 161
Campbell, D.T., 211
Cantor, G., 100–103, 105, 108, 112, 113, 115
Caputo, J., 264
Carroll, S., 162

Cartwright, N., 204
Casanova, J., 332
Cassirer, E., 54, 59
Cauchy, 100, 101
Causal closure, 34, 35, 40, 43–45, 212
Causal (dis)continuity, 179
Causal power, 19, 23, 35, 38, 41–47, 173, 176, 177, 184–187, 190, 191, 238
Causality, 23, 28, 35, 38, 46, 63, 173, 178–180, 183, 184, 197, 212–216, 219, 235, 238
Causation, 19, 20, 38, 43, 46, 172, 174, 175, 178, 179, 184, 185, 187, 191, 192, 196, 197, 208, 211–214, 218–220, 225, 233, 235, 238, 266, 282
Chambers, R., 155, 159
Chance, 4, 126, 152, 157, 160, 161, 166, 304
Chaplin, J., 4, 258
Characters, 6, 35, 53, 67, 99, 121, 137, 152, 172, 204, 232, 258, 297, 307
Chase, G., 104–106
Chester, T., 298
Cheyne, C., 186, 187
Christian ethic, 306
Christian philosophy, 3, 7, 20, 56, 106, 115, 124–126, 224, 263, 267
Christian theology, 21, 22, 106, 151, 158–160, 247
Christian worldview, 145, 224, 241, 243, 250
Churchill, J., 43
Cioran, S., 323
Clayton, P., 3, 24, 205, 211–215, 220, 232, 234
Clemenceau, L., 241
Clement of Alexandria, 241
Clifford, R., 265
Clouser, R., 69, 199, 236
Coates, R., 326
Co-emergence, 24, 216, 217, 220, 238–240
Coherence, 2, 8, 13, 15, 16, 18, 21, 58, 61, 63, 87, 90, 104, 106, 109–111, 113, 132, 160, 175, 186, 190, 196, 206, 240, 242, 246, 247, 262, 263
Coletto, R., 245
Collins, J., 266
Comperz, T., 52
Concentration, 8, 17, 125, 241
Concentration point, 241
Concept, 1, 54, 77, 98, 138, 163, 186, 204, 212, 232, 260, 261, 266, 289, 321
Concept vs. idea, 14
Contingence, 206, 207
Contingency, 5, 22, 28, 142, 145, 160, 166, 205, 206, 209, 223, 224

Continuity, 9, 62, 68, 102, 159, 174, 206
Continuum, 21, 102, 103, 108, 112, 113
Conway Morris, S., 164, 166
Cook, H., 147, 199
Cooper, J., 249, 264, 282
Cooper, L., 140
Copan, P., 187
Correlated electron systems, 22, 140–142
Cosmic order, 8, 14, 15, 28, 52, 53, 56–59, 226, 227
Cosmology, 18, 223, 324
Craig, W.L., 187
Crane, T., 41
Craver, C.F., 204
Creation, 1, 37, 52–65, 67, 106, 119, 140, 153, 204, 239, 258, 289–299, 301, 318
Creation order, 1–29, 52, 67–93, 147, 177, 204–227, 241, 258, 289–299, 301
Creator, 2, 3, 19, 24, 25, 33, 68, 69, 72, 73, 86, 91, 109, 143, 145, 172, 173, 175, 176, 189, 190, 208, 224, 227, 239, 241, 244, 259
Crenshaw, J., 260
Crowe, F., 282
Crystal structure, 123, 138–140, 147
Cunningham, C., 128, 131
Curl, R., 139
Cuvier, G., 155

D

Darwin, C., 3, 4, 64, 126, 128, 130, 151, 153, 155, 156, 162, 198, 205
Darwin, F., 198
Davies, P.C.W., 211
Davies, P.S., 204
Dawkins, R., 22, 137, 157, 158, 160, 161, 164
de Bruyn Kock, P., 60
de Buffon, C., 154
de Chardin, T., 156, 159
De Ridder, J., 1, 48, 57, 60, 69, 78, 105, 173, 176, 179, 180, 298, 301
Deacon, T., 193, 195–197, 216
Dedekind, R., 108, 115
Deism, 3, 224, 225
Delbourgo, J., 152
Democritus, 136
Dengerink, J., 174
Dennett, D., 160
Derrida, J., 292
Descartes, R., 54, 58, 72, 111, 234
Desmond, W., 318
Determinism, 123, 212
Dew, N., 152

DeWeese, G., 278
d'Holbach, P.-H.T., 154
Diachronic emergence, 146, 147, 180, 181, 196
Diagonal proof, 112, 113
Diderot, D., 154
Dieguez, S., 232
Differentiation, 17, 63, 146, 193–196, 215, 298, 321, 323, 330, 332, 333
Direction, 17, 28, 44, 55, 60, 76, 80, 85, 129, 143, 146, 147, 156–158, 162, 196, 197, 213, 220, 223, 232, 248, 258, 259, 262–264, 267, 277, 279–283, 295, 296
Disclosure, 3, 6, 177, 180, 208, 224, 259, 268, 277
Discontinuity, 62, 68–70, 72, 74–77, 79, 81–84, 89–91, 174, 178–180, 189, 193, 195, 196, 208, 214
Discreteness, 9, 21, 105, 108, 115, 207
Distinctness of levels, 214
Diversity, 7, 8, 14–16, 18, 19, 58, 65, 69, 73, 78, 85, 87–89, 109, 125, 127, 129, 132, 158, 160, 166, 173, 175, 177, 187, 189, 192, 206, 207, 236, 245, 248
Divine agency, 171
Divine economy, 319, 320
Dobzhansky, 120
Doctrine of the three estates, 290, 291, 293
Dods, M., 152
Dooyeweerd, H., 1, 52, 53, 56–61, 63, 69, 78, 86–88, 90, 92, 98, 109, 112, 114, 115, 119–122, 133, 136, 138, 146, 172–199, 205–210, 212, 223, 224, 227, 232–250, 283, 294, 296, 298
Dooyeweerd on the heart, 17–18
Dooyeweerdian philosophy, 92, 136, 137
Douma, J., 4, 295
Dowe, P., 191
Driesch, H., 206
Dubiel, H., 290
Duffy, S., 278
Dummett, M., 112, 114, 115
Dunbar, R., 166, 167
Dupré, J., 19, 42–46
Dynamic, 6, 17, 24, 52, 60, 62, 63, 74, 77, 80, 127, 133, 136, 137, 142–144, 152, 173, 184, 190, 209, 211, 216–221, 261, 262, 264, 272, 282, 283
Dynamical systems (theory), 24, 219
Dyson, F., 70

E

Eccles, J., 120, 241
Echeverria. E., 3, 258

Echolocation, 165
Eckhart, M., 327
Elementary sequence, 101
El-Hani, C.N., 234
Emergence, 1, 119, 136, 151, 172, 205, 232, 274
Emptiness, 24, 71, 238, 249
Encapsis, 22, 145–147
Encaptic relation, 146, 195
Entity, 8, 9, 11, 12, 87, 173, 175, 179–181, 184, 185, 190, 191, 193–197, 234, 313
Epigeneticists, 153
Ericksen, R., 302–305, 310, 311
Eschatology, 68, 69, 319, 321, 324, 325, 327, 332
Euclid, 99, 100
Euler, 99
Evans, R., 302, 305
Everyday experience, 8, 13, 16, 68, 175, 178
Everyday life, 28, 291
Evolution, 3, 57, 69, 119, 146, 151, 172, 204, 232
Evolutionary emergence, 172–174, 176, 178–184, 195, 197–199
Evolutionary theory, 3, 4, 7, 22, 28, 155, 157, 160, 161, 204–210, 215, 223–225
Evtuhov, C., 319, 320, 328
Excluded antinomy, 59, 60, 109, 179, 185, 198

F
Factual side, 57, 63, 112, 176, 181, 184, 207, 209, 210, 214
Fern, R., 98
Feuerbach, L., 327
Fichte, J., 327
Finnis, J., 36
Fisher, R.A., 156
Fitness landscapes, 163
Florenskii, P., 318, 323, 324, 327–329, 331
Foster, M., 243
Foucault, M., 282, 292
Four mandates, 26, 302, 311–314
Fox, M., 269, 271
Fractional quantum Hall effect (FQHE), 141
Fraenkel, A., 98, 100, 107, 108, 113
Free will, 35, 38, 46, 63, 145, 326
Freedom, 18, 35, 38, 45, 52, 55, 58, 129, 130, 133, 147, 235, 236, 240, 260, 291, 309, 310, 326
Frege, G., 103, 108, 112
Fretheim, T., 274, 278

G
Galileo, G., 54
Gallagher, S., 219
Garfield, J., 238
Garfinkel, A., 40
Garrett, D., 271
Geertsema, H., 20, 25, 28, 67, 83, 92, 120, 124, 185, 188, 189, 199, 210, 223, 297
Gegenstand, 240, 244–246, 248, 294
Gehlen, A., 290
Geim, A.K., 139
Gender symbolism, 274
Germany, 302–308, 312, 313, 319
Gerrish, B.A., 260
Gippius, Z., 331
Glas, G., 1, 204, 232, 239, 301
Gödel, K., 58, 108, 109, 111
Godfrey-Smith, P., 204
Gollerbakh, E., 331
Goudzwaard, B., 259, 279
Gould, S., 157, 166
Gousmett, C., 209
Grant, J., 273
Great Chain of Being, 153, 155
Gregersen, N., 237
Grene, M., 198
Griffioen, S., 259, 279, 296–298
Ground idea, 57–59
Ground motive, 19, 57, 58, 120, 242
Grünbaum, A., 112
Gunton, C., 244

H
Hadot, P., 264, 280, 282
Haldane, J., 198
Hamann, J.G., 290, 298
Hamilton, 107
Hamlet, 142
Hanson, N.R., 144
Harrison, P., 2, 152
Hart, H., 174, 190, 260, 267, 279, 281, 282
Hatton, P., 266, 268, 271
Hauerwas, S., 259
Hawking, 135
Heart, 3, 8, 16–19, 37, 47, 82, 90, 137, 205, 239–242, 244, 245, 247, 258, 268, 270, 277, 279, 305, 307
Hegel, G., 27, 290, 325–327, 329
Heim, K., 271
Heine, E., 100, 101
Heisenberg's uncertainty principle, 144
Henderson, R., 3, 187, 189

Hendry, R., 19, 34, 42
Hersh, R., 98, 99
Hesselink, I.J., 260
Heyting, A., 102, 103, 113
Hilbert, D., 98, 103, 108, 111
Hiley, B.J., 191
Hirsch, E., 26, 302, 303, 305–312, 314
History, 3, 7, 11, 20–23, 26, 40, 52, 53, 58, 59, 61, 62, 75, 78, 79, 82, 89, 98–100, 104, 107, 108, 110, 113, 120, 121, 128–133, 152–162, 164–168, 174, 192, 209, 211, 212, 239, 241, 246, 248, 249, 261, 265, 268, 272–274, 297, 303–310, 312, 314, 318, 320, 321, 323–326, 328–330, 332
History as revelation, 26, 304, 305, 308–310
Hitler, A., 302, 303, 308, 311
Hobbes, T., 54
Hobson, P., 44, 45
Holding of laws, 20, 28, 206, 208, 210, 215, 222, 226
Holladay, W., 274
Holz, F., 55
Honecker, M., 289, 290
Hooykaas, R., 153
Horkheimer, M., 18
Household, 319–321, 330
Howard, D., 140
Huizinga, J., 129
Hurley, S., 219
Husserl, E., 103, 220
Huxley, J., 64, 156, 158
Huxley, T.H., 156, 161
Hyatt, A., 156
Hypostatization, 186, 239

I
Ideas, 2–8, 10, 12–23, 26, 28, 41, 52–65, 67, 68, 74–79, 83–85, 87, 88, 91, 98, 101–105, 108, 111, 113, 122, 124, 126, 130, 137, 141, 142, 145, 151–154, 156, 157, 160, 161, 163, 166, 167, 172, 174–177, 185, 190, 199, 204, 206–208, 210, 211, 216, 220, 222–226, 232, 233, 235, 237, 238, 240, 242–244, 260–264, 266, 302, 305, 306, 308, 309, 311, 313, 325, 327–331
Idionomy, 22, 145–147, 208
Immanence, 18, 27, 318, 326–330
Immanentism, 27, 326–329
Individuality structure, 8, 57, 59, 61, 78, 86–88, 119, 121, 122, 180, 181, 185, 188

Infinite totality, 21, 101, 104, 105, 113, 114
Infinity, 20, 21, 98, 102–106, 111–114, 233
Integration, 63, 88, 219, 221, 318
Intelligent design, 124, 158
Intentionality, 166, 167, 232
Interaction, 4, 8–10, 12, 34–36, 39, 41, 42, 44, 45, 47, 121–126, 136, 140, 141, 143, 152, 161, 162, 179, 183, 193–197, 207–209, 211, 216–218, 220, 224, 237
Intertextuality, 275, 276
Intratextuality, 271, 276
Intuitionism, 102, 114, 115
Intuitionistic mathematics, 99, 113
Irigaray, 264
Ivanov, V., 323

J
Jaeger, L., 24, 27, 28, 174, 206, 224, 232, 233, 245
Jakim, J., 328
Jaspers, K., 264
Jeffery, C., 165
Jeremiah, 273–276
Job, 47, 163, 260, 264, 265, 267, 272, 276, 278, 283
Joint attention, 19, 44, 45
Jones, G., 165
Judaism, 47

K
Kalsbeek, L., 7
Kamerlingh Onnes, H., 140
Kant, I., 14, 54, 55, 59, 61, 65, 111, 237, 240, 261, 281, 282, 292, 318, 327
Kattsoff, L., 111
Kauffman, S., 137, 214, 216, 225
Kelso, S., 217, 221
Kendler, K., 204
Keown, G., 274
Kepler, 153
Kim, J., 24, 196, 232, 233, 235
Kingsley, C., 158
Kitcher, P., 41, 204
Klapwijk, J., 3, 119, 122, 124, 126, 133, 145, 147, 172–174, 179, 180, 183–185, 187, 189, 199, 208, 209
Kline, M., 99
Kok, J., 242
Kondrashov, F., 163
Koons, R., 173
Kroto, H., 139

Krüger, A., 139
Kuvakin, V., 327
Kuyper, A., 7, 10, 56, 57, 242, 258, 259, 294, 296–298

L
Lady Wisdom, 265
Lama, D., 238
Lamarck, J.-B., 154, 155, 158, 159
LaMothe, K., 264
Lampe, P., 83
Lange, M., 204
Langemeijer, G., 7
Laughlin, R.B., 139, 141, 142, 145, 146
Laugwitz, D., 100
Lavoisier, A., 154
Law, 1, 33, 52, 68, 103, 120, 136, 153, 172, 204, 259, 294, 304, 321
Law as boundary, 172, 189, 207
Law of nature/natural law, 2, 4, 6, 7, 14, 19, 33, 36, 38, 54, 61, 62, 65, 121, 124, 129–131, 156, 159, 192, 206, 258, 261, 311
Lawfulness, 4, 20, 23, 24, 28, 53, 59, 65, 173, 174, 176, 204–206, 208, 214, 224–227
Law-idea, 52, 53, 206, 260
Law-order, 56–59, 65, 107, 259, 294
Laws of physics, 33–36, 39, 140, 167
Law-side, 3, 21, 54, 57, 63, 92, 93, 112, 113, 120, 121, 124, 130, 176, 181–184, 207, 210, 214, 224, 259
Law-sphere, 8, 21
Law–subject distinction, 23, 24, 208, 209, 227
Le Van Quyen, M., 217
Leekam, S., 44
Leibniz, 99, 100, 108, 186, 233
Leman, L., 174
Levin, D., 262
Lewis, D., 233
Life-forms, 25, 156, 158, 161, 162, 168, 292, 293
Limit, 17, 21, 27, 61, 71–75, 78, 83, 88, 93, 100–102, 107, 124, 139, 140, 144, 145, 157, 162, 166, 167, 180–185, 188, 193, 195, 197, 219, 222, 226, 235, 247, 248, 250, 259, 261, 275, 277, 278, 282, 320, 321, 329
Link, C., 290, 296
Linnaeus, 153, 183
Liu, W., 42
Logos, 24, 27, 52, 209, 240–245, 248, 250, 282

Longo, G., 108
Lorenzen, P., 103
Lossky, N., 331
Louth, A., 318
Love, 6, 9, 27, 35, 46, 57, 129, 168, 258, 279, 282, 299, 318, 324, 328, 329, 331
Lowe, E.J., 44
Lucas, E., 266
Luckmann, T., 55
Luhmann, 293
Luther, M., 25, 26, 289–291, 293, 303, 304, 306, 307, 313, 327
Lutheran theology, 26, 289–294, 297, 298, 312

M
Macdonald, C., 34, 35
Macdonald, G., 34, 35
Mach, 127
Mackie, J., 191
Mādhyamaka Buddhism, 238
Marriage, 25, 26, 130, 261, 265, 291, 295–298, 311, 313, 314, 331
Marx, K., 327
Mathematics, 2, 8, 19–22, 98–115, 167, 220, 221
Maturana, H., 216, 237
Maurer, W., 290
Maxwell, 144
Mayr, E., 120, 157
McGrath, A., 137
McInerny, R., 36, 282
McIntire, C.T., 190
McKane, W., 273, 274
McLaughlin, P., 204
McMullin, E., 144
McShea, D.W., 204
Meaning nucleus/nuclei, 9
Memory, 80–84, 90, 130, 159
Mental phenomena, 10–12, 221
Menzies, P., 213
Merleau-Ponty, M., 219
Metaphysics, 2, 3, 6, 19, 24, 33–48, 53, 172, 174, 213, 223, 224, 226, 239, 246, 282, 305, 320–326, 329
Middleton, R., 259, 280
Midgley, M., 137
Milbank, J., 6
Miller, K., 128
Milton, J., 243
Mind–body dualism, 11, 12, 242, 245, 246
Modal analysis, 10
Modal anticipation, 110

Modal aspect, 8–10, 12–14, 16–18, 23, 57–60, 62, 63, 78, 79, 86–88, 90, 92, 105, 109–112, 114, 115, 119–122, 125, 129, 138, 146, 173, 175, 177, 178, 181, 183, 184, 188, 189, 208, 236, 240–242, 244–248, 250, 296
Modal difference, 187
Modal function, 8, 17, 18, 60
Modal law, 60–63, 124, 182, 183, 208
Modal point of view, 10, 11
Modal retrocipation, 110
Modes, 9–12, 21, 54, 59, 60, 63, 78, 100, 103, 104, 108–111, 133, 175, 189, 195–197, 208, 212, 236
Monism, 80, 84, 85, 87, 88, 90, 92, 115, 214, 232, 234, 250, 324, 326–328
Moore Smith, G.C., 152
Moreland, J., 6
Moreno, A., 216
Morgan, D.F., 260
Mosher, A., 26, 28
Müller, A., 140
Murphy, M., 36
Murphy, N., 80–84, 86, 88, 89, 212, 234, 235
Murphy, R., 25, 265, 267, 275
Murray, M., 167
Mystery, 2, 16, 25, 46, 85, 89, 180, 246, 260, 264, 265, 267, 278, 282, 283, 325

N

Nāgārjuna, 238
Naïve experience, 23, 240, 246
National Socialism, 304, 307
Nationalism, 303, 304
Nazism, 302, 303, 313
Necessity, 1, 3, 13, 24, 28, 61, 112, 140, 152, 156, 161, 165, 166, 204, 223–225, 243, 326, 327
Necessity view of laws, 3, 204, 223
Needham, J.T., 154
Neidhardt, J., 144
Neo-Calvinism, 258, 278, 281, 297
Neo-Calvinist (philosophy), 2, 3, 17, 20, 26, 27, 258–260, 280, 294, 295, 297
Neo-Kantianism, 8
Neoplatonism, 198
Neuroscience, 19, 23, 24, 44, 45, 204, 205, 216, 221, 222, 226, 227, 232, 234, 239
New covenant, 274, 276, 277
New creation, 20, 68–79, 82, 90–92
Newton, I., 99, 100, 126, 136
Nickel, J., 106

Nicolas of Cusa, 329
Nicole, E., 249
Nietzsche, 264
Nijhoff, R., 241
Noë, A., 219
Noll, M., 145
Nominalism, 16, 20, 54, 56, 59, 65, 198, 199, 227, 261, 277
Non-/nearly/minimally/decomposable systems, 218, 219
Non-denumerability, 112, 113, 115
Non-reductionism, 12, 24, 80, 199, 212, 232–236
Non-reductionist ontology, 20, 60, 65, 109, 110, 113, 115
Non-reductive physicalism, 12, 80, 199, 234, 235
Normative structure, 6, 281
Normativity, 7, 62, 63, 258, 259, 278, 279, 298
Norming principles, 62, 63, 65
Novoselov, K.S., 139
Nowak, M., 167

O

O'Connor, T., 43, 172, 232, 234
O'Donovan, O., 295
O'Leary, D., 232
Oakeshott, W., 322
Old Testament, 25, 243, 260, 262, 267
Olthuis, J., 64, 190, 261, 283
Ontic normativity, 7, 298
Ontological monism, 234
Ontological reductionism, 19, 34, 39, 41–43, 45, 46, 137, 138, 174
Opening-up, 90, 120, 125, 131, 146, 147, 153, 224
Order, 1–7, 10, 12–16, 18–21, 23–29, 40, 41, 52–65, 67–70, 72, 73, 78, 91–93, 100, 101, 105, 107–109, 112, 114, 115, 120–122, 124, 129, 131, 132, 144–146, 152–156, 159, 163, 167, 172–179, 181–187, 189, 190, 194–199, 204–221, 223–227, 232, 236–238, 240–242, 258–262, 265–283, 289, 305, 311, 330
Orderliness, 2, 3, 53–55, 64, 65, 223, 225, 266
Orgogozo, V., 164
Origen, 241
Origin, 2, 3, 14, 15, 17, 18, 24, 27, 37, 52, 58, 102, 120, 188, 206, 239, 242, 248, 280
Ostwald, 127

P

Parmenides, 72, 76
Part-whole relation(ship) and whole-part relation(ship), 10, 12, 144, 146, 173, 195, 196
Pascal, B., 233
Paul, 68, 71, 82, 83, 90, 91, 115, 154, 244, 260, 278, 294, 296, 302–304
Pearcey, N., 145
Perdue, L., 271
Pereira, A.M., 234
Periodicity, 140
Person, 6, 12, 17, 18, 20, 23, 25, 26, 35–39, 44–48, 53, 55, 63, 80, 82–89, 91, 130, 133, 144, 204, 211, 221, 225, 232, 233, 237, 246–250, 261, 275, 283, 304–307, 311, 324, 329, 330
Personal identity, 68, 79–91
Perspective, 11, 16, 20, 24–26, 42, 57–59, 65, 68–71, 74, 85, 86, 88, 89, 91, 92, 98, 105–107, 109, 175, 178, 181, 187, 188, 196, 199, 207, 212, 219, 221, 224, 277, 283, 318, 320, 322, 325, 333
Perspectivism, 4
Peters, T., 80
Phase transitions, 136, 142, 225
Phenomenology, 221, 232, 237, 239, 240, 319, 322, 324, 325
Philo, 243, 282
Philosophy of the cosmonomic idea, 52, 122
Philosophy of the law-idea, 52, 260
Physical world, 3, 21, 23, 119, 121, 125, 126, 147, 234, 236
Physicalism, 6, 12, 18, 35, 58, 80, 234–236, 248
Physics, 3, 8, 10, 19, 22, 34, 35, 39, 42, 58, 61, 75, 78, 107, 123, 126, 127, 136–147, 166, 173, 212, 234–238, 248
Pickavance, T., 173
Pines, D., 139, 142
Planck, 127
Plantinga, A., 5, 278
Plantinga, C., 278
Plato, 52, 53, 65, 124, 266, 282, 325, 328, 329, 331
Plotinus, 75, 106
Plurality, 4, 5, 174, 324, 327
Poelwijk, F.J., 163
Polanyi, M., 198
Polkinghorne, J., 69–72, 74, 80–86, 88, 143, 199
Pollard, W., 199
Popper, K., 120, 167
Povolotskaya, I., 163
Poythress, V., 106
Preexistence of laws, 205
Preformationism, 153
(Pre)givenness, 7, 28, 107
Progress, 22, 129–131, 151–160, 222
Progressionism, 157, 160
Promise-command, 20, 26, 81, 82, 85, 88–90, 93
Propensities, 21, 119–134, 224–226
Properties, 8–11, 13, 16, 19, 21, 24, 35, 41–45, 81, 84–86, 89, 92, 120–123, 125–127, 129, 133, 137–139, 142, 143, 145, 147, 157–160, 166, 167, 183, 184, 187, 191, 195–197, 204, 207, 211, 212, 216, 218–221, 224, 233–235, 238, 239
Proverbs, 25, 258, 260, 263, 264, 266–277, 279, 281
Psychologism, 18, 58
Psychology, 11, 19, 23, 24, 44, 45, 137, 204, 205, 216, 222, 226, 227, 239, 248
Purpose, 1, 22, 23, 34, 70, 73, 119, 131, 133, 137, 151, 152, 156, 158–168, 309, 313
Putnam, H., 72

R

Radical orthodoxy, 6
Rae, S., 6
Randomness, 123, 164, 209, 217
Rasputin, 327
Rationalism, 16, 20, 29, 54, 55, 198, 248, 267, 279, 319
Rayleigh-Bénard convection, 22, 142, 143
Real numbers, 101, 103, 104, 112, 113, 115
Reddy, V., 44
Reductionism, 19, 20, 28, 34, 35, 40–43, 45–48, 120, 136–138, 141, 143, 146, 189, 213, 225, 232, 234, 235, 249
Reductive physicalism, 12, 80, 199, 234, 235
Reformational philosophy, 1–4, 6, 7, 28, 56–60, 69, 109, 145, 206, 215, 221–227, 262, 279, 301
Reformed theology, 294–297
Regularity view of laws, 3, 208
Reidemeister, K., 100
Reification, 11, 12, 16, 58, 109, 186, 238, 239, 245, 249
Relation frames, 120–122, 124, 125, 127, 129–133
Responding/answering nature, 20, 26, 81, 82, 85, 86, 88–90, 93, 145, 297
Resurrection, 20, 68, 75–77, 79–92, 249, 259

Revelation, 26, 37, 38, 71, 74, 160, 243–245, 264, 265, 280, 281, 283, 304–310, 325, 329, 332
Richards, F., 41
Right living, 25, 259, 262, 280
Roberts, J., 204
Robertson, J., 152
Robinson, A., 101
Roe, S., 153
Rosen, G., 186
Rosenberg, A., 204
Rubenstein, M.-J., 264, 282
Ruiz-Mirazo, K., 216
Ruse, M., 155–157
Russell, B., 103, 107, 108
Russell, R., 199
Russian Silver Age, 319, 323

S
Sagan, C., 137
Saghatelian, A., 174
Sailhamer, J., 260, 278
Salmon, W., 191
Salthe, S., 193, 196
Sartenaer, O., 172, 174, 189
Schaeffer, H., 25, 26, 28, 29, 289–299
Schindewolf, O., 61, 62
Schmidt, A., 331
Schnatz, H., 53
Schrieffer, J.R., 140
Schuele, A., 82
Schutz, A., 55
Schwarz, R., 290
Sciences, 2–8, 10, 11, 14–16, 18–25, 27, 28, 33–35, 40, 42, 45, 52, 55, 57, 58, 62, 63, 68–73, 77, 80, 86, 89, 91–93, 98, 100, 107, 123, 124, 131, 137, 145, 152, 154, 167, 174, 188, 204–206, 210, 211, 214, 221–227, 232–234, 236, 238, 240, 242, 291, 298, 318, 321–323, 325, 326, 329, 330, 333
Sciences of the person, 23, 204
Scotus, D., 56
Searle, J., 219
Second-person perspective, 85, 86, 88, 89, 91
Secularist scientific picture (SSP), 19, 33
Seerveld, C., 261, 271, 277
Self, 82, 84–86, 89, 175–177, 217, 240, 241, 244, 245, 282
Self-organization, 142, 208–210, 215–217, 221, 237, 239
Selye, H., 222

Set theory, 99, 100, 102, 103, 108
Shaffner, K., 204
Shapin, S., 152, 153
Shavenbaby, 164
Sikkema, A.E., 22, 28, 138, 144, 145, 212
Skillen, J., 258
Smalley, R., 139
Smith, James, 259, 277, 282
Smith, John, 282
Smith, R., 282
Smith, T., 318
Smith, W., 283
Sober, E., 186, 204
Social action, 320, 321, 324–333
Social ethics, 289, 292, 294, 296, 297, 308
Sociology, 137, 290, 318
Solov'ëv, V., 324, 328, 331
Sophia, 27, 318–321, 323–325, 327, 329–333
Sophiology, 27, 318–322, 324–333
Soskice, J., 190
Soul, 6, 37, 79–84, 86, 89–92, 234, 242, 245, 247, 249, 250, 261
Southgate, C., 167
Sovereignty, 10, 294
Spencer, H., 156
Spencer, S., 155
Spengler, O., 7
Sphere sovereignty, 10, 113, 242, 296–298
Sphere universality, 113
Stafleu, M.D., 21, 23, 28, 59, 119, 192, 207
Stebbins, G., 120
Steen, P., 190
Stegmüller, W., 99
Stern, D.L., 164
Störmer, H., 141
Straus, E., 198
Strauss, D., 4, 20, 21, 28, 62–64, 107, 108, 113, 174, 207, 208, 227, 260
Strong notion of law, 23, 204, 215, 225
Structural identity, 89–92
Structural principle, 6, 181, 184
Structural side, 88
Structure, 1, 3, 6, 7, 10, 13, 17, 20, 22, 23, 25, 42, 61, 63, 69, 70, 78, 79, 87–90, 92, 93, 105, 107, 110, 114, 115, 119, 121–124, 127, 133, 138–140, 155, 157, 162–164, 174–176, 180, 181, 183, 185, 186, 188, 205, 206, 208–210, 212, 214, 216, 218, 220, 221, 224, 225, 234, 244, 245, 259, 264, 268–278, 282, 283, 290, 293, 295–297, 306, 310
Stump, E., 19, 20, 28, 36, 38, 40, 42, 43, 146

Subject-side, 21, 57, 89, 92, 121, 130, 176, 181, 183, 189, 206–208, 214, 224
Substance dualism, 83–88, 90–92, 236, 239, 244–247
Successive infinite, 20, 102, 105, 111, 113–115
Suggate, A., 298
Summa Theologiae (*ST*), 36
Superconductivity, 140, 141, 145
Supervenience, 232, 233, 235
Sweetman, R., 260, 261, 280, 282, 283
Symmetry breaking, 142, 146, 147, 196
Synthesis, 52, 53, 128, 156, 178, 179, 241, 243–245, 261, 332
Synthesis thinking, 261

T
Tataryn, M., 326
Taylor, C., 5, 332
Tempier, B., 145
Temple, F., 158
Temporal becoming, 3, 172, 205, 206
Temporality, 78, 173, 176, 181
Tertullian, 241
Thaxton, C., 145, 174
Theism, 19, 47, 156
Theological ethics, 289, 292
Theology, 2, 21, 22, 24, 26, 38, 64, 67, 98, 104–106, 151, 158–161, 215, 223, 243, 247, 249, 250, 258, 262, 263, 281, 289–299, 301–314, 318
Theoretical thinking, 14, 19, 58
Third Reich, 290, 298, 301, 302
Thom, R., 108
Thompson, E., 24, 204, 205, 211, 215–221, 237, 239, 274
Thompson, J.A., 274
Time, 2, 33, 52, 68, 98, 120, 141, 153, 172, 208, 232, 274, 296, 301, 319
Time-order, 23, 112–115, 133
Timmis, S., 298
Tipler, F., 70
Tolstoi, L., 331
Top-down causation, 43, 46, 211–214, 218–220, 225, 235, 238
Torrance, 144
Transcendence, 17, 18, 27, 120, 175, 178, 244, 318, 326–329
Transcendental, 13–16, 18, 23, 26, 28, 58–60, 65, 175, 176, 181, 186, 192, 206, 207, 215, 240, 283, 292, 298, 320–325
Transcendental framework, 16, 205, 206
Transcendental idea/order, 13–15, 58, 175, 176, 319
Transcendental-empirical method, 59, 60, 65
Tremble, A., 153
Triblè, P., 274
Trust, 2, 9, 71, 90, 129, 263, 283
Trustworthiness, 28, 223
Tsui, D., 141
Type laws, 23, 60–63, 121, 124, 176, 177, 180–188, 190–192, 208, 209

U
Uffelmann, D., 321, 332
Ulrich, H., 289, 290, 292–296
Unity, 15, 18, 19, 45, 57, 58, 80, 81, 85, 87, 99, 104, 106, 173, 175, 178, 189, 190, 206, 236, 240–242, 244, 245, 247, 296, 313, 319, 323, 325, 327, 328, 330
Universality, 3, 20, 54, 59–61, 63, 65, 99, 113, 126, 141, 142, 145, 277, 328, 329
Upward causation and downward causation, 196, 197
Ur-Offenbarung, 304

V
Vaganova, N., 320
Van den Bercken, W., 318
Van der Hoeven, J., 4, 207
Van der Kooi, K., 290, 296
van der Meer, J., 23, 28, 120, 121, 146, 172
van der Walt, B.J., 261, 267
Van Dunné, J., 8
Van Egmond, A., 296
Van Fraassen, B., 233, 234
Van Holst, R., 199
van Inwagen, P., 34, 46, 47
Van Kessel, J., 27, 29, 318, 323, 332
Van Leeuwen, R., 269, 272, 277
Van Riessen, 206
Van Til, C., 54
Van Woudenberg, R., 190, 241, 244, 295–298
VanDrunen, D., 258
Varela, F., 24, 216, 221, 237, 238
Vasquez, M., 137
Vaughan, J., 239
Veenhof, C., 57
Verburg, M., 120, 190
Verhoogt, J., 259, 279
Viedma, C., 196
Volk, 26, 302–314
Volk theology, 26, 304–308, 310–312

Index 347

Vollenhoven, D., 56, 260
Vollmert, B., 174
von Neumann, 108
von Rad, G., 264–266, 283
von Schelling, F., 324
von Weizsäcker, C.F., 54

W
Wallace, A.R., 153
Walsh, B., 186, 259, 280
Waltke, B., 266, 270, 272
Wang, H., 58, 111
Wannenwetsch, B., 25, 289, 292, 293, 296, 298
Ward, L., 160
Waters, B., 298
Wearne, B., 120
Weber, A., 55, 237, 323
Weeks, S., 266
Weierstrass, K., 100, 101, 103, 108
Weinberg, S., 136–138
Weinreich, D.M., 162
Welker, M., 69, 72
Welshon, R., 196
Weyl, H., 98, 101, 102, 105, 111
Whole-part causation and part-whole causation, 196
Wijsbegeerte der wetsidee, 58, 176, 260
Wilkins, J., 153, 186
Wilkinson, D., 69–72, 74–79
William of Occam, 57
Wilson, E.O., 157
Wilson, J., 196
Wippermann, W., 305

Wisdom, 25, 27, 28, 206, 225, 232, 243, 258, 260–268, 271, 273, 274, 276–278, 280–283, 318, 320, 321, 325, 331, 333
Wisdom literature, 25, 260, 262, 264, 266, 267, 273, 279
Wittgenstein, L., 238
Wolfson, H., 243
Woltereck, J., 206
Woltereck, R., 198
Wolters, A.M., 3, 295
Wolterstorff, N., 5, 259, 278, 281
Woltjer, 242
Wong, H.Y., 172, 232, 234
Wong, Y., 158
World soul, 323, 325, 330, 332
Worldviews, 3, 17, 20, 23, 24, 28, 39, 40, 46–48, 73, 85, 92, 145, 205, 206, 210, 223–227, 241, 243, 244, 250, 322
Wright, N.T., 70, 278

Y
Yett, D., 268
Yoder, J., 259
Young, W., 246
Yourgrau, P., 109, 111

Z
Zander, L., 318, 328
Zen'kovskii, V., 319
Zermelo, 103, 108
Zuidema, S.U., 261
Zwaan, J., 57
Zwahlen, R., 318, 320, 324
Zylstra, U., 186, 199

CPSIA information can be obtained
at www.ICGtesting.com
Printed in the USA
LVHW081409090619
620638LV00008B/497/P